Deer and People

edited by

Karis Baker, Ruth Carden
and Richard Madgwick

WIND*gather*
PRESS

Windgather Press is an imprint of Oxbow Books

Published in the United Kingdom in 2015 by
OXBOW BOOKS
10 Hythe Bridge Street, Oxford OX1 2EW

and in the United States by
OXBOW BOOKS
908 Darby Road, Havertown, PA 19083

© Windgather Press and the individual contributors 2015

Paperback Edition: ISBN 978-1-909686-54-0
Digital Edition: ISBN 978-1-909686-55-7

A CIP record for this book is available from the British Library

Printed in Malta by Gutenberg Press

For a complete list of Windgather titles, please contact:

United Kingdom
Oxbow Books
telephone (01865) 241249
Fax (01865) 794449
Email: oxbow@oxbowbooks.com
www.oxbowbooks.com

United States of America
Oxbow Books
telephone (800) 791-9354
Fax (610) 853-9146
Email: queries@casemateacademic.com
www.casemateacademic.com/oxbow

Oxbow Books is part of the Casemate group

Front cover: Red deer, photo by R. Carden. Inset, top to bottom: Queen Elizabeth I at
a stag hunt, British Library, London/The Bridgeman Art Library; red deer teeth, photo
by E. Stephan; Bronze Age scene from Boregtiin Gol, Mongolia, photo by R. Kortum;
worked antler and bone, photo by E. Gál.
Back cover: Drawing by J. Cotton

Contents

Post-Medieval Hunting in the UK

Deer Management

Preface

Deer have been central to human cultures throughout time and space: whether as staples to hunter-gatherers, icons of empire, or the focus of sport. Their social and economic importance has seen some species transported across continents, transforming landscape as they went with the establishment of menageries and parks. The fortunes of other species have been less auspicious, some becoming extirpated, or being in threat of extinction, due to pressures of over-hunting and/or human-instigated environmental change. In spite of their diverse, deep-rooted and long standing relations with human societies, no multi-disciplinary volume of research on cervids has until now been produced.

This volume draws together research on deer from wide-ranging disciplines and in so doing substantially advances our broader understanding of human-deer relationships in the past and present. Themes include species dispersal, exploitation patterns, symbolic significance, material culture and art, effects on the landscape and management. The temporal span of research ranges from the Pleistocene to the modern day and covers Europe, North America and Asia.

Papers derived from an international conference of the same title, hosted at the University of Lincoln in collaboration with the British Deer Society and from a session at the 11th International Conference of Archaeozoology in Paris. The volume has been produced in conjunction with the Dama International project, funded by the Arts and Humanities Research Council (standard grant no. AH/I026456/1). The editors are grateful to the authors for providing such wide-ranging and stimulating contributions and to the many anonymous reviewers who gave up their time to assist with the volume's production. We also thank the staff at Windgather/Oxbow, especially Tara Evans, Julie Gardiner and Claire Litt for their assistance and efficiency in bringing this volume together.

Our gratitude is most richly deserved by Naomi Sykes, without whom this volume would not have been possible. She deserves editorship far more than we three, but refused to have her name on the volume. Not only has Naomi been a driving force behind cervid studies in archaeology throughout her career, but she has fostered interdisciplinary networks to make research more relevant to modern cervid ecology and management. She continues to inspire researchers to follow in her path.

This volume is dedicated to the memory of Carol Gelvin-Reymiller, formerly of the University of Alaska, who contributed to this volume. Carol sadly passed away before the volume came to print.

Karis Baker, Ruth Carden and Richard Madgwick

Contributors

Martyn G. Allen
Department of Archaeology, University of Reading,
RG6 6AB

Richard Almond
Independent Scholar (Medieval History and Art
History), Richmond, North Yorkshire

Roger de Bayle des Hermes†
Museum National d'Histoire Naturelle, Laboratoire
de Préhistoire, 1, rue René Panhard, 75013 Paris,
France

Mandy de Belin
The Centre for English Local History, University
of Leicester, LE1 7QR

Katherine Boyle
McDonald Institute for Archaeological Research,
University of Cambridge, CB2 3ER

Gabriele Carenti
Dipartimento di Scienze della Natura e del Territorio,
Università degli Studi di Sassari, 07100 Sassari, Italy

Laurent Chiotti
Department of Prehistory, National Museum
of Natural History, UMR, 7194 Paris, France

John Clark
Department of Archaeological Collections
and Archive, Museum of London, EC2Y 5HN

Laurent Crépin
Department of Prehistory, National Museum
of Natural History, UMR 7194, Paris, France

Dorothée G. Drucker
Department of Geosciences, University of Tübingen,
Tübingen, Germany

William Fitzhugh
Arctic Studies Center, Smithsonian Institution,
Washington DC, USA

Erika Gál
Archaeological Institute of the Hungarian Academy
of Sciences, Budapest 1014, Hungary

Carol Gelvin-Reymiller†
Department of Anthropology, University of Alaska,
Fairbanks, Alaska 99775–7720, USA

Elisabetta Grassi
Dipartimento di Scienze della Natura e del Territorio,
Università degli Studi di Sassari, 07100 Sassari, Italy

Derek Hall
Centre for Economic History and Politics, University
of Stirling, FK9 4LA

Ellen Hambleton
School of Applied Sciences, Bournemouth University,
BH12 5BB

Kerry Harris
Faculty of Humanities, University of Southampton,
SO17 1BF

Dominique Henry-Gambier
PACEA, UMR 5199, University of Bordeaux, Talence,
France

Matilda Holmes
Freelance Zooarchaeologist, Leicestershire

Richard Kortum
Department of Philosophy & Humanities, East
Tennessee State University, Johnson City, Tennessee,
USA

John Langton
School of Geography and the Environment,
University of Oxford, OX1 3QY

Kenneth Lymer
Wessex Archaeology, Salisbury, Wiltshire, SP4 6EB

Kevin Malloy
Anthropology Department, University of Wyoming,
Larami, Wyoming 82071, USA

Mark Maltby
School of Applied Sciences, Bournemouth University, BH12 5BB

Anna M. De Marinis
ISPRA Institute for Environmental Protection and Research, Ozzano dell'Emilia, Italy

Stefano Masala
Dipartimento di Scienze della Natura e del Territorio, Università degli Studi di Sassari, 07100 Sassari, Italy

Marco Masseti
IUCN SSC Deer Specialist Group *and* Laboratories of Anthropology and Ethnology, Department of Evolutionistic Biology, University of Florence, Italy

Allan D. McDevitt
School of Biology and Environmental Science, University College Dublin, Belfield, Dublin, Ireland

Roland Nespoulet
Department of Prehistory, National Museum of Natural History, UMR 7194, Paris, France

Richard Oram
Department of History and Politics, University of Stirling, FK9 4LA

Konstantinia Papastergiou
Department of Environmental Protection, Municipality of Rhodes, Greece

Delphine Remy
Université Paul Valéry-Montpellier III UMR 5140 'Archéologie des societies méditerranéennes', Route de Mende, 34, 199 Montpellier cedex 5, France

Karoline Schmidt
Freelance Scientist, Zandergasse 7, Perchtoldsdorf A, 2380, Austria

Dale Serjeantson
School of Humanities, University of Southampton, SO17 1BJ

Spencer Gavin Smith
Royal Commission on the Ancient and Historical Monuments of Wales, Aberystwyth, SY23 1NJ

Elisabeth Stephan
Landesamt für Denkmalpflege im Regierungspräsidium Stuttgart, Konstanz, Germany

Nikos Theodoridis
Decentrated Administration of Aegean, Directorate General of Forest and Agriculture Piraeus, Greece

Katerina Trantalidou
Ministry of Culture, Ephorate for Palaeo-anthropology-Speleology of Southern Greece, Department of History, Archaeology and Social Anthropology, University of Thessaly, Greece

Carole Vercoutère
Department of Prehistory, National Museum of Natural History, UMR 7194, Paris, France

Cristiano Vernesi
Department of Biodiversity and Molecular Ecology, Research and Innovation Centre, Fondazione E. Mach, S. Michele all'Adige, Trento, Italy

Barbara Wilkens
Dipartimento di Scienze della Natura e del Territorio, Università degli Studi di Sassari, 07100 Sassari, Italy

Fay Worley
English Heritage, Fort Cumberland, Fort Cumberland Road, Portsmouth, PO4 9LD

Frank E. Zachos
Naturhistorisches Museum Wien, Mammal Collection, 1010 Vienna, Austria

Deer Dispersal and Interactions
with Humans

Genetic Analyses of Natural and Anthropogenic Movements in Deer

Allan D. McDevitt and Frank E. Zachos

Introduction

The advent of molecular and genetic techniques has allowed naturalists to gain a greater understanding of both past and present movements (in terms of gene flow) in a whole multitude of organisms. Frequently used molecular markers to address these questions are sequences of the mitochondrial DNA (mtDNA) and nuclear microsatellite DNA. MtDNA has a number of favourable properties that make it especially suitable for intraspecific studies, such as a high mutation rate (so that conspecific individuals may carry different variants or alleles), the absence of recombination (which results in an effectively clonal inheritance from female only to her offspring; Freeland 2005) and, in general, a lack of both pseudogenes and repetitive DNA. Contrary to the haploid (uniparentally inherited) mtDNA, nuclear microsatellite loci are diploid markers (biparentally inherited: offspring receive a maternal and a paternal allele) and thus may show homozygous and heterozygous genotypes. This is important in, amongst other things, analyses of inbreeding. Microsatellites further show very high mutation rates making it possible to individually identify animals within populations and to infer pedigrees. Both mtDNA and microsatellites have been used in a variety of studies on different deer species. In this paper we shall focus on their application in addressing questions of historical biogeography ('phylogeography') and human-mediated translocations.

Phylogeography of deer species in Europe and North America

Phylogeography is defined as the 'field of study concerned with the principles and processes governing the geographical distributions of genealogical lineages, especially at intraspecific level' (Avise *et al.* 1987). This, put simply, is the study of the genetic relationships between populations or individuals and their placement in a geographical context. From its inception, studies have relied upon mtDNA because of its above-mentioned properties. These DNA sequences are used to construct phylogenetic trees or networks (Figure 1.1) using

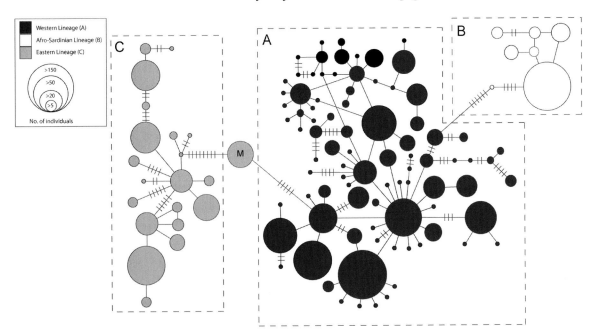

FIGURE I.I. Median-joining phylogenetic network based on parsimony of red deer (*Cervus elaphus*) in Europe. In the network, circles represent haplotypes and size corresponds to the number of individuals having that particular haplotype. Vertical lines represent mutational steps when greater than one. The network is split into the three main lineages (A, B and C) known from previous studies of European red deer (see main text for details). The intermediate haplotype 'M' between lineages A and C belongs to the Mesola red deer. Figure adapted from Carden *et al.* (2012).

various methods such as Neighbour-joining, Bayesian sampling and parsimony to name a few.

Phylogeography can then be combined with coalescent theory which dictates that the current alleles found amongst individuals in a population can be traced back through time to a point at which they coalesce back to an individual ancestral allele in their most recent common ancestor (MRCA; Rosenberg and Nordborg 2002). Further information can be gathered by applying a 'molecular clock' (usually independently calibrated with the fossil record or geological evidence; Freeland 2005) to estimate the time in calendar years when these genealogical lineages split. Therefore, the field of phylogeography can make powerful inferences about the effects of past events in shaping broad-scale genetic structure.

Phylogeographic studies of terrestrial animals have revealed overarching trends of temperate species contracting their ranges to more southern regions during the glacial cycles, followed by rapid expansion into de-glaciated regions once the ice subsided (Taberlet *et al.* 1998; Hewitt 2000). Fossil evidence had previously pointed towards these southern regions, with skeletal remains persisting through the last glacial maximum (LGM, Sommer and Nadachowski 2006). However, fossil data alone cannot reveal the subsequent expansion and spread of animals from these regions and molecular tools have been pivotal in addressing these issues (Sommer *et al.* 2008; 2009).

There have been generally consistent patterns emerging of deer phylogeography in both Europe and North America, undoubtedly the best-studied regions and the ones on which we shall focus in this paper. Beginning with Europe, several refugial areas for deer species have been described. These are the Iberian, Italian

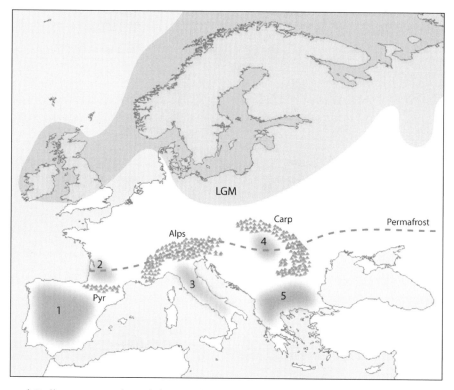

FIGURE 1.2. Extent of the ice sheet during the LGM (grey shade) and position of the permafrost limit (dashed line). Clustered grey triangles represent the major mountain ranges (from left to right); Pyr = Pyrenees, Alps = Alps and Carp = Carpathians. Glacial refugia in Europe during the LGM for red deer (and other mammals) are indicated in grey (after Taberlet *et al.* 1998; Sommer and Nadachowski 2006; Sommer *et al.* 2008). 1. Iberia; 2. Dordogne, France; 3. Italy; 4. Carpathian Basin; 5. the Balkans.

and Balkan peninsulas of the Mediterranean, the Dordogne region of France and the Carpathian Mountains (Figure 1.2). Regions further east have been proposed, although these remain poorly studied, not just with regards to deer, but for many organisms (Korsten *et al.* 2009).

Red deer (*Cervus elaphus*) form three distinct mtDNA lineages in Europe (Fig. 1.1; Skog *et al.* 2009; Niedziałkowska *et al.* 2011; Zachos and Hartl 2011). A 'western' lineage (known as lineage 'A' in Skog *et al.* 2009) occupies Iberia, France, Germany, eastern Central Europe, Scandinavia and the majority of the British Isles (Zachos and Hartl 2011; Carden *et al.* 2012). This is proposed to have arisen from an Iberian/southern French refuge during the LGM and subsequently colonised western Europe (Sommer *et al.* 2008; Zachos and Hartl 2011) and spread as far eastwards as Poland and Belarus (Niedziałkowska *et al.* 2011). The Balkans is proposed to have been another refugium, and this lineage occupies much of eastern and central Europe (lineage 'C'; Skog *et al.* 2009; Niedziałkowska *et al.* 2011). It is likely that there was a Carpathian refugium for red deer during the LGM (Sommer and Nadachowski 2006; Sommer *et al.* 2008) with possible gene flow occurring between there and the Balkans. This has possibly blurred the signal of the impact of deer from this refugium on the recolonisation of Europe but from our understanding of other mammalian species, it may have only been a local spread (McDevitt *et al.* 2012). The Mesola red deer in Italy are believed to be the last surviving population of the Italian lineage (Zachos *et al.* 2009 and references therein) and are almost intermediate between the western and Balkan lineages (Figure 1.1; Niedziałkowska *et al.* 2011). There is another lineage present in Europe, known as the Afro-Sardinian lineage found in Sardinia, Corsica, North Africa, Scotland and Ireland (lineage 'B'; Figure 1.1) but we shall address the distribution of this lineage later when dealing with anthropogenic translocations.

The identification of refugia and postglacial recolonisation patterns is more complex in roe deer (*Capreolus capreolus*) in Europe, with many recent studies attempting to address these questions (see Sommer *et al.* 2009 and references therein). Most studies point to western, eastern and central-northern lineages (Randi *et al.* 2004; Sommer *et al.* 2009). Similar to red deer, the western lineage arose in Iberia/southern France, and the eastern lineage is proposed to have originated in the Balkan region. However, unlike the red deer, the western lineage was relatively confined to Iberia, with the possibility of multiple refugia within this peninsula (Royo *et al.* 2007). Multiple genetic groups also exist within the Italian peninsula (Lorenzini and Lovari 2006) and it has been proposed that this region housed multiple refugia for a variety of species (Vega *et al.* 2010). Like the other peninsulas, the eastern lineage has remained relatively confined to the Balkan region (Sommer *et al.* 2009). It seems that unlike the red deer, the north central-northern lineage has recolonised most of Europe. Considering both fossil and genetic data, a Carpathian or further eastern origin for this lineage seems likely (Sommer *et al.* 2009) but it is clear that further genetic data is required to fully resolve this.

We certainly know a great deal about the history of red and roe deer (two of the best-studied mammals in terms of phylogeography) in Europe but we know much less about other ungulate species in Europe. Putative refugia and recolonisation processes are more difficult to address for species like the fallow deer (*Dama dama*). Although fossil data points to the Balkans as being the main refugium for the species, no widespread genetic study of the species has been undertaken but it can safely be assumed that humans have transported this species around the whole of Europe (Masseti *et al.* 2008) making inferences of natural genetic signatures problematic. Very limited data has been published on the moose (*Alces alces*) in Europe, with the proposal that both European and North American populations arose in Eurasia (Hundertmark *et al.* 2002) based on a small number of samples. A detailed study of Siberian moose has found haplotypes grouping with both European and North American lineages present in the same territory (Moskvitina *et al.* 2011). A more detailed study, using both mtDNA and microsatellites, is currently underway with detailed sampling from throughout the species' range in Europe (M. Niedziałkowska *et al.* unpublished data). Likewise, data on reindeer (*Rangifer tarandus*) in Europe is limited but a larger scale study is also underway (G. Yannic *et al.* unpublished data). Like the moose, it has been proposed that most of the current genetic diversity present in Europe and North America arose from a single refugium in Eurasia, possibly in eastern Siberia (Flagstad and Røed 2003). However, another distinct genetic lineage is present in the more western part of its Eurasian range

(Scandinavia) so it is likely that the species occupied more westerly refugia also. This is further supported by the presence of the species in Ireland and Britain up to the Younger Dryas glaciation (Yalden 1999; Carden *et al.* 2012).

Like the situation in Europe, several overarching patterns exist in the identification of refugia and postglacial recolonisation processes in North America. In most cases, two general models have emerged. During the LGM, the majority of what is now modern-day Canada and the most northerly parts of the United States of America were covered by ice. This led to fauna retreating south or moving into the ice-free regions of Beringia or eastern Siberia during the onset of major glaciations. Several deer species certainly originated from either Beringia or further west. Moose seem to have originated from one of these refugia alone, given the low genetic diversity and star-like pattern of haplotypes observed in mtDNA, consistent with survival and subsequent expansion from a single refugium (Hundertmark *et al.* 2003). North American elk or wapiti (known as red deer in Eurasia but sometimes designated as a subspecies, *C. elaphus canadensis* or even full species, *Cervus canadensis*) are proposed to have originated from Eurasia and crossed the Beringian landbridge to North America (Polziehn *et al.* 1998). Several subspecies within North America are still described but it does appear that they arose from a single lineage (Polziehn *et al.* 1998), even if this inference was based on a limited number of samples. The picture in North America is clearly influenced by the widespread extirpation of the species on a continental scale in the early 1900s and subsequent human translocations to augment herds with declining numbers (see next section).

One of the mitochondrial lineages of caribou (known as reindeer in Eurasia) in North America probably originated from either Beringia or eastern Siberia (see above; Flagstad and Røed 2003; McDevitt *et al.* 2009a). However, the other lineage present in North America (largely associated with the woodland subspecies *R. t. caribou*; McDevitt *et al.* 2009a) definitely arose from another refugium, probably south of the ice sheet in North America. These lineages separated prior to the LGM but have since come into secondary contact, possibly first occurring along the ice-free corridor along the eastern slopes of the Rocky Mountains around 14,000 years ago (McDevitt *et al.* 2009a; Weckworth *et al.* 2012). This has led to the now widespread introgression of mtDNA of certain subspecies into others and an investigation into the genetic variation of caribou on a continental scale (as well as Eurasia) is currently underway (see above). Mule deer (*Odocoileus hemionus*) also show evidence of survival south of the ice sheet but with an apparently more complex pattern of persistence in multiple southerly refugia (Latch *et al.* 2009). There are taxonomic issues among subspecies of *O. hemionus* but it is generally divided into mule deer (numerous subspecies designations) and black-tailed deer (*O. h. columbianus*). Several distinct mtDNA lineages are present on the continent, showing multiple putative refugia for mule deer. The black-tailed deer however belongs to a single mtDNA lineage and is proposed to have persisted in a refugium in the Pacific north-west during the LGM (Latch *et al.* 2009).

Deer translocations

Deer, as iconic game animals of significant financial and symbolic value, have been translocated for probably more than a millennium in Europe (see Hartl *et al.* 2003 and references therein). This is also evidenced by the introduction of five non-native species of deer to Europe (white-tailed deer *Odocoileus virginianus*, sika *Cervus nippon*, axis deer *Axis axis*, Reeves' muntjac *Muntiacus reevesi* and Chinese water deer *Hydropotes inermis*) and the translocation of red deer to, among other places, Argentina and New Zealand; fallow deer, although originally present also in Europe, owe much of their present distribution on the continent to deliberate introductions by humans (Linnell and Zachos 2011).

Molecular markers can be very useful, and at times even indispensable, in identifying an animal's origin. Red deer in particular have been moved around extensively by humans in the last few centuries (Zachos and Hartl 2011). In North America, because of human-caused extirpations and very low numbers remaining in much of their former range, humans have translocated animals across the continent on hundreds of recorded occasions during the last 100 years (Polziehn *et al.* 1998). These events are recorded in the DNA of current North American populations, with proposed subspecies and genetic lineage designations blurred between isolated regions (Polziehn *et al.* 1998). Based on results derived from microsatellite genotypes of the species in Europe, Frantz *et al.* (2006) were able to confirm the illegal introduction of four non-native red deer into a hunting population in Luxembourg, and Kuehn *et al.* (2004), again using molecular data, were able to show that, after the species' extirpation in the seventeenth century, eastern Switzerland was recolonised by red deer from Liechtenstein. In Ireland, molecular data corroborated historical data in detecting translocations from Scotland and its islands to the north-west and east of the country within the last 200 years and more recent introductions from English Parks to the west of Ireland (Carden *et al.* 2012). In tandem, this data also demonstrated translocations occurring between disconnected regions within Ireland over the last few decades (McDevitt *et al.* 2009b; Carden *et al.* 2012).

It is somewhat surprising that the extensive translocation of red deer has not blurred the continental phylogeographical pattern in Europe of three allo- or parapatric lineages (see above), which means that deer have been translocated within the distribution ranges of these lineages rather than among them, at least as far as the female sex is concerned (mtDNA is maternally inherited). However, surviving written sources and records do not indicate a male bias in translocation events (see e.g. Beninde 1937; Niethammer 1963). There is one clear exception to this and it is the occurrence of the Afro-Sardinian mtDNA lineage on the Scottish island of Rum and in the east of Ireland (Nussey *et al.* 2006, Carden *et al.* 2012). How this lineage came to be found extensively in Rum is not exactly known but it was obviously brought by humans at some point. Multiple introductions to the island took place between 1840 and 1920 from mainland Britain (Nussey *et al.* 2006). British deer parks certainly contained red deer from various continental regions at the time (Whitehead 1960). The presence of the lineage in the east of Ireland may be more straight-forward. The translocation of unknown numbers of deer from the Scottish island of Islay took place during the 1800s (Whitehead 1960) and there are no other documented introductions of this species to the east of Ireland. It is therefore plausible that this Afro-Sardinian lineage is also present on other Scottish islands due to translocations occurring within them, and this lineage was then subsequently brought to the east of Ireland (Carden *et al.* 2012). In Poland and Belarus, genetic data has shown concordance with known translocations of red deer after the species were nearly extirpated in the region in the eighteenth and nineteenth centuries (Niedziałkowska *et al.* 2012). This analysis also showed that the native red deer stock still persists in one of the forests in the region. Non-admixed populations of red deer (in terms of translocations) can still be found in parts of the Carpathians (Feulner *et al.* 2004) and the Scottish mainland (Pérez-Espona *et al.* 2009, 2013).

We highlighted previously the importance of combining fossil and genetic data in order to understand natural recolonisation processes occurring after the LGM. The combination of zooarchaeology and genetic techniques is now enlightening us of the role of ancient peoples in determining current deer distributions. It is becoming more apparent that not only did modern humans transport deer species between locations but that ancient peoples did so also (Carden *et al.* 2012). Interestingly, two of the taxonomically acknowledged red deer subspecies seem to owe their existence to human interference: *C. e. corsicanus* was probably introduced to the Tyrrhenian islands from mainland Italy, and *C. e. barbarus* has been hypothesised to go back to introductions of Tyrrhenian red deer to North-Africa (Hajji *et al.* 2008; Zachos and Hartl 2011). Using a combination of fossil, morphological and genetic data Hajji *et al.* (2008 and references therein) clarified the origin of red deer (*C. e. corsicanus*) on the Tyrrhenian islands Sardinia and Corsica, which, as a result, are believed to have been introduced to the islands during the Late Neolithic (Sardinia) and just before the beginning of Classical Antiquity (Corsica).

The analysis of fossil and contemporary genetic data independently certainly tells us much but it is really the advent of techniques allowing us to sequence and analyse ancient DNA (aDNA) that allows us to look more directly at the relationships between ancient and modern populations of deer (Carden *et al.* 2012). Although ancient samples are undoubtedly scarce and sequencing aDNA is obviously much more challenging than sequencing contemporary DNA, the insights gained are well worth the effort and can tell us about both natural and human-induced processes over time. In terms of deer movements, little to date has been done on natural processes but this would be expected to change as the methodology becomes more widely used. Using modern and ancient caribou samples from western Canada, Kuhn *et al.* (2010) observed a partial replacement of caribou genetic lineages occurring approximately 1,000 years ago, consistent with a volcanic eruption occurring in the region at the time. In addition, recent aDNA analysis of pre-LGM red deer bones from Crimea in the Black Sea has revealed that this may have been a north-eastern refugium for red deer, and possibly other mammalian species (Stankovic *et al.* 2011).

Temporal shifts in genetic lineages or haplotypes also tell us about the close relationships between humans and deer throughout ancient and modern times and are indicative of changing management strategies (moving from hunting to animal husbandry for instance). In reindeer in Scandinavia, none of the haplotypes associated with modern domestic reindeer were found in the Late Stone and Iron Ages, indicating a distinctive shift from a hunting to herding economy in the region (Bjørnstad *et al.* 2012). Importantly, the use of aDNA in comparison to modern individuals gives us insights into the movements of ancient peoples and the wild animals they brought with them. Ireland has remained a tricky problem in terms of how species reached it and the origin of red deer on the island has been a contentious issue in particular (Yalden 1999). Modern introductions in historical times are well known (see above) but the species has a good fossil record in Ireland also. Carden *et al.* (2012) analysed modern samples from throughout the species range in the British Isles and Europe and compared them to aDNA from fossils in Ireland from approximately 2,000 up to 30,000 years ago. It was found that one population in the south-west of the country was of Neolithic origin and was brought there from Britain. Their appearance in Ireland at the same time as domestic animals indicates a special, cultural significance beyond acting merely as a food source. Fallow deer are also known to have been extensively translocated during the last millennium (Masseti *et al.* 2008, Linnell and Zachos 2011). Sykes *et al.* (2011) analysed ancient samples from Britain from approximately 2,000 years ago and compared them to a limited dataset from modern individuals. They found that the ancient fallow deer most closely resembled modern individuals from Italy, consistent with the likelihood of a Roman origin.

Conclusions

Genetic methods have revolutionised the fields of ecology and evolution. Molecular characterisation of extant populations yields insights into demographic dynamics, historical biogeography and the distribution and partitioning of genetic diversity. Apart from merely academic results, this kind of data is also informative for conservation biology and, particularly in deer, management of harvested stocks and populations. Genetic signatures of human influences can help disentangle natural from anthropogenic processes, and the recent advent of ancient DNA analysis and its synthesis with palaeontology and archaeology have been especially fruitful, deepening our understanding of how present distribution patterns and genetic structuring have occurred and opening up an interdisciplinary research field of its own at the same time.

10 *Allan D. McDevitt and Frank E. Zachos*

References

Avise, J. C., Arnold, J., Ball, R. M., Bermingham, E., Lamb, T., Neigel, J. E., Reeb, C. A. and Saunders, N. C. (1987) 'Intraspecific phylogeography: The mitochondrial DNA bridge between population genetics and systematics', *Annual Review of Ecology and Systematics* 18, 489–522.

Beninde, J. (1937) *Zur Naturgeschichte des Rothirsches*, Paul Parey, Hamburg and Berlin (reprinted in 1988).

Bjørnstad, G., Flagstad, Ø., Hufthammer, A. K. and Røed, K. H. (2012) 'Ancient DNA reveals a major genetic change during the transition from hunting economy to reindeer husbandry in northern Scandinavia', *Journal of Archaeological Science* 39, 102–108.

Carden, R. F., McDevitt, A. D., Zachos, F. E., Woodman, P. C., O'Toole, P., Rose, H., Monaghan, N. T., Campana, M. G., Bradley, D. G. and Edwards, C. J. (2012) 'Phylogeographic, ancient DNA, fossil and morphometric analyses reveal ancient and modern introductions of a large mammal: the complex case of red deer (*Cervus elaphus*) in Ireland', *Quaternary Science Reviews* 42, 74–84.

Flagstad, Ø. and Røed, K. H. (2003) 'Refugial origins of reindeer (*Rangifer tarandus* L.) inferred from mitochondrial DNA sequences', *Evolution* 57, 658–670.

Feulner, P. G., Bielfeldt, W., Zachos, F. E., Bradvarovic, J., Eckert, I. and Hartl, G. B. (2004) 'Mitochondrial DNA and microsatellite analyses of the genetic status of the presumed subspecies *Cervus elaphus montanus* (Carpathian red deer)', *Heredity* 93, 299–306.

Frantz, A. C., Tigel Pourtois, J., Heuertz, M., Schley, L., Flamand, M. C., Krier, A., Bertouille, S., Chaumont, F. and Burke, T. (2006) 'Genetic structure and assignment tests demonstrate illegal translocation of red deer (*Cervus elaphus*) into a continuous population', *Molecular Ecology* 15, 3191–3203.

Freeland, J. (2005) *Molecular Ecology*, John Wiley and Sons Ltd. Chichester, UK.

Hajji, G. M., Charfi-Cheikrouha, F., Lorenzini, R., Vigne, J.-D., Hartl, G. B., Zachos, F. E. (2008) 'Phylogeography and founder effect of the endangered Corsican red deer (*Cervus elaphus corsicanus*)', *Biodiversity and Conservation* 17, 659–673.

Hartl, G. B., Zachos, F. and Nadlinger, K. (2003) 'Genetic diversity in European red deer (*Cervus elaphus* L.): anthropogenic influences on natural populations', *Comptes Rendus Biologies* 326, 37–42.

Hewitt, G. M. (2000) 'The genetic legacy of the Quaternary ice ages', *Nature* 405, 907–913.

Hundertmark, K. J., Shields, G. F., Udina, I. G., Bowyer, R. T., Danilkin, A. A. and Schwartz, C. C. (2002) 'Mitochondrial phylogeography of moose (*Alces alces*): late Pleistocene divergence and population expansion', *Molecular Phylogenetics and Evolution* 22, 375–387.

Hundertmark, K. J., Bowyer, R. T., Shields, G. F. and Schwartz, C. C. (2003) 'Mitochondrial phylogeography of moose (*Alces alces*) in North America', *Journal of Mammalogy* 84, 718–728.

Kuhn, T. S., McFarlane, K. A., Groves, P., Mooers, A. Ø. and Shapiro, B. (2010) 'Modern and ancient DNA reveal recent partial replacement of caribou in the southwest Yukon', *Molecular Ecology* 19, 1312–1323.

Kuehn, R., Haller, H., Schroeder, W. and Rottmann, O. (2004) 'Genetic roots of the red deer (*Cervus elaphus*) population in eastern Switzerland', *Journal of Heredity* 95, 136–143.

Latch, E. K., Heffelfinger, J. R., Fike, J. A. and Rhodes, O. E. (2009) 'Species-wide phylogeography of North American mule deer (*Odocoileus hemionus*): cryptic glacial refugia and postglacial recolonisation', *Molecular Ecology* 18, 1730–1745.

Linnell, J. D. C. and Zachos, F. E. (2011) 'Status and distribution patterns of European ungulates: genetics, population history and conservation'. In *Ungulate Management in Europe: Problems and Practices*, eds R. Putman, M. Apollonio and R. Andersen, CUP, Cambridge, Mass, 12–53.

Lorenzini, R. and Lovari, S. (2006) 'Genetic diversity and phylogeography of the European roe deer: the refuge area theory revisited', *Biological Journal of the Linnean Society* 88, 85–100.

Masseti, M., Pecchioli, E. and Vernesi, C. (2008) 'Phylogeography of the last surviving populations of Rhodian and Anatolian fallow deer (*Dama dama dama* L., 1758)', *Biological Journal of the Linnean Society* 93, 835–844.

McDevitt, A. D., Mariani, S., Hebblewhite, M., DeCesare, N. J., Morgantini, L., Seip, D., Weckworth, B. V. and Musiani, M. (2009a) 'Survival in the Rockies of an endangered hybrid swarm from diverged caribou (*Rangifer tarandus*) lineages', *Molecular Ecology* 18, 665–679.

McDevitt, A. D., Edwards, C. J., O'Toole, P., O'Sullivan, P., O'Reilly, C. and Carden, R. F. (2009b) 'Genetic structure of, and hybridisation between, red (*Cervus elaphus*) and sika (*Cervus nippon*) deer in Ireland', *Mammalian Biology* 74, 263–273.

McDevitt, A. D., Zub, K., Kawałko, A., Oliver, M. K., Herman, J. S. and Wójcik, J. M. (2012) 'Climate and refugial origin influence the mitochondrial lineage distribution of weasels (*Mustela nivalis*) in a phylogeographic suture zone', *Biological Journal of the Linnean Society* 106, 57–69.

Moskvitina, N. S., Nemoikina, O. V., Tyuten'kov, O. Y. and Kholodova, M. V. (2011) 'Retrospective evaluation and modern state of the moose (*Alces alces* L.) population in West Siberia: ecological and molecular-genetic aspects', *Contemporary Problems of Ecology* 4, 444–448.

Niedziałkowska, M., Jędrzejewska, B., Honnen, A.-C., Otto, T., Sidorovich, V. E., Perzanowski, K., Skog, A., Hartl, G. B., Borowik, T., Bunevich, A. N., Lang, J. and Zachos, F. E. (2011) 'Molecular biogeography of red deer *Cervus elaphus* from eastern Europe: insights from mitochondrial DNA sequences', *Acta Theriologica* 56, 1–12.

Niedziałkowska, M., Jędrzejewska, B., Wójcik, J. M. and Goodman, S. J. (2012) 'Genetic structure of red deer population in north-eastern Poland in relation to the history of human interventions', *Journal of Wildlife Management* 76, 1264–1276.

Niethammer, G. (1963) *Die Einbürgerung von Säugetieren und Vögeln in Europa*, Paul Parey, Hamburg and Berlin.

Nussey, D. H., Pemberton, J., Donald, A. and Kruuk, L. E. B. (2006) 'Genetic consequences of human management in an introduced island population of red deer (*Cervus elaphus*)', *Heredity* 97, 56–65.

Pérez-Espona, S., Pérez-Barberia, F. J., Goodall-Copestake, W. P., Jiggins, C. D., Gordon, I. J. and Pemberton, J. M. (2009) 'Genetic diversity and population structure of Scottish Highland red deer (*Cervus elaphus*) populations: a mitochondrial survey', *Heredity* 102, 199–210.

Pérez-Espona, S., Hall, R. J., Pérez-Barberia, F. J., Glass, B. C., Ward, J. F. and Pemberton, J. M. (2013) 'The impact of past introductions on an iconic and economically important species, the red deer of Scotland', *Journal of Heredity* 104, 14–22.

Polziehn, R. O., Hamr, J., Mallory, F. F. and Strobeck, C. (1998) 'Phylogenetic status of North American wapiti (*Cervus elaphus*) subspecies', *Canadian Journal of Zoology* 76, 998–1010.

Randi, E., Alves, P. C., Carranza, J., Milošević-Zlatanović, S., Sfougaris, A. and Mucci, N. (2004) 'Phylogeography of roe deer (*Capreolus capreolus*) populations: the effects of historical subdivisions and recent nonequilibrium dynamics', *Molecular Ecology* 13, 3071–3083.

Rosenberg, N. A. and Nordborg, M. (2002) 'Genealogical trees, coalescent theory and the analysis of genetic polymorphisms', *Nature Reviews Genetics* 3, 380–390.

Royo, L. J., Pajares, G., Alvarez, I., Fernández, I. and Goyache, F. (2007) 'Genetic variability and differentiation in Spanish roe deer (*Capreolus capreolus*): a phylogeographic reassessment within the European framework', *Molecular Phylogenetics and Evolution* 42, 47–61.

Skog, A., Zachos, F. E., Rueness, E. K., Feulner, P. G. D., Mysterud, A., Langvatn, R., Lorenzini, R., Hmwe, S. S., Lehoczky, I., Hartl, G. B., Stenseth, N. C. and Jakobsen, K. S. (2009) 'Phylogeography of red deer (*Cervus elaphus*) in Europe', *Journal of Biogeography* 36, 66–77.

Sommer, R. S. and Nadachowski, A. (2006) 'Glacial refugia of mammals in Europe: evidence from fossil records', *Mammal Review* 36, 251–265.

Sommer, R. S., Zachos, F. E., Street, M., Jöris, O., Skog, A. and Benecke, N. (2008) 'Late Quaternary distribution dynamics and phylogeography of the red deer (*Cervus elaphus*) in Europe', *Quaternary Science Reviews* 27, 714–733.

Sommer, R. S., Fahlke, J. M., Schmölcke, U., Benecke, N. and Zachos, F. E. (2009) 'Quaternary history of the European roe deer *Capreolus capreolus*', *Mammal Review* 39, 1–16.

Stankovic, A., Doan, K., Mackiewicz, P., Ridush, B., Baca, M., Gromadka, R., Socha, P., Weglenski, P., Nadachowski, A. and Stefaniak, K. (2011) 'First ancient DNA sequences of the Late Pleistocene red deer (*Cervus elaphus*) from the Crimea, Ukraine', *Quaternary International* 245, 262–267.

Sykes, N. J., Baker, K. H., Carden, R. F., Higham, T. F. G., Hoelzel, A. R. and Stevens, R. E. (2011) 'New evidence for the establishment and management of the European fallow deer (*Dama dama dama*) in Roman Britain', *Journal of Archaeological Science* 38, 156–165.

Taberlet, P., Fumagalli, L., Wust-Saucy, A.-G. and Cosson, J.-F. (1998) 'Comparative phylogeography and post-glacial colonization routes in Europe', *Molecular Ecology* 7, 453–464.

Vega, R., Amori, G., Aloise, G., Celleni, S., Loy, A. and Searle, J. B. (2010) 'Genetic and morphological variation in a Mediterranean glacial refugium: evidence from Italian pygmy shrews, *Sorex minutus* (Mammalia, Soricomorpha)', *Biological Journal of the Linnean Society* 100, 774–787.

Weckworth, B. V., Musiani, M., McDevitt, A. D., Hebblewhite, M. and Mariani, S. (2012) 'Reconstruction of caribou evolutionary history in Western North America and its implications for conservation', *Molecular Ecology* 21, 3610–3624.

Whitehead, G. K. (1960) *The Deerstalking Grounds of Great Britain and Ireland*, Hollis and Carter, London.

Yalden, D. W. (1999) *The History of British Mammals*, Poyser, London.

Zachos, F. E., Hajji, G. M., Hmwe, S. S., Hartl, G. B., Lorenzini, R. and Mattioli, S. (2009) 'Population viability analysis and genetic diversity of the endangered red deer population from Mesola, Italy', *Wildlife Biology* 15, 175–186.

Zachos, F. E. and Hartl, G. B. (2011) 'Phylogeography, population genetics and conservation of the European red deer *Cervus elaphus*', *Mammal Review* 41, 138–150.

Historic Zoology of the European Fallow Deer, *Dama dama dama*: Evidence from biogeography, archaeology and genetics

Marco Masseti and Cristiano Vernesi

Introduction

Within the genus *Dama* two extant varieties are recognised: the European fallow deer (*Dama dama dama*) and the Mesopotamian/Persian fallow deer (*Dama dama mesopotamica*). Current knowledge suggests that the post-glacial diffusion of this genus was limited almost exclusively to the northern Mediterranean and the Near East (Uerpmann 1981; Masseti 1996; Chapman and Chapman 1997), where despite the contiguity of their native geographical distribution, these species were not sympatric. Fallow deer of the Taurus range belong to the European form (Danford and Alston 1880; Harrison, 1968), while the original range of the Mesopotamian variety is reported east of the *Nur Dağlari*, the ancient *Amanus*; a mountain range in the Hatay Province of south-eastern Turkey (Figure 2.1). This paper examines the historic zoology of the European fallow deer by reviewing biogeographic evidence from archaeology and genetics.

Distribution of the European fallow deer during the Late Pleistocene

At the beginning of the Late Pleistocene, European fallow deer resided within continental Europe and commonly occur in faunal assemblages from the last interglacial (*c.*130,000–70,000 BP) of mid–latitude Europe (Butzer 1972; Leonardi and Petronio 1976; Stuart 1991). The general opinion previously held was that European fallow deer had been ousted from most of the European mainland before the last interglacial, and that its post-Pleistocene natural occurrence in Western Europe was unproven.

In effect, during the last glacial period, European fallow deer retreated into the southernmost areas of its former distribution, the north-eastern

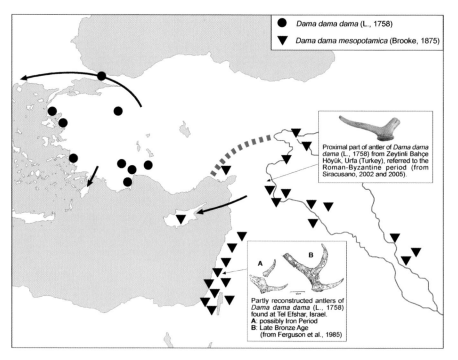

FIGURE 2.1. Past and present distributions of the two subspecies of fallow deer in the Near East based on Uerpmann (1987), Vogler (1997), Masseti (1999; Masseti 2002), and Davis (2003). Arrows indicate the westward routes of exportation of the two deer. The red hatched line indicates the *Nur Dağlari*, a mountain range of south-eastern Anatolia which divides the coastal region of Cilicia from inland Syria, also apparently marking a biogeographic barrier. Ancient introductions of European fallow deer in the natural distributional range of *D. d. mesopotamica* are also indicated.

Mediterranean shores and especially Asia Minor, but it did not subsequently return as far north as its previous range (Uerpmann 1987; Masseti and Rustioni 1988; Stuart 1991).

Current knowledge suggests that the European fallow deer survived in the late Upper Pleistocene in southern Italy, Sicily, and the southern Balkan Peninsula, still persisting also in southern Anatolia (Uerpmann 1987; Masseti and Rustioni 1988; Stuart 1991; Masseti 1996; Vogler 1997; Davis 2003; Starkovich and Stiner 2010; Atici 2011; Trantalidou and Masseti this volume). As far as is presently known, several artistic Epipalaeolithic representations of European fallow deer from southern Italy and Sicily constitute the latest chronological evidence for the persistence of this deer in Italy during the Late Glacial Maximum (*c*.13,000–10,000 BP) (Masseti and Rustioni 1988; Masseti 1996). After this, fallow deer appeared for the first time in the Early Neolithic assemblages, dated between the sixth and fifth millennium BC, of Rendina situated near Melfi in southern Italy (Bökönyi 1977). These fallow deer may represent either the remains of a native population that survived in small refuges of the Italian peninsula, or perhaps, the earliest evidence of the importation of the fallow deer in Italian prehistory. The fact that the Rendina fallow deer were discovered alongside elements of a Near Eastern origin, such as the Asiatic mouflon (*Ovis orientalis*) Gmelin, 1774, and the Bezoar goat (*Capra aegagrus*) (see Masseti 1997), may support the latter scenario (Masseti 1996). Indeed, from this period onwards nearly all European populations of *D. d. dama* are derived from human movement (see next section). The only geographical area where European fallow deer has persisted as a native form is reputed to be the coastal wooded plains of southern Anatolia (Uerpmann 1987), where the borders of its historical range started in the Izmir region and followed the coastal mountains to the south-east (Borovali 1986). According to Chapman and Chapman (1997), by 1934 fallow deer were still thought to survive in the northern and southern foothills of the western Taurus range (see also Banoğlu 1952). Today, only a small population of fallow deer (<100 individuals) (see Arslangündoğlu *et al.* 2010; Albayrak *et al.* 2012) remain in the Turkish wildlife reserve of Düzlerçami in Termessos National Park, Antalya. This population is believed to represent the last autochthonous stock of wild *D. d. dama* in Asia Minor.

Fallow deer and Europe: the role of man

Of all the deer of the world, fallow are perhaps the species whose current distribution has been most influenced by man. Although it has become extirpated in most of its former range, man has taken this species in its semi-domesticated form into most of Europe, the Near East, North and South America, northern and southern Africa, Australia and New Zealand (Chapman and Chapman 1980).

According to archaeological evidence, fallow deer were imported beyond their natural distribution in the early Neolithic (Masseti 1998; Masseti 1999; Guilaine *et al.* 2000). European fallow deer were recorded from the early Neolithic sites of the Aegean, such as Aghios Petros in the Sporades islands, and Knossos on Crete (cf. Masseti 1998; Masseti 1999). There is also evidence for the importation of the European fallow deer from the late sixth to early fifth millennium BC into the island of Rhodes (see Halstead 1987; Halstead and Jones 1987), possibly from Asia Minor (Masseti *et al.* 2008). European fallow deer, in much or all of Europe, are considered exotic animals (Masseti 1996). Hence archaeological remains of European fallow deer can tell us about peoples' migratory and trading patterns in the antiquity. The Romans, for example, exported European fallow deer beyond their original homeland to the territories of western Europe, such as Great Britain (Sykes 2007; Sykes *et al.* 2011) and the Iberian peninsula (Davis and MacKinnon 2009). There is historical evidence confirming that, at least between the fourteenth and seventeenth century, European fallow deer were still captured in Asia Minor, especially in Rhodes, and taken to western Europe to embellish palaces and game parks (Masseti 1996). In the sixteenth century, at the time of Suleiman the Magnificent, the immense palatial complex of the Topkapi, in Istanbul, was still covered by huge wooded areas where herds of fallow deer '*roamed amongst the cypresses and along the shady paths*' (Dash 1999).

Population genetics of the European fallow deer

The extant fallow deer of western Europe are generally regarded as being characterised by a low biochemical variability and lack of polymorphism (Randi and Apollonio 1988; Randi *et al.* 2001). However, a recent molecular survey of mitochondrial DNA (mtDNA) sequence variation in individuals from Turkey, Rhodes, Italy, continental Greece and Hungary revealed substantial diversity and peculiar phylogeographical characteristics (Masseti *et al.* 2006; Masseti *et al.* 2008). This analysis identified three main clusters of mtDNA haplotypes: one representing Turkish individuals, the second representing individuals from Rhodes (which exhibited an 80 base pair insertion, distinct to this cluster) and the third representing all other individuals (including some from Rhodes, which lacked the 80 base pair insertion (see Figure 2.2). The results from the mtDNA analyses show that there is no clear connection between individuals from Turkey and all other locations. The probable foundation of the Rhodian fallow deer population by humans in Neolithic times has resulted in the chance preservation of a significant mitochondrial genetic variability of the species up to the present. The finding that many of the individual fallow deer from Rhodes possess a unique mitochondrial lineage is significant. It is likely that this insertion, found exclusively in animals from Rhodes, derived directly from the first deer introduced into the island during the Neolithic. Preserving these mtDNA lineages is therefore equivalent to sustaining original wild variants, most likely introduced from Anatolia, that are nowadays present only on the island of Rhodes (Masseti *et al.* 2008). The survival of this population represents not only the first documented instance of '*ad hoc* conservation', but it is also significant historically, archaeologically and ecologically.

A recent analysis of ancient specimens from the UK Roman period confirmed that mtDNA sequences cluster together with modern individuals from Italy, Hungary and Greece (Sykes *et al.* 2011). A more recent study revealed a close relationship between mtDNA haplotypes from modern German populations and Turkey (Ludwig *et al.* 2012) (see Figure 2.2) which may reflect the foundation of the German populations about 2,000 years ago by founders of Turkish origin.

Medieval fallow deer from Sardinia

Other than the extant fallow deer of Rhodes, the population of the island of Sardinia, which was extirpated at the end of the 1960s (Reihard and Schenk 1969; Schenk 1976; Carenti *et al.,* this volume), represented the only other population with anthropochorous stock to have persisted on Mediterranean islands since antiquity. As far as is presently known, the first European fallow deer in Sardinia do not pre-date the sixth to eighth century AD. On this island, fallow deer increased in abundance from the fourteenth century onwards (Delussu 1996; Wilkens and Delussu 2003). The analysis of mtDNA data from five Medieval Sardinian samples has shown that these samples are characterised by two mtDNA haplotypes which group with haplotypes from Italy, Hungary and continental Greece (see Figure 2.2).

The occurrence of extant European fallow deer in Italy is probably directly

FIGURE 2.2. Network of mtDNA haplotypes based on sequenced individuals from Turkey, Rhodes (Greece), Italy, Hungary, continental Greece, and Sardinia (Italy). The label 3012 refers to the ancient mtDNA haplotype found in two Middle Age specimens from Sardinia. The location of the archaeological site which yielded these specimens is indicated by ▲ (Geridu, Sorso, north-western Sardinia), while * denotes the location of the preserve of San Rossore, near the town of Pisa. The latter is the supposed gate of entrance of Sardinian fallow deer in continental Italy, according to historical documents.

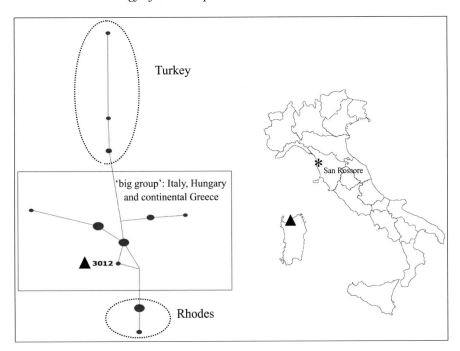

related to the island of Sardinia. The mainland Italian population (comprised within the 'big group' see Figure 2.2) is believed to have originated from the translocations of individuals from the historical preserve of San Rossore, Pisa (Tuscany, Italy) (Perco 1987; Focardi and Toso 1991; Masseti *et al.* 1997). San Rossore itself was established during the Middle Ages with animals entering Pisa harbour upon payment of customs duty (R. Arch. di Stato di Pisa. *Breve gabellarum porte Degathie de maris.* AD 1362; Simoni 1910; Apollonio and Toso 1988; Masseti 1996; Masseti 2003). The merchandise which normally entered Pisa through the gate called '*Porta Degathie de maris*' was of overseas origin, mainly imported from Sicily and/or Sardinia. Simoni (1910) invoked the latter island as the most likely origin of the San Rossore fallow deer population. Our data can be interpreted as being concordant with an origin of mainland Italian fallow deer from Sardinia although other hypotheses, which take into account different historical documents, should be considered.

The fifteenth century Italian artistic representations of the European fallow deer may suggest a Near Eastern origin (e.g. School of Gentile Bellini, c.1513–1516, *Reception of the Venetian embassy in Cairo.* Paris, Musée du Louvre; and/or the cycle of canvas (1502–1507) by Vittore Carpaccio at the Scuola di San Giorgio degli Schiavoni, Venice).

Alternatively, the origin of the Italian fallow deer may be related to the Moors. Commenting on the encomiastic poem *Ambra* written in 1485 by Angiolo Poliziano for Lorenzo the Magnificent, Targioni-Tozzetti (1773) suggested other sources for the origin of the fallow deer population of Tuscany: '*Atque aliud nigris missum (quis credat ?) ab Indis ruminat ignotas armentum*

discolor herbas' ('*And another ungulate with the coat of various colour, arrived here [who can believe it?] from the black Indies, is ruminating herbs unknown to it*') (Poliziano 1485, *Ambra*). The 'black Indies' might be identified in some of the countries ruled at that time by the Moors, such as North Africa, the Near East or even the Iberian Peninsula.

The Moors were also known to have helped spread the practice of keeping European fallow deer in large enclosed areas (deer parks) during the Middle Ages (Zeuner 1963). Around this time, the distribution and popularity of fallow deer increased because of its connections with hunting and providing entertainment for European kings and their courts (cf. Cummins 1988; Masseti 1996). In various courts, in Austria, Spain, and Medici's Italy, one of the favourite activities was the 'fallow deer race' (Lamberini 1975; Rubio Aragonés 1996; Masseti 2002), which could be performed with the assistance of dogs. An illustration of this can be found in the famous painting '*La caceria del tabladillo*' by Juan Buaptista Martinez del Mazo (*c*.1612–1667), currently housed in the Prado Museum in Madrid (cat. no. 2571). The painting portrays the courtly pursuit, as it was practised in the gardens of the royal palace of Aranjuez, at the time of Phillip IV. The race involved constraining a number of fallow deer and forcing them to run along straight courses which were bordered by fabric boundaries or other provisional structures and thick hedges. At the end of the race the hunters lay in wait to put the ungulates to the sword, to the amusement and delight of the noble women watching the spectacle from a raised platform (*tabladillo*) (Figure 2.3). In the Medici villa of Poggio a Caiano, close to Florence (Tuscany, central Italy) during the period of the Grand-dukes, fallow deer were purposely bred for this past time and there existed an entire apparatus designed for the performance of the 'fallow deer races' (ASF, RR Possessioni, n. 3556,

FIGURE 2.3. Detail of *La caceria del tabladillo*, painting by Juan Bauptista Martinez del Mazo (1612 *c.*–1667), Prado Museum, Madrid (cat. no. 2571) (courtesy Prado Museum, reproduced by permission).

relazione del luglio 1715 indirizzata agli Ill. Sigg. Deputati delle Possessioni di SAR; Père Labat 1730; Lamberini 1975; Masseti 2002; 2003). It seems that the tradition of 'fallow deer races' originated in antiquity. In fact, according to Pliny the Elder *(Naturalis historia*, VIII, 149–150*)*, they were already performed by the king of *Albania* or *Albani* (cf. Borghini *et al.* 1983), a country located west of the Caspian Sea (cf. McEvedy 1967), and even by Alexander the Great, on his way to India.

Concluding remarks

It would appear that the artificial diffusion of the European fallow deer beyond its natural homeland begun in prehistoric times with the transfer of the extant Rhodian population from Anatolia (Masseti *et al.* 2008). In this case, we can hypothesise that humans unknowingly preserved a remarkable portion of the original genetic diversity of the source population. As far as the more recent diffusion of the species in the western world is concerned, it cannot be ruled out that the popularity of the European fallow deer for aristocratic leisure pursuits began in Sicily with the Arabs and the Normans and spread then from the twelfth century on to Great Britain (possibly through the Normans from Sicily: Rackham 1997; Sykes 2007; Masseti 2009), Italy, Spain, France and the remaining European countries. It went on to become the most widespread cervid in the parks and preserves of Europe.

Acknowledgements

In the course of this project we have been fortunate to have the help of many friends and colleagues.

We should like to express special thanks to Francesca Alhaique, Beni Levent Atici, Norma Chapman, Barton Mills, Simon Davis, Matteo Girardi, Clarice Innocenti, Giovanni Siracusano, Naomi Sykes, and Barbara Wilkens for their contributions in material and information.

References

Albayrak, T., Giannatos, G. and Kabasakal, B. (2012) 'Carnivore and ungulate populations in the Beydaglari Mountains (Antalya, Turkey): border region between Asia and Europe', *Polish Journal of Ecology* 60, 419–428.

Apollonio, M. and Toso, S. (1988) 'Analisi della gestione di una popolazione di daini e delle sue conseguenze sui parametri demografici e biometrici', in *Atti del I Convegno Nazionale dei Biologi della Selvaggina*, eds M. Spagnesi and S. Toso *Supplementi alle Ricerche di Biologia della Selvaggina* 14, 525–539.

Arslangündoğlu, Z., Kasparek, M., Saribaşk, H., Kaçar, M. S., Yöntem, O. and Şahin, M. T. (2010) 'Development of the population of the European fallow deer, *Dama dama dama* (Linnaeus, 1758), in Turkey', *Zoology in the Middle East* 49, 3–12.

Atici, L. (2011) *Before the Revolution: Epipalaeolithic Subsistence in the Western Taurus Mountains, Turkey*, BAR International Series 2251, Archaeopress, Oxford

Banoğlu, N. A. (1952) *Turkey: A Sportsman's Paradise*, The Press, Broadcasting and Tourism Department, Ankara.

Bökönyi, S. (1977) 'The Early Neolithic fauna of Rendina: a preliminary report', *Origini* 82, 345–350.

Borghini, A., Giannarelli, E., Marcone, A. and Ranucci, G. (1983) *Gaio Plinio Secondo. Storia naturale. II. Antropologia e zoologia. Libri 7–11*. Giulio Einaudi editore, Torino.

Borovali, Ö. (1986) 'Turkey', in *The SCI record book of trophy animals*, Tucson, AZ, Safari Club International, 528–530.

Butzer, K. W. (1972) *Environment and Archaeology: An Ecological Approach to Prehistory*, 2nd edition, Methuen, London.

Chapman, D. I. and Chapman, N. G. (1997) *Fallow deer*, Coch-y-bonddu Books, Machynlleth.

Chapman, N. G. and Chapman, D. I. (1980) 'The distribution of fallow deer: a worldwide review', *Mammal Review* 10, 61–138.

Cummins, J. (1988) *The Hound and the Hawk: The Art of Medieval Hunting*, Weidenfeld and Nicolson, London.

Danford, C. G. and Alston, E. R. (1880) 'On the Mammals of Asia Minor', Part II, *Proceedings of the Scientific Meetings of the Zoological Society of London* 1880, 50–64.

Dash M. (1999) *Tulipomania. The Story of the World's Most Coveted Flower and the Extraordinary Passions it Aroused,* Crown Publisher, New York.

Davis, S. (2003) 'The zooarchaeology of Khirokitia (Neolithic Cyprus), including a view from the mainland', in *Le Néolithique de Chypre*, eds J. Guilaine and A. Le Brun (avec la collaboration de O. Daune-Le Brun), Actes du Colloque International organisé par le Département des Antiquités de Chypre et l'École Française d' Athenes (Nicosie 17–19 May 2001), 253–268.

Davis, S. and MacKinnon, M. (2009) 'Did the Romans bring fallow deer to Portugal?', *Environmental Archaeology* 14, 15–26.

Delussu, F. (1996) 'Il villaggio medievale di Geridu (Sorso, SS). Campagne di scavo 1995/1996: relazione preliminare. I resti faunistici', *Archeologia Medievale* 23, 530–533.

Focardi, S. and Toso, S. (1991) 'Daino' in *I Cervidi: biologia e gestione*, eds M. Spagnesi and S. Toso, Istituto Nazionale di Biologia della Selvaggina, *Documenti tecnici*, 8, 35–56.

Guilaine, J., Briois, F., Vigne, J. -D. and Carrère, I. (2000) 'Découverte d'un Néolithique précéramique (fin 9e, début 8e millénaires cal. BC), apparenté au PPNB ancien/moyen du levant nord', *Comptes Rendu de l'Académie des Sciences Paris, Sciences de la Terre et des planètes* 330, 75–82.

Halstead, P. (1987) 'Man and other animals in later Greek Prehistory', *Annual of the British School at Athens* 82, 71–83.

Halstead, P. and Jones, G. (1987) 'Bioarchaeological remains from Kalythies Cave, Rhodes', in *The Neolithic period in the Dodecanese* (in Greek), ed. A. Sampson, Ministry of Culture, Athens, 135–152.

Harrison, D. L. (1968) *The mammals of Arabia. Vol. II. Carnivora – Artiodactyla – Hyracoidea*, Ernest Benn Ltd, London.

Lamberini, D. (1975) 'Le Cascine di Poggio a Caiano – Tavola', *Prato. Storia e Arte* 16, 3–77.

Leonardi, G. and Petronio, G. (1976) 'The fallow deer in European Pleistocene', *Geologoca Romana* 15, 1–67.

Ludwig, A., Vernesi, C., Lieckfeldt, D., Lattenkamp, E. Z., Wiethölter, A. and Walburga, L. (2012) 'Origin and patterns of genetic diversity of German fallow deer as inferred from mitochondrial DNA', *European Journal of Wildlife Research* 58, 495–501.

Masseti, M. (1996) 'The post-glacial diffusion of the genus Dama Frisch, 1775, in the Mediterranean region', *Supplemento alle Ricerche di Biologia della Selvaggina* 20, 7–29.

Masseti, M. (1997) 'The prehistoric diffusion of the Asiatic mouflon, *Ovis gmelini* Blyth, 1841, and the Bezoar goat, *Capra aegagrus* Erxleben, 1777, in the Mediterranean area beyond their natural distributions', in *Proceedings of the Second International Symposium on Mediterranean Mouflon 'The Mediterranean mouflon: management and conservation'*, ed. E. Hadjisterkotis, Game Fund of Cyprus/IUCN Species Survival Commission, Caprinae Specialist Group, Nicosia, 1–19.

Masseti, M. (1998) 'Holocene endemic and anthropochorous wild mammals of the Mediterranean islands', *Anthropozoologica* 28, 3–20.

Masseti, M. (1999) 'The fallow deer, *Dama dama* L., 1758, in the Aegean region', *Contributions to the Zoogeography and Ecology of the Eastern Mediterranean Region* 1, Supplement, 17–30.

Masseti, M. ed. (2002) *Island of deer. Natural history of the fallow deer of Rhodes and of the vertebrates of the Dodecanese (Greece),* City of Rhodes, Environmental Organization, Rhodes (Greece).

Masseti, M. (2003) *Fauna toscana: Galliformi non migratori, Lagormorfi e Artiodattili,* ARSIA (Agenzia Regionale per lo Sviluppo e l'Innovazione nel settore Agricolo-forestale)/Regione Toscana, Firenze.

Masseti, M. (2009) 'In the gardens of Norman Palermo, Sicily (twelfth century AD)', *Anthropozoologica* 44, 7–34.

Masseti, M. and Rustioni, M. (1988) 'Considerazioni preliminari sulla diffusione di *Dama dama* (Linnaeus, 1758) durante le epoche tardiglaciale e postglaciale nell'Italia mediterranea', *Studi per l'Ecologia del Quaternario* 10, 93–119.

Masseti, M., Pecchioli, E. and Vernesi C. (2008) 'Phylogeography of the last surviving populations of Rhodian and Anatolian fallow deer (*Dama dama dama* L., 1758)', *Biological Journal of the Linnean Society* 93, 835–844.

Masseti, M., Cavallaro, A., Pecchioli, E. and Vernesi, C. (2006) 'Artificial occurrence of the fallow deer, *Dama dama dama* (L., 1758), on the island of Rhodes (Greece): insight from mtDNA analysis', *Human Evolution* 21, 167–175.

Masseti, M., Pecchioli, E., Romei, A., Tilotta, G., Vernesi, C. and Chiarelli, B. (1997) 'RAPD (Random Amplified Polymorphic DNA) fingerprinting analysis of some Italian populations of fallow deer *Dama dama*', *Italian Journal of Zoology* 64, 235–238.

McEvedy, C. (1967) *The Penguin Atlas of Ancient History.* Penguin Books, Harmondsworth (UK).

Perco, F. (1987) *Ungulati,* Carlo Lorenzini editore, Udine.

Rackham, O. (1997) *The History of the Countryside: The Classic History of Britain's Landscape, Flora and Fauna,* Phoenix, London.

Randi, E. and Apollonio, M. (1988) 'Low biochemical variability in European fallow deer (*Dama dama*): natural bottlenecks and the effect of domestication', *Heredity* 67, 405–410.

Randi, E., Mucci, N., Claro-Hergueta, F., Bonnet, A. and Douzery, E. J. P. (2001) 'A mitochondrial DNA control region phylogeny of the Cervinae: speciation in *Cervus* and implications for conservation', *Animal Conservation* 4, 1–11.

Rubio Aragonés, M. J. (1996) *La caza y la Casa Real. Una visión de la caza através de los Reyes de España.* Ayuntamento de Badajoz, Badajoz.

Schenk, H. (1976) 'Analisi della situazione faunistica della Sardegna. Uccelli e mammiferi' in *SOS fauna. Animali in pericolo in Italia*, F. Pedrotti ed. Edizione WWF, Camerino, 465–556.

Simoni, D. (1910) *San Rossore nella storia*, Leo S. Olschki editore, Firenze.

Starkovich, B. M. and Stiner, M. C. (2010) 'Upper Paleolithic animal exploitation at Klissoura Cave 1 in southern Greece: Dietary trends and mammal taphonomy', *Eurasian Prehistory* 7, 107–132.

Stuart, A. J. (1991) 'Mammalian extinctions in the Late Pleistocene of Northern Eurasia and North America', *Biological Reviews of the Cambridge Philosophical Society (London)* 66, 453–562.

Sykes, N. J. (2007) The Norman Conquest: A Zooarchaeological Perspective. BAR International Series 1656, Archaeopress, Oxford.

Sykes, N. J., Baker, K. H., Carden, R. F., Higham, T. F. G., Hoelzel, A. R. and Stevens, R. E. (2011) 'New evidence for the establishment and management of the European fallow deer (*Dama dama dama*) in Roman Britain', *Journal of Archaeological Science* 38, 156–165.

Targioni-Tozzetti, G. (1773) *Relazioni d'alcuni viaggi fatti in diverse parti della Toscana, per osservare le produzioni naturali e gli antichi monumenti di essa*, volume V. Firenze.

Uerpmann, H. P. (1981) 'The major faunal areas of the Middle East during the late Pleistocene and early Holocene', *Préhistoire du Levant*, Paris, Centre Nationale de la Recherche Scientifique, 99–106.

Uerpmann, H. P. (1987) *The Ancient Distribution of Ungulate Mammals in the Middle East*. Beihefte zum Tübinger Atlas des Vorderen Orients, Reihe A 27. Dr. Ludwig Reichert Verlag, Wiesbaden.

Vogler, U. (1997) Paunenhistorische Untersuchungen am Sirkeli Höyük/Adana, Türkei (4.–1. Jahrtausend v. Chr.). Unpublished PhD dissertation. Ludwig-Maximilians-Universitat, München.

Wilkens, B. and Delussu, F. (2003) 'Wild and domestic mammals in holocenic Sardinia', in *The New Panorama of Animal Evolution*, eds A. Legakis, S. Sfenthourakis, R. Polymeni and M. Thessalou-Legaki, Proceeding of the 18th International Congress of Zoology, Pensoft, Sofia – Moskow, Athens, 303–308.

Zeuner, F. E. (1963) *A History of Domesticated Animals*, Hutchinson, London.

Human–Deer Interactions in Sardinia

*Gabriele Carenti, Elisabetta Grassi, Stefano Masala
and Barbara Wilkens*

Introduction

The Sardinian red deer (*Cervus elaphus corsicanus*) and the European fallow deer (*Dama dama dama*) are not native to Sardinia, both having arrived on the island in antiquity. The Sardinian red deer represents an endangered population, which is somewhat distinct from other continental European red deer (*Cervus elaphus*) populations (see Hmwe *et al.* 2006), whilst the European fallow deer is known to have been introduced to Sardinia during the Medieval period (Wilkens 2012a, 113). The natural historian Francesco Cetti wrote in 1774 that fallow deer were abundant across the island with at least 3,000 being culled annually. Unfortunately, this species became extinct in 1968 following intense hunting and deforestation. This paper will review evidence from the archaeozoological record to provide insight into the natural and cultural history of these two species in Sardinia.

The origins of the Sardinian-Corsican red deer

Phylogenetic and phylogeographic reconstructions of the Sardinian red deer, carried out by different scholars, using different methodologies, have found close relationships with the North African red deer (Hartl *et al.* 1995; Ludt *et al.* 2004, 1075; Zachos and Hartl 2011); with the red deer from continental Italy (currently present in Gran Bosco della Mesola) and Spain (Hmwe *et al.* 2006, 691; Hajji *et al.* 2008, 669; Zachos *et al.* 2003) and even Bulgarian red deer, which are phylogenetically very similar to the Eastern subspecies (Hartl *et al.* 1995), even though the latter hypothesis has not been confirmed by later studies (Hajji *et al.* 2008, 660).

By integrating genetic, palaeontological and historical/archaeological data Hajji *et al.* (2008, 669) propose an anthropic introduction of the red deer on the island. These authors propose an arrival from the Italian peninsula on the basis of morphological similarities observed by Vigne (1988, 231; 1999, 314) in Neolithic red deer from Southern Italy (mainly from zooarcheological data from the site of Torre Sabea). Contrastingly, Wilkens (1989) has noted substantial

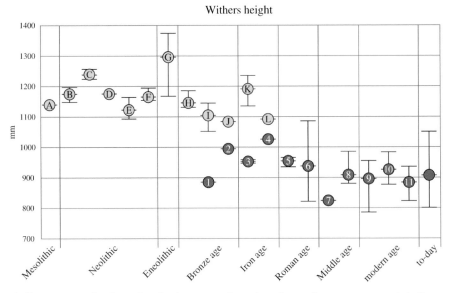

FIGURE 3.1. Withers height: comparison between Italian and Sardinian deer heights.

differences in both individual size and antler shape between material from Central-Southern Italian localities and Sardinia.

In terms of animal height (measured at the withers), it appears that mainland red deer have never been smaller than 1 m (average 110–120 cm, with a peak in the Eneolithic at Conelle) whereas Sardinian red deer are rarely larger than 1 m (Wilkens 1989, 72; Figure 3.1). Sardinian red deer also have smaller antlers, with the palm of the distal part being less evident and rarer compared to its mainland Italian relative (see Figure 3.2). This phenotypic differentiation must have evolved with time and since the red deer were first introduced to the island. Evidence for this comes from size differences, which become more pronounced over time (Wilkens 1989, 72).

Historical and archaeological analysis of the arrival of the red deer in Sardinia

Even though the most widely accepted hypotheses point towards the arrival of the Sardinian red deer in the Neolithic, it is still an open question whether red deer could have reached Sardinia naturally. An earlier (pre-Neolithic) presence in Sardinia and Corsica cannot be excluded with certainty due to the discovery of *Cervus elaphus rossii* in Corsica in the Middle Pleistocene (Pereira 2001). Another early find of *Cervus elaphus* in Sardinia dates to 8040±80 BP from a site characterised by domestic fauna and pottery (Sanges 1987, 826). It is not however until the Neolithic that Sardinian red deer appear more frequently in the archaeological record: e.g. finds from Early and Middle Neolithic Grotta Corbeddu (Sanges 1987); from Middle Neolithic Su Coloru (Masala 2011) and from Late Neolithic Grotta Verde (Wilkens 2012b) (see Table 3.1).

During the Bronze and Iron Ages the Nuragic civilisation developed on the island. The Nuragic economy appears to have been heavily based on agriculture

FIGURE 3.2. Italian and Sardinian deer antler: 1. Neolithic antler from S. Maria in Selva (Italy); 2. Sardinian antler from Nuraghe S. Antine (Bronze Age) and; 3. from Sorso (Roman Age).

and caprine, bovine and pig husbandry. It appears that hunting was of secondary importance and, when it occurred, it focused on the Sardinian red deer and the wild boar (*Sus scrofa meridionalis*) (Wilkens 2003a, 187). All forms of fauna held a symbolic importance to the Nuragic people who represented different animal species, and also red deer, in bronze artefacts (Lo Schiavo and Manconi 1999–2000). Use of the lost wax technique spread on the island during this period, with the so-called 'bronzetti', or bronze figurines, being produced for use mainly as votive offerings in sacred areas.

With the arrival of the Phoenicians, and subsequently the Punics, the exploitation of red deer through hunting became increasingly important and widely practised, particularly in the interior sites of Sardinia. The Sulcis area provides an example of the exploitation of red deer connected with the knowledge and control of the territory (Carenti and Wilkens 2006; Carenti 2012a): the inland sites, placed on the mining and agricultural routes, specialised in deer hunting for the production of meat and hides, while the coastal towns, which functioned as ports, provide evidence of antler working (Wilkens 2008, 252–254). This specialisation may have been inherited from the Nuragic populations (e.g. Palmavera, eleventh to ninth century BC, Figure 3.4) since in the Archaic period the production of antler handles was highly advanced, as is exemplified at Sulky (eighth century BC, Carenti and Unali 2013), and continued throughout the Punic and Roman periods (Figure 3.4). In these contexts deer may also have been of ritual significance: for instance, an antler from a juvenile Sardinian red deer was recovered from a votive deposit in Nuraghe Sirai (seventh to sixth century BC) and astragali were deposited in a burial at Monte Sirai (sixth century BC, Carenti and Guirguis in press).

The Roman conquest of the island and its annexation as a province brought no significant change in deer exploitation, despite the extensive urbanisation of

NR	Site	Period	Domestic mammals	Other wild mammals	Cervidae	Cervus elaphus	Dama dama	Total	References
1	Su Coloru (Perfugas)	Neolithic	286	8		2		296	Masala 2011
2	Grotta Verde (Alghero)	Neolithic	2109	270		5		2384	Wilkens 2012b
3	Grotta Corbeddu (Oliena)	Neolithic	X	X		X			Sanges 1987
4	Monte d'Accoddi (Sassari)	Eneolithic	2573	36		5		2614	Wilkens 2003b
5	Flumenelongu (Alghero)	Nuragic	97	21		6		124	Masala in Wilkens 2012a, 90–91
6	Madonna del Rimedio (Oristano)	Middle Bronze	274	1		2		277	Santoni and Wilkens 1996
7	Bruncu Madugui (Gesturi)	Middle Bronze	1076			74		1150	Fonzo 1987
8	Nuraghe Miuddu (Birori)	Final Bronze	591	4		39		634	Delussu 1997
9	Sant'Imbenia (Alghero)	Middle/Final Bronze	2037	21		169		2227	Manconi 2000
10	Nuraghe Arrubiu (Orroli)	Middle/Late Bronze	2231	8		131		2370	Fonzo 2003
10	Nuraghe Arrubiu (Orroli)	Final Bronze	861	9		38		908	Fonzo 2003
10	Nuraghe Arrubiu (Orroli)	Final Bronze/Iron	393	5		62		460	Fonzo 2003
11	S. Anastasia (Sardara)	Iron	102			27		129	Fonzo 1987
12	Nuraghe Is Paras (Isili)	Broze/Iron	1992	109		218		2319	Wilkens 2003b
13	S. Antonio (Siligo)	Iron	738	7		7		752	Wilkens 2003b
14	Genna Maria (Villanovaforru)	Iron	252	11		90		353	Fonzo and Vigne 1993
15	Sant'Antine (Torralba)	Final Bronze	1628	308	53	77		2066	Masala in Wilkens 2012a, 89
26	Nora (Pula)	Iron	395		1	26		422	Sorrentino 2009
16	Tharros (Oristano)	Phenician-Punic	621	2		17		640	Farello 2000
17	Nuraghe Sirai (Carbonia)	Phenician-Punic	1041	29		538		1608	Carenti 2005
18	Monte Sirai (Carbonia)	Phenician-Punic	453	4		426		883	Carenti and Wilkens 2006
19	Cronicario (S. antioco)	Phenician-Punic	273	5		36		314	Wilkens 2008
20	Santu Pedru (Alghero)	Phenician-Punic	495			253		748	Masala in Wilkens 2012a, 97–98
26	Nora (Pula)	Punic	59		2	4		65	Sorrentino 2009
5	Flumenelongu (Alghero)	Roman	197	41		27		265	Masala in Wilkens 2012a, 102
21	Purissima (Alghero)	Roman	876	13		313		1202	Masala 2012
22	Nuraghe Mannu (Cala Gonone)	2nd–4th AD	840	77		52		969	Delussu 2000
23	S'Imbalconadu (Olbia)	2nd–1st BC	172	2		32		206	Manconi 1997
24	Turris Libissonis (Porto Torres)	2nd BC	2974	61		55		3090	Delussu 2005
25	Su Cuguttu (Olbia)	2ndAD	379	1		9		389	Manconi 1996
26	Nora (Pula)	Roman	1188		1	78		1267	Sorrentino 2009
27	Santa Filitica (Sorso)	Vandalic	696	6	4	11		717	Wilkens 2003b
27	Santa Filitica (Sorso)	Bizantine	884	10	7	16		917	Wilkens 2003b
28	Villa Sant'Imbenia (Alghero)	Early Middle Age	1383	8	10	34	1	1436	Grassi in Baldino et al. 2008, 113–116

NR	Site	Period	Domestic mammals	Other wild mammals	Cervidae	Cervus elaphus	Dama dama	Total	References
29	Tergu Santa Maria	Early Middle Age	438	5	2	19		464	Wilkens in Baldino *et al.* 2008, 116–119
29	Tergu Santa Maria	Middle Age	383	3	1	5		392	Wilkens in Baldino *et al.* 2008, 122–123
29	Tergu Santa Maria	Middle Age	205	12	1	2	2	222	Wilkens in Baldino *et al.* 2008, 122–123
30	Sassari Cappuccine	12th AD	849	3	15	4	9	880	Carenti 2009
30	Sassari Piazza Duomo	12th–13th AD	379	8		2	3	392	Wilkens in Baldino *et al.* 2008, 131
31	San Nicola di Trullas (Semestene)	14th AD	367	1	3	2	4	377	Carenti in Pandolfi *et al.* 2007, 190–194; Carenti 2010a
32	S.Maria di Seve (Banari)	14th AD	23	1	15	12	24	75	Wilkens and Delussu 2000
33	Ardu (Sassari)	14th AD	828	1	23	8	29	888	Grassi in Rovina and Grassi 2006, 169–171
34	Olmedo	14th AD	96	2	14	21	3	136	Wilkens in Baldino *et al.* 2008, 127–128
35	Maddalena (Alghero)		984	1	3	11	15	1014	Wilkens in Baldino *et al.* 2008, 128–129
30	Sassari via Satta	14th AD	3156	21	37	40	34	3288	Wilkens in Baldino *et al.* 2008, 131–132
30	Sassari Castello	14th AD	1057	3	19	1	15	1095	Grassi 2012a
36	Taniga (Sassari)		46			1		47	Wilkens in Baldino *et al.* 2008, 133
37	Geridu (Sorso)	14th–15th AD	5088	8	87	59	120	5362	Delussu 1996; Grassi 2004
38	Monteleone Roccadoria		3482	2	262	99	130	3975	Grassi and Secchi in Baldino *et al.* 2008, 136–139
30	Sassari Largo Pazzola	15th AD	613		7	1	2	623	Grassi 2012a
29	Tergu Santa Maria	Post-Medieval	681	4	5	11	6	707	Wilkens in Baldino *et al.* 2008, 144
34	Olmedo	16th AD	290	2	52	40	46	430	Wilkens in Baldino *et al.* 2008, 145
39	Bosa Castello		272	2		2		276	Delussu 2000
40	Saccargia (Codrongianus)		9119	22	384	114	1165	10804	Baldino 2005
41	Castelsardo		3616	13	41	5	16	3691	Carenti 2010b
30	Sassari Cappuccine	17th AD	2345		37	24	27	2433	Carenti and Grassi in Baldino *et al.* 2008, 152
30	Sassari vicolo Canne		663	2	5	4	5	679	Orgolesu in Baldino *et al.* 2008, 153–154
30	Sassari Palazzo Ducale		224	2	2			226	Wilkens 2001
30	Sassari Infermeria San Pietro		648	2		3	1	654	Cambule in Wilkens 2012a, 133–134
30	Sassari Castello	16th–17th AD	3916	4	49	21	15	4005	Grassi 2012a
35	Alghero bastioni S. Giacomo	16th–17th AD	3183	22	61	16	42	3324	Carenti 2012b
30	Sassari Castello	18th–19th AD	1826	2	3	6	1	1838	Grassi 2012a
26	Nora (Pula)	Medieval–Post medieval	354			2		356	Sorrentino 2009

TABLE 3.1. List of sites by period and NISP of domestic mammals versus cervids and other wild mammals. The first column shows the identification numbers of the sites (NR), which are represented in Figure 3.3.

the territory. As before, red deer are better represented in assemblages from rural settlements located in the interior of the island than those of an urban nature. The only variation in this pattern comes from the so-called 'sacred well' of *La Purissima* (Alghero), where the abundance of postcranial skeletal deer remains and the scarce remains of antlers are in contrast with the considerable presence of sheep horn-cores. This would seem to be related to ritual activity (Masala 2012).

Few data are available for the Early Middle Ages but it is observed that in the north of the island deer hunting was of secondary importance although one particular exception is provided by the deposits from the coastal villa of Santa Filitica (Rovina *et al.* 2011). Numerous finished artefacts and waste products were recovered from this site suggesting that a notable specialisation in antler working occurred during the Early Middle Age (Vandal and Byzantine phases) (Figure 3.4) (Rovina *et al.* 2011, 256–261).

The Medieval period (twelfth to fifteenth century AD) witnessed the introduction of the European fallow deer, which appears alongside the Sardinian red deer in all the contexts examined, often in greater numbers than the latter. To this day, how and why European fallow deer were introduced into the island is unknown. Throughout the Medieval period records from towns, villages and monasteries testify to a constant exploitation of cervids

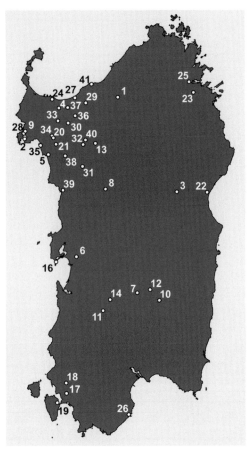

FIGURE 3.3. Location map with the sites mentioned in the text, numbers are explained in Table 3.1.

(Baldino *et al.* 2008), but one, which was always marginal in comparison to livestock husbandry. An exception is provided by the animal remains that blocked the well of the cloister of the Santa Maria di Seve monastery (Banari, province of Sassari) from where whole antlers from at least two Sardinian-Corsican red deer and eight fallow deer were recovered (Wilkens and Delussu 2000). The origins of the remains are as of yet unknown. Both shed antlers and those cut from culled animals are present, some with traces of working, while the post-cranial bones are generally whole and do not present butchering marks.

During the Modern Era (sixteenth to nineteenth century AD) the frequencies of cervids amongst zooarchaeological assemblages diminish considerably, reaching minimal percentages in urban contexts: e.g. Bosa, Castelsardo, Sassari and Alghero (see Baldino *et al.* 2008). Some sites however remain an exception. For example, cervid finds from the village of Olmedo between the fourteenth and sixteenth centuries are numerous: notably, during the fourteenth century the majority of finds are of Sardinian red deer (Wilkens 2012a, 123), however, by the sixteenth century both cervid species are represented in equal proportions (Wilkens 2012a, 126; Baldino *et al.*, 2008, 145). The high concentration of cervids is probably due to geographical location: deer were already frequent in the area (Nuraghe Talia) during the Roman period (Wilkens 2012a, 107). Furthermore, information left

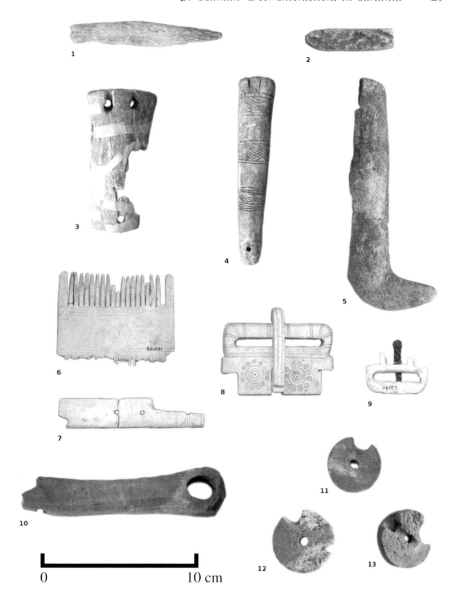

FIGURE 3.4. Antler artefacts: 1. spatula, Grotta verde (Neolithic); 2. spatula, Monte d'Accoddi (Eneolithic); 3. knife handle, Nuraghe Palmavera (Bronze Age); 4–5. knifes handles, S. Antioco (Iron Age); 6–7. combs, 8–9. buckles, S. Filitica (Middle Age); 10 part of a horse or donkey trapping; 11–13. crossbow nuts, Sassari castle (Modern Age).

by the natural historian F. Cetti (1774) suggests that by the eighteenth century Sardinian red deer were restricted to certain parts of the island e.g. the Nurra plain. At Saccargia, on the other hand, there is a clear prevalence of fallow deer over Sardinian red deer (Baldino 2005; Baldino *et al.* 2008, 147–149). An increased abundance of both Sardinian red deer and fallow deer is also noted at Sassari castle with one particular context, the fill of an external ditch (area 1500–1600 AD), being used as a refuse site for a craft area (Grassi 2012a; Grassi 2012b). These cervid remains are mostly fragments of frontal bone, detached from the base of the antler, and fragments of antler that show clear signs of working. Finished products are also present, such as crossbow nuts and what are probably parts of horse or donkey trappings (Figure 3.4) (Grassi 2012a).

The zooarchaeological record from more recent periods is scarce, however, events, which took place in the twentieth century AD, are known. In 1939 Italian law protected both the Sardinian red deer and the fallow deer, furthermore, the Sardinian red deer was listed as endangered by the IUCN (Baille and Groombridge 1996, 226). However by 1968 uncontrolled hunting, deforestation and illegal hunting led to the complete extinction of the fallow deer and a drastic reduction in population size for the Sardinian red deer (Puddu and Viarengo 1993, 245). Due to rigorous protection, the Sardinian red deer population has increased notably and it is currently present in five areas and twelve corrals in various parts of the island managed by the *Ente Foreste della Sardegna* (Puddu *et al.* 2009, fig. 1; 2011). Some fallow deer were reintroduced in the late 1980s from the mainland (Puddu *et al.* 2009), and today these may be found in seven areas which host captive and free-ranging exemplars (Apollonio 2003).

Conclusions

The data currently available concerning red and fallow deer in Sardinia are anything but abundant and definitive. Undoubtedly, obtaining further Sardinian red deer material from excavations, particularly from the Neolithic, would provide valuable information. One such potential site is Grotta Verde, near Alghero, which is currently being excavated and some cervid remains, probably red deer, have been recovered (Wilkens *pers. comm.*). Future work on Sardinian red deer is required to elucidate relationships with red deer found elsewhere in the Mediterranean. Such work could focus efforts on both further archaeological investigation and ancient DNA. On this last point, it is important to underline the fact that the genetic data available today relates to modern day exemplars and greater use of DNA extracted from ancient samples would be highly desirable, especially in light of the large number of deer samples available from the Palaeolithic.

References

nell'antichità. Atti del Convegno dei Giovani Archeologi, Sassari 27–30 settembre 2006, eds M. G. Melis and Muros, 563–565.

Carenti, G. (2010a) 'I resti faunistici. Primi dati archeozoologici', in *San Nicola di Trullas. Archeologia architettura paesaggio*, eds A. Boninu and A. Pandolfi, Sassari, 129–140.

Carenti, G. (2010b) 'Le faune e l'alimentazione, in Castelsardo'. *Archeologia di una fortezza dai Doria agli Spagnoli* ed. M. Milanese, *Sardegna Medievale 2*, Sassari, 95–98.

Carenti, G. (2012a) 'Lo sfruttamento del cervo sardo nel Sulcis. Controllo del territorio ed espressione di potere', in *L'Africa Romana. Trasformazione dei paesaggi del potere nell'Africa settentrionale fino alla fine del mondo antico*. Atti del XIX convegno di studio. Sassari 16–19 dicembre 2010, eds M. B. Cocco, A. Gavini, A. Ibba, Roma, Carocci Editore, 2945–2952.

Carenti, G. (2012b) 'L'area 20100 del bastione S. Giacomo ad Alghero: le faune tra XVI e XVII secolo', in *Atti del VI Convegno Nazionale di Archeozoologia*, eds J. De Grossi Mazzorin, D. Saccà and C. Tozzi, Parco dell'Orecchiella, San Romano in Garfagnan, Lucca, Italy, 291–296.

Carenti, G. and Guirguis, M. (in press) 'La necropoli di Monte Sirai tra il tardo arcaismo e la prima età punica: sepolture infantili, riti funerari, studi archeozoologici', in *L'archeologia funeraria in Sardegna. Società dei vivi, comunità dei morti: un rapporto [ancora?] difficile*, Atti del Convegno di Studi, Sanluri 8–9 Aprile 2011.

Carenti, G. and Unali, A. (2013) 'Ancient trade and crafts in Sardinia: The evidence from Sulcis', in *Identity and Connectivity: Proceedings of the 16th Symposium on Mediterranean Archaeology*, eds L. Bombardieri, A. D'Agostino, G. Guarducci, V. Orsi, S. Valentini, Florence, Italy, 1–3 March 2012, BAR International Series S2581, Archaeopress, Oxford, 723–732.

Carenti, G. and Wilkens, B. (2006) 'La colonizzazione fenicia e punica e il suo influsso sulla fauna sarda', *Sardinia, Corsica et Baleares Antiquae* 4, 173–186.

Cetti, F. (1774) *I quadrupedi di Sardegna*, Piattoli, Sassari.

Delussu, F. (1996) 'I resti faunistici', in *Il villaggio medievale di Geridu (Sorso, SS). Campagne di scavo 1995/1996: relazione preliminare*, ed. M. Milanese, *Archeologia medievale* 23, 530–533.

Delussu, F. (1997) 'Le faune dell'Età del Bronzo del Nuraghe Miuddu (NU)', *Rassegna di Archeologia*, 14, 189–204.

Delussu, F. (2000) 'Lo stato attuale degli studi sulle faune oloceniche della Sardegna centro-settentrionale', in *Atti del 2° Convegno Nazionale di Archeozoologia* (Asti 1997), Forlì, 183–192.

Delussu, F. (2005) 'Produzione e consumo dei prodotti animali nell'ambito dell'economia di Turris Libisonis (Porto Torres – SS) in età imperiale', in *Atti 3° Convegno Nazionale di Archeozoologia* (Siracusa, 2000), eds I. Fiore, G. Malerba and S. Chilardi, Roma, 379–407.

Farello, P. (2000) 'Reperti faunistici punici da Tharros (OR), Sardegna', in *Atti del 2° Convegno Nazionale di Archeozoologia* (Asti 1997), Forlì, 293–300.

Fonzo, O. (1987) 'Reperti faunistici in Marmilla e Campidano nell'età del Bronzo e nella prima età del Ferro', in *Un millennio di relazioni tra i Paesi del Mediterraneo*, Atti del 2° Convegno di Studi, Selargius-Cagliari 27–30 novembre 1986, 233–242.

Fonzo, O. (2003) 'L'ambiente e le sue risorse: la caccia e l'allevamento del bestiame', in *La vita nel nuraghe Arrubiu*, eds T. Cossu, F. Campus, V. Leonelli, M. Perra and M. Sanges, Quartu Sant'Elena, 113–133.

Fonzo, O. and Vigne, J-D. (1993) 'Reperti osteologici', in *Genna Maria II, 1 Il deposito votivo del mastio e del cortile*, eds C. Lilliu, L. Campus, F. Guido, O. Fonzo and J.-D. Vigne, Cagliari, 161–173.

Grassi, E. (2004) *Il villaggio di Geridu. Analisi archeozoologica per lo studio dell'economia agro-pastorale nel medioevo sardo*. Unpublished MSc dissertation. University of Sassari.

Grassi, E. (2012a) *L'economia a Sassari dal medioevo all'età moderna. Contributo archeozoologico*. Unpublished PhD dissertation. University of Sassari.

Grassi, E. (2012b), 'Faunal remains from Sassari (Sardinia, Italy). An urban archaeozoological case study', in *Proceedings of the General Session of 11th ICAZ International Conference*, Paris 23–28 agoust 2010, ed. C. Lefèvre, BAR international series 2354, Archaeopress, Oxford, 127–136.

Hartl, G. B., Nadlinger, K., Apollonio, M., Markov, G., Klein, F., Lang, G., Findo, S. and Markowski, J. (1995) 'Extensive mitochondrial-DNA differentiation among European red deer (*Cervus elaphus*) populations: implications for conservation and management', *Zeitschrift für Säugetierkunde* 60, 41–52.

Hajji, G. M., Charfi-Cheikrouha, F., Lorenzini, R., Vigne, J.-D., Hartl, G. B. and Zachos, F. E. (2008) 'Phylogeography and founder effect of the endangered Corsican red deer (*Cervus elaphus corsicanus*)', *Biodiversity and Conservation* 17, 659–673.

Hmwe, S. S., Zachos, F. E., Eckert, I., Lorenzini, R., Fico, R. and Hartl, G. B. (2006) 'Conservation genetics of the endangered red deer from Sardinia and Mesola with further remarks on the phylogeography of *Cervus elaphus corsicanus*', *Biological Journal of the Linnean Society* 88, 691–701.

Lo Schiavo, F. and Manconi, F. (1999–2000) 'Animals in nuragic Sardinia', *Accordia Research Papers* 8, 101–132.

Ludt, C., Schroeder, W., Rottmann, O. and Kuehn, R. (2004) 'Mitochondrial DNA phylogeography of red deer (*Cervus elaphus*)', *Molecular Phylogenetics and Evolution* 31, 1064–1083.

Manconi, F. (1996) 'Olbia. Su Cuguttu 1992: i reperti faunistici', in *Da Olbìa a Olbia. 2500 anni di storia di una città mediterranea*, eds A. Mastino and P. Ruggeri, Sassari, 447–460.

Manconi, F. (1997) 'I resti animali', in *Una fattoria di età romana nell'agro di Olbia*, ed. A. Sanciu, Sassari, 213–221.

Manconi, F. (2000) 'La fauna dell'età del Ferro degli scavi 1988 e 1990 del nuraghe S. Imbenia di Alghero (Sassari)', in *Atti del 2° Convegno Nazionale di Archeozoologia* (Asti 1997), Forlì, 267–277.

Masala, S. (2011) 'La fauna della grotta Su Coloru: scavi 1996–98', *Aidu Entos. Archeologia e beni culturali* 5–6, 4–17.

Masala, S. (2012) 'I resti faunistici rinvenuti nel tempio a pozzo della *Purissima* presso Alghero (SS)', in *Atti del VI Convegno Nazionale di Archeozoologia*, eds J. De Grossi Mazzorin, D. Saccà and C. Tozzi, Parco dell'Orecchiella, San Romano in Garfagnan, Lucca, Italy, 227–234.

Pandolfi, A., Fiori M., Padua G., Carenti G., Angius L. and Petruzzi E. (2007) 'San Nicola di Trullas a Semestene. Chiesa e monastero', in *Committenza, scelte insediative e organizzazione patrimoniale nel medioevo*, eds L. Ermini Pani, *De re monastica* 1, 167–206.

Pereira, E. (2001) '*Cervus elaphus rossii* (Mammalia, Artiodactyla), a new endemic subspecies from the middle Pleistocene in Corsica', *Palaeovertebrata* 30, 189–213.

Puddu, F. and Viarengo, M (1993) *Animali di Sardegna. I Mammiferi*, Carlo Delfino Editore, Sassari.

Puddu, G., Maiorano, L. Falcucci, A., Corsi, F. and Boitani L. (2009) 'Spatial-explicit assessment of current and future conservation options for the endangered Corsican Red Deer (*Cervus elaphus corsicanus*) in Sardinia', *Biodiversity and conservation* 18, 2001–2016.

Rovina, D. and Grassi, E. (2006) 'Il villaggio medievale di Ardu', in *Vita e morte dei villaggi rurali tra medioevo ed età moderna*, ed. M. Milanese, *Quaderni del centro di documentazione dei villaggi medievali abbandonati*, 2, Firenze, 161–172.

Rovina, D., Garau, E., Mameli, P. and Wilkens B. (2011) 'Attività produttive nell'insediamento romano e altomedievale di Santa Filitica (Sorso – SS)', *Erentzias*, 1, 245–268.

Sanges, M. (1987) 'Gli strati del Neolitico antico e medio nella grotta Corbeddu di Oliena (Nuoro). Nota preliminare', in *Il Neolitico in Italia*, Atti della XXVI Riunione Scientifica dell'Istituto Italiano di Preistoria e Protostoria, Firenze 7–10 Novembre 1985, 825–830.

Santoni, V., Wilkens, B. (1996) 'Il complesso nuragico *del Rimedio* di Oristano', *Quaderni della Soprintendenza Archeologica di Cagliari-Oristano*, 13, 29–43.

Skog, A., Zachos, F. E., Rueness, E. K., Feulner, P. G. D., Mysterud, A., Langvatn, R., Lorenzini, R., Hmwe, S. S., Lehoczky, I., Hartl, G. B., Stenseth, N. C. and Jakobsen, K. S. (2009) 'Phylogeography of red deer (*Cervus elaphus*) in Europe', *Journal of Biogeography* 36, 66–77.

Sorrentino, C. (2009) 'Il materiale osteologico animale', in *Nora. Il foro romano. Storia di un'area urbana dall'età fenicia alla tarda antichità. 1997–2006*. Volume II.2: I materiali romani e gli altri reperti, eds J. Bonetto, G. Falezza, A. R. Ghiotto, Padova, 891–903.

Vigne, J.-D. (1988) *Les mammifères post-glaciaires de Corse, étude archéozzologique*, CNRS, Paris.

Vigne, J.-D. (1999) 'The large "true" Mediterranean islands as a model for the Holocene human impact on the European vertebrate fauna? Recent data and new reflections' in *The Holocene History of the European Vertebrate Fauna. Modern Aspects of Research*, ed. N. Benecke, Verlag Marie Leihdorf, Rahden, 295–322.

Vigne, J.-D. (2003) 'Les restes de vertébrés du site de Torre Sabea', in *Torre Sabea, un établissement du Néolithique ancien en Salento*, eds J. Guilaine and G. Cremonesi, Collection de l'École Française de Rome 315, 251–278.

Wilkens, B. (1989) 'Il cervo dal Mesolitico all'età el Bronzo nell'Italia centro-meridionale', *Rassegna di Archeologia* 8, 63–95.

Wilkens, B. (2001) 'I resti faunistici del Palazzo Ducale di Sassari nel quadro della Sardegna postmedievale', in VV. AA. *Dal mondo antico all'età contemporanea. Studi in onore di Manlio Brigaglia*, Roma, 325–345.

Wilkens, B. (2003a) 'La fauna sarda durante l'Olocene: le conoscenze attuali', *Sardinia, Corsica et Baleares Antiquae*, 1, 181–197.

Wilkens, B. (2003b) *Archeozoologia. Manuale per lo studio dei resti faunistici dell'area mediterranea*, CD-Rom, Schio.

Wilkens, B. (2008) 'I resti faunistici dell'US 500', in *Il cibo nel mondo fenicio e punico d'occidente. Un' indagine sulle abitudini alimentari attraverso l'analisi di un deposito urbano di* Sulky *in Sardegna*, eds L. Campanella, Fabrizio Serra Editore, Pisa–Roma, 249–259.

Wilkens, B. (2012a) *Archeozoologia. Il Mediterraneo, la stroia, la Sardegna*, Edes, Sassari.

Wilkens, B. (2012b) 'Resti faunistici dai livelli neolitici della grotta Verde di Capo Caccia (Alghero)', in *Atti del VI Convegno Nazionale di Archeozoologia*, eds J. De Grossi Mazzorin, D. Saccà and C. Tozzi, 125–129.

Wilkens, B. and Delussu, F. (2000) 'Resti ossei dal convento di S. Maria di Seve (Banari – SS)', *Archeologia Medievale* 27, 311–313.

Zachos, F., Hartl, G. B., Apollonio, M. and Reutershan, T. (2003) 'On the phylogeographic origin of Corsican red deer (*Cervus elaphus corsicanus*): evidence from microsatellites and mitochondrial DNA', *Mammalian Biology* 68, 284–298.

Zachos, F. E. and Hartl, G. B. (2011) 'Phylogeography, population genetics and conservation of the European red deer *Cervus elaphus*', *Mammal Review* 41, 138–150.

Enduring Relationships: Cervids and humans from Late Pleistocene to modern times in the Yukon River basin of the western Subarctic of North America

Carol Gelvin-Reymiller[†]

Introduction

A brief review is given here of cervid and human relationships within the interior region of the Western Subarctic of North America, drained by the Yukon River and its largest tributary, the Tanana River. The review is based on archaeological and ethnographic records as well as modern data, representing cervid-human material associations, which span the last 14,300 years. The prehistory and history of Alaska's interior and the Yukon Territory (Yukon) of farthest western Canada (Figure 4.1) are replete with evidence indicating cervids were key contributors as sources of food and to nearly all classes of material culture from clothing and shelter, to weaponry and other hand tools (cf. Helm 1981). The region is the eastern extent of Beringia, noted for archaeological sites interpreted as among the earliest evidence of human presence in North America (Goebel and Buvit 2011). In addition, direct contact in the interior of Alaska between Euroamericans and Indigenous people was among the latest on the continent (AD 1840s–1880s) (McKennan 1969; Helm *et al.* 1975). Reduction in use of cervid materials has occurred since that time, but strong associations to deer species have continued in spite of material replacement.

Moose (*Alces alces*), caribou (*Rangifer tarandus*) and wapiti (*Cervus canadensis*) are the three cervid species (Courtois *et al.* 2003; Ludt *et al.* 2004; Hundertmark and Bowyer 2004) with which modern Western Subarctic societies are most closely associated. The details of ancient prehistoric connections to these deer species (and their direct ancestors) are of course linked to challenging issues of preservation of faunally derived materials. Cold climate parameters create several uniquely favorable contexts for preservation, including permafrost, loess deposits, and highland ice patches, but the majority of organic archaeological materials, especially hide and antler, unfortunately do not survive. This contrasts to sites on Subarctic and Arctic coastal planes (not reviewed here), which can have well-preserved organic assemblages.

Today, complex relationships (multi-faceted) between humans and cervid species persist across the Subarctic, and a full discussion of modern game management policies and their effects are beyond the scope of this paper. The focuses of this paper are concerned with the (i) social and economic contexts and examples of modern cervid-based material culture in the interior, followed by a (ii) short review of cervid and human ecology relevant to the Yukon River basin and (iii) an abridged archaeological review related to the *Cervidae*.

Modern associations

Moose (referred to as elk in Europe) and caribou (reindeer in Europe) are both culturally and economically important to today's residents of the Western Subarctic (Fulton and Hundertmark 2004). Wild populations of *c.*200,000 moose and *c.* one million caribou exist in Alaska, many of which inhabit Arctic and Subarctic ecozones (ADFG 2009; ADFG 2011a; Environment Yukon 2011). Wapiti (red deer in Europe) have been reintroduced to Alaska and are hunted on Afognak and Raspberry Islands, and a small number are farmed in Interior Alaska near Delta Junction. Wapiti are hunted in western Canada, and along with the *c.*100 reintroduced population in southern Yukon, are the largest northern extant wild herd (Long 2003). Domestic reindeer, introduced at the end of the nineteenth century, occupy Arctic ecozones outside the interior region.

Cervid harvest totals are low compared to temperate North America and to Scandinavia (7,500 moose and 25,000 caribou harvested in Alaska in 2010), but cervid harvests are actually high per capita (ADFG 2011a). Despite fewer human inhabitants in the region, moose and caribou are at the center of volatile issues on both sides of the Alaska–Canada border. Debates and research are focused principally on impacts to wildlife by resource development and due to climate change, on concerns for caribou calving grounds, on methods and means of controlling predators (wolves (*Canis lupus*) and bears (*Ursus* sp.)), and on the equity of hunting regulations (Boertje *et al.* 1996; Bertram and Vivion 2002; Tesar 2007; Nuttall *et al.* 2010). At present, deer species and other wildlife in Alaska are considered public property, regulated by licensing and open seasons from several days to weeks in length. In the Western Subarctic, cervids are viewed as critical sources of food and materials; regulatory policies overtly specify that game management practices are designed foremost to provide sources of food to residents (Health Canada 2007; ADFG 2011b).

For Indigenous Dene, or Na-Dene (Athapaskan) speakers of the region, cultural functions are not complete without caribou or moose as food. Alaskan hunting regulations allow the taking of moose for memorial feasts known as potlatches, and Canadian treaty agreements enshrine traditional harvest rights for Yukon First Nations (Environment Yukon 2003; ADFG 2011b).

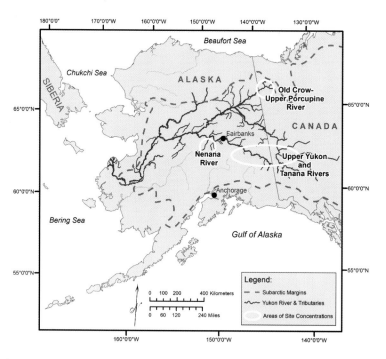

FIGURE 4.1. Map of the Western Subarctic. Polygons show areas of archaeological site concentrations with cervid materials (base map source: www.alaskamapped.org).

Significantly, jewelry and leather/other goods exchanged during feasts are part of local economies as well as markers of cultural identity (Duncan and Carney 1997; Simeone 2002). The endurance of connections with cervids is corroborated by ethnographic information about Dene of interior Alaska, which includes caribou or caribou tail members of sibs within the general dichotomous clan structure (Guédon 1974; Andrews 1975; Jette and Jones 2000). In addition, Ahtna Athapaskans consider the caribou themselves within the clan system (Simeone 2006). Cervid products persist in traditional uses; scapulae are scraped against shrubs when hunting to imitate antler thrashing by bull moose, metapodials are still used in hide processing, vestigial (rudimentary) metapodials are used as beaded leather hairpiece pins and decorated awls, and cervid antler continues as a popular material for knife and other tool handles.

Evident today in the Western Subarctic are also the influences of historic Euroamerican immigrants to the area, especially the display of 'racks' or sets of cervid antlers on buildings as signs of admiration for the animals. Antlers and bones have been pressed into service as useful replacement materials, especially in rural areas, and as materials in modern artwork. A moose antler outboard motorboat propeller is a particularly good example of expediency (Figure 4.2). In addition, moose have become cultural icons in Alaska's few urban communities (primarily Anchorage and Fairbanks). This is seen in the popularisation of these animals especially as imagery on public display. Likewise, the Canadian north is synonymous with caribou, its image appearing on national currency.

FIGURE 4.2. Circa 1930 motorboat propeller (UA140-1080) cut from moose antler brow palm and main beam, with pedicle incorporated into tip of left prop blade (University of Alaska Museum of the North [UAMN], Ethnology and History Collection; author photos).

Cervidae and human ecology in the Yukon river basin

The environment of the region today is a boreal mosaic consisting of mixed forests on intermontane plateaus and bottomlands, and taiga meadows or tundra on uplands (Nowacki *et al.* 2001). The entire region is affected strongly by the extremes of solar cycles and precipitation regimes, which in part stimulate animal migration. Paleoenvironmental reconstructions indicate grasses and herbaceous plants favoring wapiti and bison were dominant until deciduous tree species favoring browsers (moose) populated the region after about 11,000 years ago; conifers (*Picea spp.*) moved variably into river valleys beginning at 8000–9000 years ago (Lloyd *et al.* 2006). The subsequent boreal forests, with upland lichen tundra, favored caribou. The understanding of altered cervid, bovid, and other animal habitat ratios over time, as well as associated predator-prey relationships, including those involving human hunters, continues to be refined by genetic and other research (Yesner 1996; Hundertmark and Bowyer 2004; Guthrie 2006; Potter 2008a; Schmidt *et al.* 2009).

Linguistic datasets and 'traditional ecological knowledge' (TEK) are interesting adjuncts to prehistory and ecology, recording terms related to deer-human coexistence (Cruikshank 1981; Kari 1990; Berkes 1999). Some terms have poetic meaning such as the Koyukon Athapaskan term for the brow tine or 'shovel' of caribou antlers, which translates literally as 'that which blocks the sun' (Jetté and Jones 2000). Slightly more cognates for caribou than for moose appear in datasets compiled for the Koyukon language of western Alaska with implications for species representations in material culture, at least for the recent past. Scarcity of moose during the Late Holocene in western Alaska is also indicated by TEK, genetic studies, and other research (Kay 1997; Hundertmark and Bowyer 2004), but the relationship of species scarcity or abundance to human predation is not well understood for deep prehistory.

Numerous burned and calcined bone fragments at archaeological sites which illustrate marrow extraction, critical to maximising fat content in a cold climate diet, attest to the importance of cervid-sized mammals, even though many fragments are difficult to assign to species beyond 'large mammal'. However, species shifts suggested by paleoecological changes are generally corroborated by the archaeology of the region, as reported in compilations of successfully identified fauna from dated components (Potter 2007; Potter 2008b).

Archaeological highlights through time: cervid bone and antler

Broad archaeological patterns across the region have been identified, suggesting prehistoric technologies were relatively consistent, with several notable changes at around 6,000 to 5,000 years ago and 1,300 to 1,000 years ago. Causation of changes are still under debate and have potential links to shifts in cervid and human subsistence relationships, and to area-wide events such as tephra falls (Workman 1978), with implications for human spatial distribution and population sizes (Holmes 2001; Mason and Bigelow 2008; Potter 2008a). The archaeological sites with cervid remains in the Yukon River basin include all periods from the Late Pleistocene to the Late Holocene and into the historic era.

Late Pleistocene and Early Holocene

Currently, stratified sites with components dating from about 14,300 to 11,500 years ago represent the earliest cervid-human associations, with many located in the middle and upper Tanana River valley. In sites buried up to nearly two meters in depth, cervid bone and teeth are preserved by cool, well-drained and stable calcareous loess (neutral to alkaline pH) (Zazula 2006). Earliest recorded cervid associations are at Swan Point and Broken Mammoth (Holmes 2001) near the Shaw Creek Flats, and the Little John site in the Yukon near Scottie Creek (Easton *et al.* 2011). At Bluefish Caves, a site in the Yukon on a feeder stream of the Porcupine River (Yukon River tributary), cervid and other fauna have been dated to at least 13,000 years ago, and some considerably older (Morlan 2003). A majority of identifiable cervid faunal material from these early sites is wapiti, though moose and caribou are present. Associated lithic technologies include bifacial tools and/or microblades and this is variable by component and by site; assemblages are sometimes classified into the Denali (microblade) or Nenana (non-microblade) complexes (see Graf and Bigelow 2011).

Other well-known sites which date to the Early Holocene (11,500 to 6,000 years ago) and have cervid faunal material are Dry Creek (Hoffecker 2001) in the Nenana River valley, and Pelly Farm (MacNeish 1964) and KaVn-2

(Heffner 2008) on upper Yukon River tributaries. The Gerstle River site in the Tanana River valley (Potter 2006) held extensive well preserved wapiti and bison assemblages which allowed detailed butchering and processing analyses. At many of these early sites, antler preservation is poor, but a few antler slotted tools have survived from this early time period in cave sites outside the Yukon River basin in the Kuskokwim River drainage and on the Seward Peninsula (Vinson 1993; Ackerman 2011), and from ice patch contexts in the Yukon (Hare *et al.* 2004).

Middle Holocene

The introduction of side-notched points is evident in lithic tool assemblages, termed the Northern Archaic tradition, throughout the region by about 6,000 to 5,000 years ago. This change has been interpreted (and debated) as indicative of cultural adjustments and/or population movements, possibly related to the expanding spruce forests and associated faunal changes, including an increased focus on caribou herds (Dumond 2001; Esdale 2008; Mason and Bigelow 2008). However, microlithic technology, basically a composite tool kit linked to use of cervid antler and other mammal bone, continued across parts of the region throughout the Middle Holocene (Potter 2008a). As in earlier sites, formal organic tools are rare and cervids are represented mainly by fragmented bone; the exceptions to this are organic artifacts found on ice patches, discussed below.

Late Holocene

Archaeological sites near caribou migration routes, especially at topographic constrictions, river and lake crossings typify the Late Holocene (<2,000 years ago) throughout the general Athapaskan culture area; caribou fences were still in use in the Yukon River basin at the time of Euroamerican contact (Murie 1935). Pole fences were often positioned at ecotones between boreal forests and uplands, some with snare insets of babiche (rawhide) and some utilising inuksuk (rock cairns); the Ketchumstock area in the Yukon-Tanana Uplands and the Old Crow area of northern Yukon are examples (Warbelow *et al.* 1975; Andrews 1980). Archaeological sites representing seasonal caribou interception on lakes include alpine sites at the margins of the Yukon and Tanana River basins such as Paxson (Yesner 1980) and Taye lakes (Workman 1978), and at Lake Minchumina and Hahanudan Lake, used by the inland Ipiutak, a culture known for caribou bone and antler craftsmanship (Clark 1977; Holmes 1983). Lowland sites such as Klo-kut and Rat Indian Creek on the Porcupine River in Canada (Morlan 1973; LeBlanc 1984) also exemplify caribou interception subsistence strategies.

For the capture of moose, heavy snares, fence lines and tracking techniques were reportedly used as well as methods using boats in lakes and rivers (Dall 1898; Nelson 1973; Yerbury 1975). At Quartz Lake (Gelvin-Reymiller 2011), Healy

Lake (Cook 1989), and Dixthada on Lake Mansfield (Shinkwin 1979) along the Tanana River, moose bone is prevalent in artifact assemblages, and sites are located near wetland areas favored by moose, especially during summer. These sites are peripheral to major caribou habitat in the Yukon-Tanana uplands, so appear to maximise access to both cervid species. In the higher Nenana River valley to the west, the Nenana River Gorge site (Plaskett 1977) held substantial organic preservation, including both moose and caribou bone.

In the Late Holocene, bone and antler technologies of the region are particularly noteworthy (see LeBlanc (1984), Morlan (1973), and Shinkwin (1979) for breadth of bone and antler industries in this time period, and Corbin (1975) for antler reduction techniques). This prevalence of cervid material is due partly to relationships of preservation and time, but also reflects increased human populations and increased reliance on cervids as indicated by well-developed drive systems for mass procurement, as mentioned above. Ingenuity, as well as aesthetic sensibilities, are represented in this region during the Late Holocene by objects (some decorated with incised lines and inset materials) ranging from fish spears, lures, and nets to war clubs, shovels, and spoons, all created from cervid antler, bone, sinew or hide.

An interesting example of cervid bone tool technology from Late Holocene sites is the use of moose and caribou metapodials. The reduction of these bone elements have intrigued a number of researchers in the region (Yesner and Bonnichsen 1979; Binford 1981; LeBlanc 1984; Morrison 1986) and has prompted discussions of cultural exchange and technological processes (though not to the degree of bone and antler projectile point technology). Reduction techniques of metapodials illustrate nearly all methods used in general for crafting bone and antler during this time period, though regional differences as well as differences between cervid species have been noted (Morrison 1986). Techniques included 'groove and snap,' indirect percussion with wedges, direct percussion, chopping, sawing, and abrading. The nature of cervid leg bones – long, straight and with relatively thick cortex – is generally concluded to be responsible for their widespread use (noted also in other regions of the world). Metapodials were commonly crafted into hide working tools, projectile points, spear heads, and knives by both Arctic and Subarctic inhabitants, though antler was the preferred material for most projectile points and objects other which required more complex forming. Hide working tools made from moose and caribou metatarsals as 'end-of-the-bone' fleshers and sagittal split 'beamers' persisted into historic times (Clark 1981, 119), and ethnographies record decorated fleshers as among a woman's valued possessions (Jarvenpa and Brumbach 1983). Flesher grips and hafting techniques suggest standardised methods of construction; examples are retention of the small tarsals (naviculo-cuboid and fused external-middle cuneiform) for hand grips, and shifts from lithic to metal bits (Figure 4.3). The dependency of humans on cervid hides for clothing, shelter, cordage, and containers suggests tools for hide-working, whether lithic, organic, or both, were of extreme importance, especially for life in environments with lengthy winter seasons requiring protection from cold.

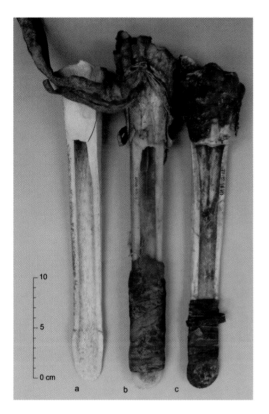

FIGURE 4.3. Circa 1900 moose metatarsal end-of-bone hide working tools, anterior surfaces removed: (L to R) a. notched working edge weathered, gnawed (UA755-2); b. notched working edge intact; grip incorporates tarsals, suspensory ligament, leather strap (UA68-56-4); c. metal working edge hafted w/cloth, leather; retained tarsals as grip (UA86-13-12) (UAMN Ethnology and History Collection; author photo).

Organic artifacts also illustrate another major technological change at 1,300 years ago, a shift from throwing dart (atlatl) technology to the bow and arrow. Significantly, artefacts found over the last decade on Alaskan and Canadian ice patches above 1,500 metres in elevation record this change. Caribou are central to these finds; layers of caribou dung and accumulated snowfall (now melting) have preserved wooden arrow shafts and antler arrowheads, as well as throwing dart elements. These remarkable organic artifacts extend back to 9,000 years and are intriguing glimpses into prehistoric human reliance on cervids. After about AD 700, materials used for arrowheads were mainly antler, while the earlier throwing dart heads were most often lithic tipped (Hare *et al.* 2004; Dixon *et al.* 2005). Other research suggests bow and arrow technology utilising cervid antler was present in eastern Beringia much earlier (Ackerman 2011). Connections to population movements, technological advantages or other causal factors for adoption of the bow continue to be explored (Morrisey 2009), as does the apparent loss of microlithic technology several centuries later at about 1,000 years ago in parts of the same region (Clark 1981; Potter 2008b).

Information shared by Dene elders such as the Reverend David Salmon and Belle Herbert have added valuable insights into the use of cervid materials as tools in protohistoric and historic times (Herbert 1988; Frink and Weedman 2005; O'Brien 2011). Rev Salmon describes a technique for using the anterior surface of a moose metatarsal for use as a heavy bear spear, and also relates the style and workmanship of moose leg bone arrowheads which made Han Gwich craftsman 'famous in their day' (O'Brien 2011, 55). Similarity of skeletal element morphology among *Cervidae* may have encouraged technological standardisation over time, as seen, for example, in hide processing tools and arrowheads. Antler as a relatively standard material, in size and shape, would also have encouraged the development of technologies over time geared specifically for utilising, and thus depending upon, material of that form. Though cervid antler was (and is) unquestionably a versatile and widely used material, certain skeletal elements, such as metapodials, apparently have unique qualities which encourage their continued use through time and across the western Subarctic in spite of possible technological requirements or constraints, and thus illustrate well the enduring cervid-human material connections.

Concluding remarks

As in other geographic regions, specific knowledge of the relationships of prehistoric humans and cervids in regard to material culture, technology and economies develops as knowledge of the archaeological past increases, and as palaeoenvironments are better understood. Our thinly populated region of the globe has many areas which have not yet been systematically archaeologically surveyed. Organic artifacts, as well as recent stable isotope and DNA research, indicate human and cervid coexistence in the Yukon River basin begins at least by the end of the Late Pleistocene epoch and continues throughout the Holocene. Our connections to modern cervids are also prosaic; on present day archaeological sites, moose and caribou trampling can expose or obscure artefacts and features.

Continued public access to cervids is considered by many current residents, both urban and rural alike, as essential. This attachment to moose and caribou today is the consequence of an amalgamation of influences from historic Euroamerican cultures and Indigenous cultures expressed uniquely in North America. The general prehistoric reliance on cervids is, of course, not exclusive to this geographic location but attests to the nearly worldwide significance of *Cervidae*. In the Yukon River basin, moose and caribou meat maintains its importance as food, and the hides, bone and antler have persisted in material culture, continuing as integral to cultural identity.

Acknowledgments

Thanks to Angela Linn, Ethnology and History Collections Manager, UAMN; Jim Whitney, Archaeology Collections Manager, UAMN. I also thank Naomi Sykes, Steve Ashby and Ben Elliott for their welcome to the Deer and People Conference and material culture session at Lincoln, UK, September 2011. Thanks to the Alaska Department of Fish and Game and to my fellow archaeologists working in the western Subarctic for discussions, especially in regard to fauna and human ecology. And finally, thanks to Ruth Carden and the anonymous reviewers who provided the opportunity to improve this paper.

References

Ackerman, R. (2011) 'Microblade assemblages in South-western Alaska: An early Holocene adaptation', in *From the Yenisei to the Yukon: Interpreting Lithic Assemblage Variability in Late Pleistocene/Early Holocene Beringia*, eds T. Goebel and I. Buvit, Texas A&M University Press, College Station, 255–269.

ADFG (2009) Alaska Department of Fish and Game, *Caribou Management Report of survey-inventory activities 1 July 2006–30 June 2008*, ed. P. Harper, Division of Wildlife Conservation, Juneau. http://www.adfg.alaska.gov/static/home/library/pdfs/wildlife/mgt_rpts/09_caribou.pdf, accessed 06.2011.

ADFG (2011a) Alaska Department of Fish and Game, *Interior Alaska Moose News*

Summer/Fall, ed. D. D. James (Region III Regional Supervisor), Division of Wildlife Conservation Region III, Fairbanks, http://www.adfg.alaska.gov/static/species/speciesinfo/moose/pdfs/interior_moose_news_fall_2011.pdf, accessed 07.2011.

ADFG (2011b) Alaska Department of Fish and Game, *2011–2012 Alaska Hunting Regulations governing, subsistence and commercial uses of Alaska's wildlife*, 52, http://www.adfg.alaska.gov/static/regulations/wildliferegulations/pdfs/general.pdf accessed 07.2011.

Andrews, E. F. (1975) *Salcha: an Athapaskan Band of the Tanana River and its Culture*, MA Thesis, Department of Anthropology, University of Alaska, Fairbanks.

Andrews, E. F. (1980) *Native and Historic Accounts of some Historic Sites in the Tananacross-Ketchumstock Area*, Cemetery and Historic Sites Committee, Doyon Limited, Fairbanks.

Berkes, F. (1999) *Sacred Ecology: Traditional Ecological Knowledge and Resource Management*, Taylor and Francis, London.

Bertram, M. R. and Vivion, M. T. (2002) 'Moose mortality in eastern interior Alaska', *Journal of Wildlife Management* 66, 747–756.

Binford, L. R. (1981) *Bones: Ancient men and Modern Myths*, Academic Press, NY.

Boertje, R. D., Valkenburg, P. and McNay, M. E. (1996) 'Increases in moose, caribou, and wolves following wolf control in Alaska', *Journal of Wildlife Management* 60, 474–489.

Clark, D. W. (1977) *Hahanudan Lake: An Ipiutak-related occupation of Western Interior Alaska*, Mercury Series 71, National Museum of Man, Ottawa.

Clark, D. W. (1981) 'Prehistory of the Western Subarctic', in *Handbook of North American Indians 6 Subarctic*, ed. J. Helm, Smithsonian Institution, Washington, DC, 107–129.

Cook, J. P. (1989) 'Historic archaeology and ethnohistory at Healy Lake, Alaska', *Arctic* 42, 108–116.

Corbin, J. E. (1975) *Aniganigaruk: A Study in Nunamiut Eskimo Archaeology*. Unpublished PhD dissertation. Department of Anthropology, Washington State University, Pullman.

Courtois, R., Bernatchez, L., Ouellet, J. and Breton, L. (2003) 'Significance of caribou (*Rangifer tarandus*) ecotypes from a molecular genetics viewpoint', *Conservation Genetics* 4, 393–404.

Cruikshank, J. (1981) 'Legend and Landscape: Convergence of oral and scientific traditions in the Yukon Territory.' *Arctic Anthropology* 18, 67–93.

Dall, W. H. (1898) *The Yukon Territory, the Narrative of W. H. Dall, Leader of the Expedition to Alaska in 1866–1868,* Downey, London.

Dixon, E. J., Manley, W. F. and Lee, C. M. (2005) 'The emerging archeology of glaciers and ice patches: examples from Alaska's Wrangell–St Elias National Park and Preserve', *American Antiquity* 70, 129–143.

Dumond, D. E. (2001) 'The archaeology of Eastern Beringia: Some contrasts and connections', *Arctic Anthropology* 38, 196–205.

Duncan, K. C. and Carney, E. (1997) *A Special Gift: The Kutchin Beadwork Tradition*, University of Alaska Press, Fairbanks.

Easton, N. A., Mackay, G. R., Young, P. B., Schnurr, P., and Yesner, D. R. (2011) 'Chindadn in Canada: Emergent evidence of the Pleistocene transition in Southeast Beringia as revealed by the Little John Site, Yukon', in *From the Yenisei to the Yukon: Interpreting Lithic Assemblage Variability in Late Pleistocene/Early Holocene Beringia*, ed. T. Goebel and I. Buvit, Texas A&M University Press, College Station, 289–307.

Environment Yukon (2003) *Hunting and Fishing Rights and Responsibilities of First Nation People*, Government of Yukon, Whitehorse, http://www.env.gov.yk.ca/fishing/documents/hunting_fishing_firstnation.pdf, accessed 06.2011.

Environment Yukon (2011) *Yukon hunting regulations summary*, Government of Yukon, Whitehorse, http://www.env.gov.yk.ca/mapspublications/documents/hunting_regs2011_en.pdf, accessed 06.2011.

Esdale, J. (2008) 'A current synthesis of the Northern Archaic', *Arctic Anthropology* 45, 3–38.

Frink, L. and Weedman, K. (2005) *Gender and Hide Production*, Altamira Press, Lanham, MD.

Fulton, D. C. and Hundertmark, K. (2004) 'Assessing the effects of a selective harvest system on moose hunters' behaviors, beliefs, and satisfaction', *Human Dimensions of Wildlife* 9, 1–16.

Gelvin-Reymiller, C. (2011) *Shaw Creek Flats East Project: Quartz Lake Sites XBD-00362, 00361, 00159, 00155 Second Year Report-2010*, Field Archaeology Permit 2009–01 (File 3420–2009), report on file with State of Alaska Office of History and Archaeology, State Historic Preservation Officer, Anchorage, Alaska.

Goebel, T. and Buvit, I. (2011) *From the Yenisei to the Yukon: Interpreting Lithic Assemblage Variability in Late Pleistocene/Early Holocene Beringia*, Texas A&M University Press, College Station.

Graf, K. E. and Bigelow, N. H. (2011) 'Human response to climate during the Younger Dryas chronozone in central Alaska', *Quaternary International* 242, 434–451.

Guédon, M. F. (1974) *People of Tetlin, Why are you Singing?* Mercury Series 9, National Museum of Man, Archaeological Survey of Canada, Ottawa.

Guthrie, R. D. (2006) 'New carbon dates link climatic change with human colonization and Pleistocene extinctions', *Nature* 441, 207–209.

Hare, G. P., Greer, S., Gotthardt, R., and Farnell, R., Bowyer, V. and Schweger, C. (2004) 'Multidisciplinary Investigations of Alpine Ice Patches in South-west Yukon, Canada: Ethnographic and Archaeological Investigations', *Arctic* 57, 260–272.

Health Canada (2007) *Eating Well with Canada's Food Guide-First Nations, Inuit, and Metis.* http://www.hc-sc.gc.ca/fn-an/food-guide-aliment/fnim-pnim/index-eng.php, accessed 07.2011

Heffner, T. (2008) 'The role of glacial lakes in the pre-contact human history of south-west Yukon territory: a late drainage hypothesis', *Northern Review* 29, 85–104.

Helm, J. ed (1981) volume *Handbook of North American Indians, 6, Subarctic*, Smithsonian Institution, Washington, DC.

Helm, J., Alliband, T., Birk, T., Lawson, V., Reisner, S., Sturtevant, C. and Witkowski, S. (1975) The contact history of the Subarctic Athapaskans: an overview, in *Proceedings: Northern Athapaskan Conference, 1971, 1*, ed. A. M. Clark, Mercury Series 27, Canadian Ethnology Service, National Museums of Canada, Ottawa, 302–346.

Herbert, B. (1988) *Shandaa: In my Lifetime*, Alaska Native Language Center University of Alaska, Fairbanks.

Hoffecker, J. F. (2001) 'Late Pleistocene and Early Holocene Sites in the Nenana River Valley, Central Alaska', *Arctic Anthropology* 38, 139–153.

Holmes, C. E. (1983) 'Lake Minchumina prehistory: an Archaeological Analysis', *Aurora Monograph Series* 2.

Holmes, C. E. (2001) 'Tanana River Valley Archaeology Circa 14,000 to 9,000 BP.', *Arctic Anthropology* 38, 154–170.

Hundertmark, K. and Bowyer, R. T. (2004) 'Genetics, evolution and phylogeography of moose', *Alces* 40, 103–122.

Jarvenpa, R. and Brumbach, H. J. (1983) 'Ethnoarchaeological perspectives on an Athapaskan moose kill', *Arctic* 36, 174–184.

Jetté, J. and Jones, E. (2000) *Koyukon Athabascan dictionary*, Alaska Native Language Center, University of Alaska, Fairbanks.

Kari, J. (1990) *Ahtna Athabaskan dictionary*, Alaska Native Language Center, University of Alaska, Fairbanks.

Kay, C. E. (1997) 'Aboriginal overkill and the biogeography of moose in Western North America', *Alces* 33, 141–164.

Le Blanc, R. (1984) *The Rat Indian Creek site and the Late Prehistoric Period in the interior northern Yukon*, Mercury Series 120, National Museum of Man, Archaeological Survey of Canada, Ottawa.

Lloyd, A. H., Edwards, M. E., Finney, B. P., Lynch, J. A., Barber, V. and Bigelow, N. H. (2006) 'Holocene development of the Alaskan boreal forest', in *Alaska's Changing Boreal Forest*, eds F. S. Chapin, M. W. Oswood, K. Van Cleve, L. Viereck, and D. L. Verbyla, OUP, Oxford, 62–78.

Long, J. (2003) *Introduced Mammals of the World: Their History, Distribution and Influence*, CABI Publishing, Wallingford.

Ludt, C. J., Schroeder, W., Rottmann, O. and Kuehn, R. (2004) 'Mitochondrial DNA phylogeography of red deer (*Cervus elaphus*)', *Molecular Phylogenetics and Evolution* 3, 1064–1083.

MacNeish, R. S. (1964) 'Investigations in South-west Yukon: Archaeological excavations, comparisons and speculations', in *Papers of the Robert S. Peabody Foundation for Archaeology* 6, 199–488.

Mason, O. K. and Bigelow, N. H. (2008) 'The crucible of Early to Mid-Holocene climate in Northern Alaska: Does Northern Archaic represent the people of the spreading forest?', *Arctic Anthropology* 45, 39–70.

McKennan, R. A. (1969) 'Athapaskan groups of Central Alaska at the time of white contact', *Ethnohistory* 16, 335–343.

Morlan, R. E. (1973) *The Later Prehistory of the Middle Porcupine Drainage, Northern Yukon territory*, Mercury Series 11, Archaeological Survey of Canada, National Museum of Man, Ottawa.

Morlan, R. E. (1980) *Taphonomy and archaeology in the Upper Pleistocene of the Northern Yukon Territory*, Mercury Series 94, Archaeological Survey of Canada, National Museums of Canada, Ottawa.

Morlan, R. E. (2003) 'Current perspectives on the Pleistocene archaeology of eastern Beringia', *Quaternary Research* 60, 123–132.

Morrisey, G. (2009) *Tools and change: The shift from atlatl to bow on the British Columbia plateau*, M.A. thesis, Department of Archaeology, Simon Fraser University, Burnaby, BC.

Morrison, D. (1986) 'Inuit and Kutchin bone and antler industries in North-western Canada', *Canadian Journal of Archaeology* 10, 107–125.

Murie, O. J. (1935) *Alaska-Yukon Caribou*, United States Department of Agriculture, Government Printing Office, Washington, DC.

Nelson, R. K. (1973) *Hunters of the Northern Forest*, University of Chicago Press, Chicago.

Nowacki, G., Spencer, P., Fleming, M., Brock, T. and Jorgenson, T. (2001) *Ecoregions of Alaska*, U.S. Geological Survey Open-File Report 02–297 (map). Downloadable at: http://agdc.usgs.gov/data/usgs/erosafo/ecoreg/index.html, accessed 06.2011.

Nuttall, M., Berkes, F., Forbes, B., Kofinas, G., Vlassova, T., and Wenzel, G. (2010) 'Climate change impacts on Indigenous caribou systems of North America', in

Arctic Climate Impact Assessment, International Arctic Science Committee, CUP, 650–690.

O'Brien, T. A. (2011) *Gwich'in Athabascan Implements: History, Manufacture, and Usage according to Reverend David Salmon*, University of Alaska Press, Fairbanks.

Plaskett, D. C. (1977) *The Nenana River Gorge Site: A Late Prehistoric Athapaskan Campsite in Central Alaska*, MA thesis, Department of Anthropology, University of Alaska, Fairbanks.

Potter, B. A. (2006) *Site Structure and Organization in Central Alaska: Archaeological Investigations at Gerstle River*. Unpublished PhD dissertation. Department of Anthropology, University of Alaska, Fairbanks.

Potter, B. A. (2007) 'Models of faunal processing and economy in Early Holocene interior Alaska', *Environmental Archaeology* 12, 3–23.

Potter, B. A. (2008a) 'Radiocarbon chronology of Central Alaska: Technological continuity and economic change', *Radiocarbon* 50, 181–204.

Potter, B. A. (2008b) 'Exploring models of intersite variability in mid to late Holocene Central Alaska', *Arctic* 61, 407–425.

Schmidt, J. I., Hundertmark, K. J., Bowyer, R. T. and McCracken, K. G. (2009) 'Population structure and genetic diversity of moose in Alaska', *Journal of Heredity* 100, 170–180.

Shinkwin, A. (1979) *Dakah De'nin's village and the Dixthada Site: A contribution to Northern Athapaskan Prehistory*, National Museum of Canada, Ottawa.

Simeone, W. E. (2002) *Rifles, Blankets and Beads: Identity, History, and the Northern Athapaskan Potlatch*, University of Oklahoma Press, Norman.

Simeone, W. E. (2006) *Some Ethnographic and Historical Information on the Use of Large Land Mammals in the Copper River Basin*, US National Park Service Alaska Region Technical Report Series CRR 2006–56, Anchorage.

Tesar, C. (2007) 'What price the caribou? Canadian Arctic Resources Committee', *Northern Perspectives* 31, 1–39. www.carc.org, accessed 06.2011.

Warbelow, C., Roseneau, D. and Stern, P. (1975) 'The Kutchin caribou fences of north-eastern Alaska and the Northern Yukon', in *Studies of Large Mammals along the proposed Mackenzie Valley Gas Pipeline Route from Alaska to British Columbia*. Biological Report Series 32, JRD, Renewable Resources Consulting Services Ltd.

Workman, W. (1978) 'Prehistory of the Aishihik-Kluane Area, South-west Yukon Territory', *Mercury Series* 74, Archaeology Survey of Canada, National Museums of Canada, Ottawa.

Yerbury, J. C. (1975) *An Ethnohistorical Reconstruction of the Social Organization of Athabascan Indians in the Alaskan Subarctic and in the Canadian Western Subarctic and Pacific drainage*, M.A. thesis, Department of Political Science, Sociology and Anthropology, Simon Fraser University, Burnaby, BC.

Yesner, D. R. (1980) 'Caribou exploitation in interior Alaska: paleoecology at Paxson Lake', *Anthropological Papers of the University of Alaska*, 19, 15–31.

Yesner, D. R. (1996) 'Human Adaptation at the Pleistocene-Holocene boundary (circa 13,000 to 8,000 BP)', in *Humans at the end of the Ice Age: The Archaeology of the Pleistocene-Holocene Transition*, eds L. G. Straus, B. V. Eriksen, J. M. Erlandson, and D. R. Yesner, Plenum Press, N.Y., 255–276.

Zazula, G. (2006) *Paleoecology of Late Pleistocene Arctic Ground Squirrel Middens and Glacial Environments of the West-Central Yukon Territory*. Unpublished PhD dissertation. Department of Biological Sciences, Simon Fraser University, Burnaby, BC.

Cervid Exploitation and Symbolic Significance in Prehistoric and Early Historic Periods

Hunting, Performance and Incorporation: Human–deer encounter in Late Bronze Age Crete

Kerry Harris

Introduction

The role of deer in prehistoric Crete is somewhat enigmatic; as deliberate human importations they cannot be described as wild in the traditional sense (e.g. as independent of human influence)[1] and yet do not seem to be treated by humans in the same way as the domestic species.

There is evidence for their presence on Crete since the Neolithic, yet rarely in any great numbers. During the Late Bronze Age on the southern Greek mainland the use of hunting imagery (including deer hunting) becomes a symbolic resource for 'Mycenaean' elites,[2] yet the osteological remains of deer, especially European fallow deer (*Dama dama dama*), are often scarce (Yannouli and Trantalidou 1999). In Chania, west Crete, during a period in which the region is characterised by its 'Mycenaean' affiliations, the consumption of deer is significantly higher than anywhere else in Crete (Hallager 2001; Harris n.d.). The significance afforded to these animals and their role in the formulation and maintenance of social relationships and local and regional identities will be investigated through the zooarchaeological evidence for fallow deer[3] on Crete, particularly the zooarchaeological assemblages from Chania.

Cretan deer in an Aegean context

Although paleontological discoveries found evidence for endemic Pleistocene cervids on Crete, these are thought to have become extinct long before the start of the Holocene (de Vos 1996). There is, as yet, no evidence for red deer (*Cervus elaphus*) and fallow deer species on Crete prior to the Neolithic, thus on Crete, as on Mediterranean islands elsewhere (e.g. Masseti 1996; Carenti *et al.* this volume; Masseti *et al.* this volume), deer would have been deliberately, and with considerable effort, imported.

The earliest archaeological fallow deer specimens on Crete come from the Neolithic levels at Knossos (Jarman 1996; Isaakidou 2004). It is therefore supposed that this species may also have been brought along with sheep (*Ovis* sp.), goat (*Capra* sp.), cattle (*Bos* sp.) and pigs (*Sus* sp.) by the first settlers to Crete, probably from Anatolia, and may indicate its early introduction (but unsuccessful establishment) on Crete during the Neolithic (Isaakidou 2004, 297). By contrast, elsewhere on Aegean islands fallow deer do seem to have been successfully introduced along with the main domesticates during the Neolithic, with relatively numerous osteological fallow deer remains being recorded on Rhodes and Chios (Halstead and Jones 1987; Masetti 1999; Yannouli and Trantalidou 1999), although this is perhaps less surprising given the proximity of these islands to the Anatolian mainland unlike Crete. Nonetheless, the westward distribution of European fallow deer from Anatolia across Europe is acknowledged to be directly attributed to human transportation, albeit via the Macedonia and Thrace regions to the north rather than insular and peninsular Greece (e.g. Sykes *et al.* 2013).

With the Bronze Age a 'new wave of imports is seen in Crete' (Jarman 1996, 219 see also Wilkens 2003, 86). Deer appear to become more numerous and widespread (see Table 5.1) and it seems likely that fallow deer populations were established. This hypothesis finds some support in the recent study by King *et al.* (2008), based on modern genetic data, that a further human population influx from Anatolia to Crete occurred at the time associated with the Neolithic/Bronze Age transition. On the southern Greek mainland, however, the only (Holocene) sites from which fallow deer remains have been recovered so far are the Mycenaean (Late Bronze Age) palace site at Tiryns (only 0.04 percent of the assemblage, based on data in Yannouli and Trantalidou 1999) and a possible fragment from Asine (Nilsson 1996). During the Iron Age on Crete the presence of deer remains seems to decrease again.

The status of Cretan deer

The deliberate introduction of deer onto Crete is an issue of considerable interest. This phenomenon of wild fauna being introduced onto islands has been noted elsewhere in the Mediterranean (e.g. Vigne 1999), and is attributed to meeting the need of 'wild game' where large indigenous species are absent (*ibid.*). These animals are often described as being of 'symbolic' value in contrast to the 'economic' purposes of the domestic species (e.g. Jarman 1976; Jarman 1996; Vigne 1999). The question surrounding the status of deer on Crete as 'wild' needs to be examined, and many researchers acknowledge that a rigid wild/domestic dichotomy is overly simplistic to account for the variety of human/animal relationships in the past (see Note 1). Halstead (1987) asks 'why deer are referred to as wild when they have been deliberately imported by man' and Jarman (1976, 1996) also suggests in the case of fallow deer that, due to the effort expended to import this animal, they were at least semi-

Period	Site	Red deer	Fallow deer	Deer	Reference
Neolithic	Knossos	0.01% n 2	0.01% n 2		(Isaakidou 2004)
Bronze Age	Ayia Triada		1.4% n 10		(Wilkens 1996)
	Mochlos Chalinomouri farmhouse			1.8% n 12	(Reese 2004)
	Knossos		0.5% n 19		(Isaakidou 2004)
	Knossos (Minoan Unexplored Mansion)	0.8% n 22			(Bedwin 1984)
	Kommos			0.004% n 7	(Reese 1995)
	Karphi	6 antler fragments			(Pendlebury *et al.* 1937–38)
	Chania (Ayia Aikaterini)	2% n 65	6% n 195	4% n 113	(Harris n.d.)
	Chania (Odos Daskaloyannis)	3% n 57	12% n 227	0.8% n 16	(Harris n.d.)
	Chania (Mathioudaki)	3% n 8	5% n 18	0.3% n 1	(Harris n.d.)
	Chamalevri	present	present		(Tzedakis and Martlew 1999)
	Thronos/Kephala	present	present		(D'Agata 1997–2000)
	Tylissos	present			(Jarman 1996)
	Dictaen Cave		present		(Boyd-Dawkins 1902)
Iron Age	Kastro		0.1% n 10		(Snyder and Klippel 1999)
	Prinias		2.4 % n 11		(Wilkens 2003)
	Kommos			0.01% n 3	(Reese *et al.* 2000)
	Knossos (sanctuary of Demeter)		n 1?		(Jarman 1973)
	Thronos/Kephala			present	(D'Agata 1997–2000)

TABLE 5.1 Data for deer remains in zooarchaeological assemblages across Crete (figures relate to the number of identified specimens and the percentage this represents of the whole assemblage).

domestic rather than fully wild. Masseti (1996, 17), on the other hand, notes that the introduction of cervids onto islands in sufficient numbers to act as a breeding stock argues for a high degree of control over the animals and for their sophisticated management; however, he suggests that it is unlikely that fallow deer on the Mediterranean islands were domesticated, but were possibly released and hunted.

Discussions relating to the 'wild/domestic' status of deer in this context are often based upon the logic that a high quantity of remains equates to economic importance and, *ipso facto*, domestic status (e.g. Jarman 1976, 92). On Crete the generally low quantity of deer remains in zooarchaeological assemblages is seen as indicative of little economic value, however the 'symbolic' role of the animal (e.g. as a 'prestige object') is sometimes considered (e.g. Jarman 1996; Isaakidou

2004). A different perspective might consider that few remains does not indicate insignificance (as sometimes assumed), but rather that the nature of incorporation of deer into the human life-space was different to that of the domestic species, for example occasional encounters compared to the more regular cycles of husbandry practice. Indeed, at Kalythies, Rhodes, Halstead describes a difference in body part representation between sheep, goat, cattle and pig on the one hand and fallow deer on the other as suggesting a real difference in management, which he interprets as deer being hunted at a distance from settlement with only selected parts brought back to the site (Halstead and Jones 1987).

The role and perception of deer is almost certainly a complex and multifaceted one which likely incorporated elements of symbolism and economics; however the effort made to introduce living deer onto Crete as well as the evidence from the osteological remains suggests that in some contexts actual physical encounter with these animals was a meaningful and sought after interaction. The following part of this paper uses a zooarchaeological case-study from Late Bronze Age west Crete to explore this phenomenon through the lens of human engagement with the materiality and physicality of the animal body to provide an additional perspective to the issues raised above.

The zooarchaeological evidence from Chania and the human–deer intersection

At a number of contemporary Late Bronze Age sites in Chania, west Crete (Ayia Aikaterini, Odos Daskaloyannis, Mathioudaki) the zooarchaeological assemblage comprised the usual range of domestic species generally found on Cretan Bronze Age sites; in contrast however, the percentage of fallow deer remains was significantly higher (see Table 5.1). Both male and female deer were equally represented[4] and of a range of ages, although adult animals of four years or more predominate. The presence of whole deer is indicated by bones from all parts of the body including head, neck, podials, and limbs, with the latter – the main meat bearing elements – predominating. The bones show evidence for butchery with marks suggestive of carcass dismemberment and the filleting of meat, indicating that these were the remains of food consumption events. Gnawing marks on some of the remains suggest they were subsequently fed to dogs or scavenged once discarded. Whereas the remains of domestic species were recovered from a range of features across the sites, the majority of the deer remains were deposited in large pits along with large quantities of pottery, including food and drink serving and consumption vessels (Hallager and Hallager 2000; Hallager and Hallager 2003). The presence of articulating bones (and conjoining sherds) and lack of weathering damage suggests this material was deposited in the pits soon after the consumption event. Yet despite the high quantity of postcranial deer remains the presence of antler is scarce; some pieces that were found showed evidence for working, it is therefore supposed that some antler was removed from these contexts for use elsewhere.

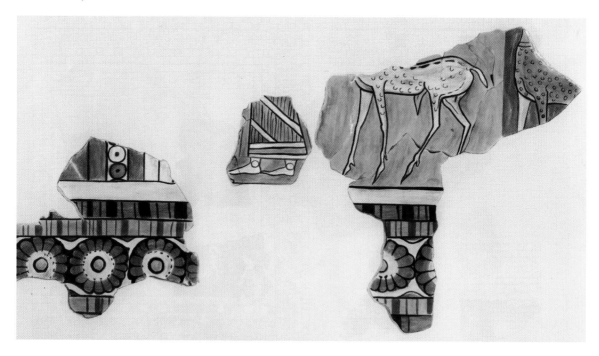

These remains speak of the entanglement of humans and deer/deer bodies in a linked sequence of highly physical and sensory practices, commencing with interaction with the live animal. Yet what was the nature of the interaction between humans and deer on Crete? The iconographic data from the Mycenaean Greek mainland portray deer in hunting narratives (Lang 1969, pl. 121) but, as noted by the aforementioned authors, the anthropogenic establishment of deer on Crete and possibly their close management for survival perhaps belies these animals as wild and hunted in the traditional sense. The significantly higher quantities of deer remains in the zooarchaeological assemblages from Chania could perhaps be evidence for the sustained maintenance of deer herds in this region. If so, there does not appear to have been a deliberate culling strategy for consistent age or sex (except perhaps an emphasis on adult animals); this might suggest killing based on a more 'chance' encounter between deer and human,[5] or perhaps the identification and selection of specific animals based on individual qualities.

Could the maintenance of deer herds on Crete be seen as a deliberate fostering of the possibility for encounter between human and deer? A creation of the potential for hunting, but perhaps not in the traditional sense, rather a *performance* of the hunt (see Marvin 2000): the heightened, physical and sensory interaction between unrestrained and dangerous sentient beings, culminating in death, perhaps even a ritualised death through sacrifice.[6] However, the intersection of human and deer does not end there, rather the zooarchaeological evidence suggests that further engagement occurs through fragmentation of the deer body: the antlers are perhaps removed (chop marks against the skull

FIGURE 5.1.
Reconstruction of a fresco fragment from Ayia Triada, Crete, depicting 'women leading fallow deer to an altar' (after Militello 1998, reproduced by permission of the Scuola Archeologica Italiana di Atene).

pedicle), the pelt is removed (cut marks on limb extremities), the body is parted (dismemberment cut marks), the flesh is stripped (filleting cut marks), the bones are given to dogs (gnawing marks). The deer is consumed.

The deer is consumed, along with other animals (and presumably other substances) in large-scale feasting events. The archaeological evidence for these events consists of sizeable pit features containing numerous serving and consumption vessels and large quantities of bone, including many articulating bones. The majority of deer remains come from these pits and a higher percentage of canid gnawing occurs on bones from these pits than from features elsewhere.

The consumption of deer marks the event as distinctive, different to the everyday. Meat is prepared in larger portion sizes at these events and the larger fragment size and lack of splintering suggests an emphasis on meat abundance but not intensive carcass exploitation. A greater participation of dogs as co-consumers perhaps further marks these events as distinctive and, given the frequent representation of hunting dogs in many of the iconographic hunt scenes (e.g. Lang 1969; Immerwahr 1990), these may well have been the dogs used in hunting.

The deer is consumed and incorporated by the participants, and the shared sensory experience of communal consumption binds and incorporates the individual participants into the wider group (e.g. Falk 1997; Hamilakis 2008). The act of gathering up and depositing the feasting remains into large pits acts as a closing performative phase, after which the pits become a material mnemonic record of the previous events (Hamilakis 2008; Hamilakis and Harris 2011). It is possible that antler (synecdochic of deer), as 'tokens of heightened moments' (Marvin 2011), were removed from these contexts for use as tools or other objects, an inherent part of the biography of which were the events that went before (e.g. Gosden and Marshall 1999). Why such practices appear to predominate in Chania, west Crete is considered below.

Hunting and the Mycenaean ideology?

Late Bronze Age Chania, *c.*1300–1100 BC, is characterised by its connections with the Mycenaean mainland; significant trade links and a common administrative system are in place (Hallager 1987; Andreadaki-Vlasaki 2000; Hallager 2005) and, in the ostentatious mainland style burial practices that also flourish in west Crete at this time (notably the 'warrior' grave), there is evidence too for a shared elite ideology, or at least the incorporation of elements of a mainland elite symbolism (Preston 2004). At a time in which the Cretan political landscape was becoming increasingly fragmented with greater variation in regional identities (e.g. Borgna 2004; Preston 2004; Smith 2005), it is possible that in west Crete a regional identity was developed that was characterised by expressions of affiliation with the Mycenaean mainland. Hunting was a prominent theme in Mycenaean high status iconography (Immerwahr 1990) and has been convincingly argued to have been an important ideological resource for Mycenaean elites as a means of generating and legitimating power and authority based, in part, on exhibition of participation in external geographical and temporal realms and demonstration of a successful confrontation with 'otherness' (Morris 1990; Hamilakis 1996; 2003).[7] The increased quantities of deer in the Chania assemblages, if we are to assume that maintenance of deer herds was associated with hunting, could perhaps be a further manifestation of affiliation with Mycenaean ideologies.

However, the issue is more complex: Hamilakis (2003) notes that the paucity of wild animals in Mycenaean zooarchaeological assemblages is at odds with the prominence of hunting in the iconographic repertoire, and as such implies that the 'deployment of the idea of hunting and its representation' counted for more than the act itself (see also Marinatos 1990; Morgan 1995; Thomas 1999). Fallow deer, particularly, are a case in point: as noted above they occur frequently in iconographic representations but osteological remains are extremely scarce. Furthermore, Cretan incorporation of Mycenaean elements was not a case of passive wholesale adoption; rather was selective, and adapted and fused with local traditions in innovative ways (Preston 2004). Here then, hunting was not so much a metaphoric or representational device but rather was performative and structured, in which confrontation with (and incorporation of) the physicality of the animal was key.

Jones (1998) notes that animals signify and embody specific aspects of the spatial and temporal landscape, are a means through which people identify themselves with such and embody the memory of a particular place. In Late Bronze Age Chania, through a fusion of 'Mycenaean' and 'Minoan' histories, the Cretan fallow deer embodies a complex paradox: the exotic, the distant, and the unfamiliar, but perhaps also tradition, ancestry, an Anatolian origin-myth. In the zooarchaeological data from Chania there is evidence that these complexities were repeatedly consumed and incorporated in a performative and possibly ritualised sequence of hunting, feasting, deposition and dispersal of animal remains. Through these practices social relationships, identities and memories could be defined and embodied on an individual, community and regional scale; a defining element of which was human interaction with deer.

Notes

1. The debate regarding the culturally specific nature of concepts such as 'wild' is extensive (e.g. see papers in Descola and Pálsson 1996) and not addressed in detail here. The stance taken by this author considers wild animals as those engaged with via extra-ordinary practices, associated with a temporality outside of the daily rhythms of husbandry practice, and which inhabit, evoke and embody different parts of the landscape to those of the domestic.
2. The use of the terms Minoan and Mycenaean here refer to Cretan or mainland-derived material culture characteristics respectively, but does not imply a correlation with ethnicity.
3. Red deer too are present but generally in lower quantities than fallow deer; they will

not be discussed in this paper as the role and perception of the different deer species may have varied. Roe deer has also been recorded, but its presence on Crete is not widely acknowledged (see Jarman 1996).

4. Based on the number of male and female pelves.

5. Here I mean that in the course of the hunt the conjunction of hunter and deer was one of chance (rather than selection of an animal dictated by its age or sex); I do not mean a chance encounter in the sense of happening upon a deer whilst in the process of another activity.

6. Figure 5.1, The Ayia Triada, Crete, fresco fragment depicts what has been interpreted as women leading deer to an altar (Militello 1998), if so it should be considered that this is not the realistic representation of a practice but the idealised perception of a docile animal subjecting itself to sacrifice.

7. Along with long-distance travel, acquisition of exotic goods and skilled crafting (after Helms 1988).

References

Andreadaki-Vlasaki, M. (2000) *The County of Khania Through its Monuments*, Ministry of Culture, Athens.

Bedwin, O. (1984) 'Appendix 2. The animal bones', in *The Minoan Unexplored Mansion at Knossos*, M. R. Popham, Thames and Hudson, Oxford, 307–308.

Borgna, E. (2004) 'Social meanings of food and drink consumption at LMIII Phaistos', in *Food, Cuisine and Society in Prehistoric Greece*, eds P. Halstead and J. C. Barrett, Oxbow, Oxford, 174–195.

Boyd-Dawkins, W. (1902) 'Remains of animals found in the Dictaean Cave in 1901', *Man* 114, 162–165.

D'Agata, A.-L. (1997–2000) 'Ritual and rubbish in Dark Age Crete: The settlement of Thronos/Kephala (Ancient Sybrita) and the Pre-Classical roots of a Greek City', *Aegean Archaeology* 4, 45–59.

de Vos, J. (1996) 'Taxonomy, ancestry, and speciation of the endemic Pleistocene deer of Crete compared with the taxonomy, ancestry, and speciation of Darwin's finches', in *The Pleistocene and Holocene Fauna of Crete and its First Settlers*, ed. D. Reese, Prehistory Press, Madison, Wisconsin, 111–124.

Descola, P. and Pálsson, G. eds (1996) *Nature and Society: Anthropological Perspectives*, Routledge, London.

Falk, P. (1997) *The Consuming Body*, Sage Publications Ltd, London.

Gosden, C. and Marshall, Y. (1999) 'The cultural biography of objects', *World Archaeology* 31, 169–178.

Hallager, B. P. (2005) 'The synchronisms Mainland-West Crete in the LMIIIA$_2$–IIIB period', in *Ariadne's Threads: Connections between Crete and the Greek mainland in Late Minoan III (LMIIIA$_2$–LMIIIC)*, A.-L. D'Agata, J. Moody and E. Williams, Scuola Archeologica Italiana di Atene, Athens, 277–292.

Hallager, E. (1987) 'The Inscribed Stirrup Jar: Implications for Late Minoan IIIB Crete', *American Journal of Archaeology* 91, 171–190.

Hallager, E. (2001) 'A waste deposit from LBA shrine in Khania (?)', in *Potnia : Deities and Religion in the Aegean Bronze Age: Proceedings of the 8th International Aegean Conference*, eds R. Laffineur and R. Hägg, Austin Université Histoire de l'Art et Archéologie de la Grèce Antique/ University of Texas at Austin, Program in Aegean Scripts and Prehistory, Liège, 175–180.

Hallager, E. and Hallager, B. P. eds (2000) *The Greek-Swedish excavations at the Agia Aikaterini Square, Kastelli, Khania 1970–1987: Results of the Excavations under the direction of Yannis Tzedakis and Carl-Gustaf Styrenius.Vol. II. Text and Plates.The Late Minoan IIIC Settlement.*, Åström, Jonsered.

Hallager, E. and Hallager, B. P. eds (2003) *The Greek-Swedish Excavations at the Agia Aikaterini Square, Kastelli, Khania 1970–1987: Results of the Excavations under the Direction of Yannis Tzedakis and Carl-Gustaf Styrenius. Vol 3, pts 1 and 2. The Late Minoan IIIB:2 Settlement*, Åström, Jonsered.

Halstead, P. (1987) 'Man and other animals in later Greek prehistory', *Annual of the British School at Athens* 82, 71–83.

Halstead, P. and Jones, G. (1987) 'Bioarchaeological remains from Kalythies Cave, Rhodes', in *I Neolithiki Periodos sta Dodekanisa*, A. Samson, Ministry of Culture, Athens, 135–152.

Hamilakis, Y. (1996) 'A footnote on the archaeology of power: zooarchaeological evidence from a Mycenaean chamber tomb at Galatas. NE Peloponnese', *Annual of the British School at Athens* 91, 153–66.

Hamilakis, Y. (2003) 'The sacred geography of hunting: wild animals, social power and gender in early farming societies', in *Zooarchaeology in Greece: Recent Advances*, eds E. Kotjabopoulou, Y. Hamilakis, P. Halstead, C. Gamble and P. Elefanti, BSA Studies 9, London, 239–248.

Hamilakis, Y. (2008) 'Time, performance and the construction of a mnemonic record: from feasting to an archaeology of eating and drinking', in *DAIS: The Aegean Feast*, eds L. Hitchcock, R. Laffineur and J. Crowley, Université de Liège, Liège, 3–22.

Hamilakis, Y. and Harris, K. (2011) 'The social zooarchaeology of feasting: the evidence from the "ritual" deposit at Nopigeia-Drapanias', in *Proceedings of the 10th International Cretological Congress (Khania, 1–8 Oct. 2006)*, eds E. Kapsomenos, M. Andreadaki-Vlazaki and M. Andrianakis, Literary Society of Khania 'Chrysostomos', Khania, 225–244.

Harris, K. (in prep) *The Social Role of Hunting and Wild Animals in Late Bronze Age Crete: Aa Zooarchaeological Analysis*. Unpublished PhD dissertation. University of Southampton..

Helms, M. W. (1988) *Ulysses' Sail: An Ethnographic Odyssey of Power, Knowledge, and Geographical Distance*, Princeton University Press, Princeton.

Immerwahr, S. A. (1990) *Aegean Painting in the Bronze Age*, The Pennsylvania State University Press, University Park and London.

Isaakidou, V. (2004) *Bones from the Labyrinth: Faunal evidence for Management and Consumption of Animals at Late Neolithic and Bronze Age Knossos, Crete*. Unpublished PhD dissertation. University College London.

Jarman, M. R. (1973) 'Chapter IX: Preliminary report on the animal bones', in *Knossos: The Sanctuary of Demeter*, ed. J. N. Coldstream, British School at Athens, 177–179.

Jarman, M. R. (1976) 'Early animal husbandry', *Philosophical Transactions of the Royal Society of London. Series B, Biological Sciences* 275, 85–97.

Jarman, M. R. (1996) 'Human influence in the development of the Cretan mammalian fauna', in *Pleistocene and Holocene Fauna of Crete and its First Settlers*, ed. D. Reese, Prehistory Press, Madison Wisconsin, 211–229.

Jones, A. (1998) 'Where eagles dare: landscape, animals and the Neolithic of Orkney', *Journal of Material Culture* 3, 301–324.

King, R. J., Özcan, S. S., Carter, T., Kalfoğlu, E., Atasoy, S., Triantaphylllidis, C., Kouvatsi, A., Lin, A. A., Chow, C.-E. T., Zhivotovsky, L. A., Michalodimitrakis, M.

and Underhill, P. A. (2008) 'Differential Y-chromosome Anatolian influences on the Greek and Cretan Neolithic', *Annals of Human Genetics* 72, 205–214.

Lang, M. L. (1969) *The Palace of Nestor at Pylos in Western Messenia. Vol II. The Frescoes*, Princeton University Press, Princeton.

Marinatos, N. (1990) 'Celebrations of death and the symbolism of the lion hunt', in *Celebrations of Death and Divinity in the Bronze Age Argolid*, eds R. Hagg and G. C. Nordquist, Stockholm, 143–148.

Marvin, G. (2000) 'Natural instincts and cultural passions: Transformations and performances in foxhunting', *Performance Research* 5, 108–115.

Marvin, G. (2011) 'Enlivened through memory: hunters and hunting trophies', in *The Afterlives of Animals. A Museum Menagerie*, S. Alberti, University of Virginia Press, Charlottesville, 346–374.

Masseti, M. (1996) 'The postglacial diffusion of the genus *Dama* Frisch, 1775, in the Mediterranean region', *Supplemento alle Richerche di Biologia della Selvaggina* 25, 7–29.

Militello, P. (1998) *Haghia Triada I. Gli Affreschi*, Scuola Archeologica Italiana di Atene, Padova.

Morgan, L. (1995) 'Of animals and men: the symbolic parallel', in *Klados: Essays in honour of J. N. Coldstream*, C. Morris, University of London, Institute of Classical Studies, London, 171–184.

Morris, C. E. (1990) 'In pursuit of the white tusked boar: aspects of hunting in Mycenaean society', in *Celebrations of Death and Divinity in the Bronze Age Argolid*, eds R. Hagg and G. C. Nordquist, Stockholm, 149–156.

Nilsson, K. M. (1996) 'Animal bones from Terrace III in the lower town of Asine', in *Asine III supplementary Studies on the Swedish Excavations 1922–1930*, eds R. Hagg, G. C. Nordquist and B. Wells, Stockholm, 111–115.

Pendlebury, H. W., Pendlebury, J. D. S. and Moneycoutts, M. B. (1937–38) 'Excavations in the Plain of Lasithi. III. Karphi: Acity of refuge in the Early Iron Age in Crete. Excavated by the students of the British School of Archaeology at Athens, 1937–1939', *British School at Athens* 38, 57–145.

Preston, L. (2004) 'A mortuary perspective on political changes in Late Minoan II-IIIB Crete', *American Journal of Archaeology* 108, 321–348.

Reese, D. (1995) 'The Minoan fauna', in *Kommos 1. The Kommos Region and Houses of the Minoan Town. Part 1, The Kommos Region, Ecology and Minoan Industries*, eds J. W. Shaw and M. C. Shaw, Princeton University Press, Princeton, 163–193.

Reese, D. (2004) 'Fauna and flora', in *Mochlos IC. Period III. Neopalatial Settlement on the Coast: The Artisans' Quarter and the Farmhouse at Chalinomouri. The Small Finds*, ed. J. S. Soles, INSTAP Academic Press, Philadelphia, 117–125.

Reese, D., Rose, M. and Ruscillo, D. (2000) 'The Iron Age fauna', in *Kommos IV. The Greek Sanctuary. Part I*, eds J. W. Shaw and M. C. Shaw, Princeton University Press, Princeton/ Oxford, 415–646.

Smith, R. A. (2005) 'Minoans, Mycenaeans and Mochlos: The formation of regional identity in Late Minoan III Crete', in *Ariadne's Threads: Connections between Crete and the Greek Mainland in Late Minoan III (LMIIIA$_2$–LMIIIC)*, eds A.-L. D'Agata, J. Moody and E. Williams, Scuola Archeologica Italiana di Atene, 2005, Athens, 185–204.

Snyder, L. M. and Klippel, W. E. (1999) 'Dark age subsistence at the Kastro Site, East Crete: Exploring subsistence change and continuity during the Late Bronze Age-Early Iron Age transition', in *Palaeodiet in the Aegean*, eds S. J. Vaughan and W. D. E. Coulson, Oxbow, Oxford, 65–83.

Sykes, N. J., Carden, R. F. and Harris, K. (2013) 'Changes in the size and shape of fallow deer – evidence for the movement and management of a species', *International Journal of Osteoarchaeology* 23, 55–68.

Thomas, N. (1999) 'The war animal: three days in the life of the Mycenaean lion', in *Polemos I: Le contexte guerrier en Egée à l'âge du Bronze: Actes de la 7e rencontre égéenne internationale, Université de Liège, 14–17 avril 1998*, ed. R. Laffineur, Univérsité de Liège, Liège, 277–312.

Tzedakis, Y. and Martlew, H. (1999) *Minoans and Mycenaeans: flavours of their time*, National Archaeological Museum, Athens.

Vigne, J.-D. (1999) 'The large "true" Mediterranean islands as a model for the Holocene impact on the European vertebrate fauna? Recent data and new reflections', in *The Holocene History of the European Vertebrate Fauna: Modern Aspects of Research*, ed. N. Benecke, Verlag Marie Leidorf GmbH, Rahden/Westf, 295–322.

Wilkens, B. (1996) 'Faunal remains from Italian excavations on Crete', in *Pleistocene and Holocene Fauna of Crete and Its First Settlers*, ed. D. Reese, Prehistory Press, Madison Wisconsin, 241–261.

Wilkens, B. (2003) 'Hunting and breeding in ancient Crete', in *Zooarchaeology in Greece: Recent Advances*, eds E. Kotajabopoulou, Y. Hamilakis, P. Halstead, C. Gamble and P. Elefanti, BSA Studies 9, British School at Athens, London, 85–90.

Yannouli, E. and Trantalidou, K. (1999) 'The fallow deer (*Dama dama* Linnaeus, 1758): Archaeological presence and representation in Greece', in *The Holocene History of the European Vertebrate Fauna: Modern Aspects of Research*, ed. N. Benecke, Verlag Marie Leidorf, Rahden, 247–281.

Archaeozoology of the Red Deer in the Southern Balkan Peninsula and the Aegean Region During Antiquity: Confronting bones and paintings

Katerina Trantalidou and Marco Masseti

Introduction

It is well known that the contribution made by wild animals to human diet and material culture declined from the Neolithic period onwards, as people came to increasingly rely on domestic species (Greenfield 1986; Bökönyi 1989, 316–7; Trantalidou 1990). Nevertheless, the social and cultural significance of wild animals endured and, in the Balkans and Aegean region, cervids continued to be hunted into the historic period.

Herein, we examine the evidence for the presence of of red deer (*Cervus elaphus*) from the Haemus peninsula and Aegean. We attempt to draw together zooarchaeological data (relative frequencies, skeletal element representation, size and morphological traits of antlers) with iconographic evidence (paintings, mainly on the Athenian vases) in order to review the dynamics of human-red deer interactions between the seventh millennium BC and the fourth century AD.

Comparisons of artistic and zooarchaeological data not only provide information about the symbolic significance of animals but they may also help to reconstruct the ancient biogeography of red deer lineages and subspecies. The red deer is known to divide into several distinct lineages and sub-species (see Skog *et al.* 2009 and Ludt *et al.* 2004 for genetic evidence and Sommer and Nadachowski 2006 for fossil record evidence). However, there is a continuing debate with regards to the division of subspecies due to the numerous and continuing data generated from molecular studies. We also discuss the synthesis of these findings and the limitations of these results if attempting to determine specific subspecies in antiquity. This novel study should be useful to other researchers attempting similar inter-disciplinary studies, as well as highlighting potential caveats and limitations in such research.

Methods

A variety of methods were employed combining both zooarchaeological and art historical techniques, which can be summarised as follows.

Frequency of red deer remains in archaeological wild animal assemblages

Zooarchaeological data were collated from 43 mainland sites (Figure 6.1, Table 6.1) located across the Balkan Peninsula. For each assemblage, data pertaining to the representation of red deer, roe deer and fallow deer were noted, together with quantification figures for the total vertebrate and total wild fauna assemblages. For reasons of comparability, only fragment counts (NISP – Number of Identified Specimens) were utilised. These data are presented in Table 6.1, which show the raw data but also the frequencies of red deer, and roe/fallow deer, both expressed as a percentage of the wild fauna total.

It is important to recognise that these data provide only generalised indications of deer representation as they are limited by cultural and natural processes of deposition, taphonomy, recovery, reporting or simply small sample size. Wherever possible attempts have been made to mitigate these factors; for instance specimens assumed to represent votive deposits, trophies or debris from craft working have been excluded from the study. However, in some cases it was not possible to account for inter-site variations in the dataset. Therefore, the data should be viewed as providing only broad brush evidence for diachronic shifts in red deer representation.

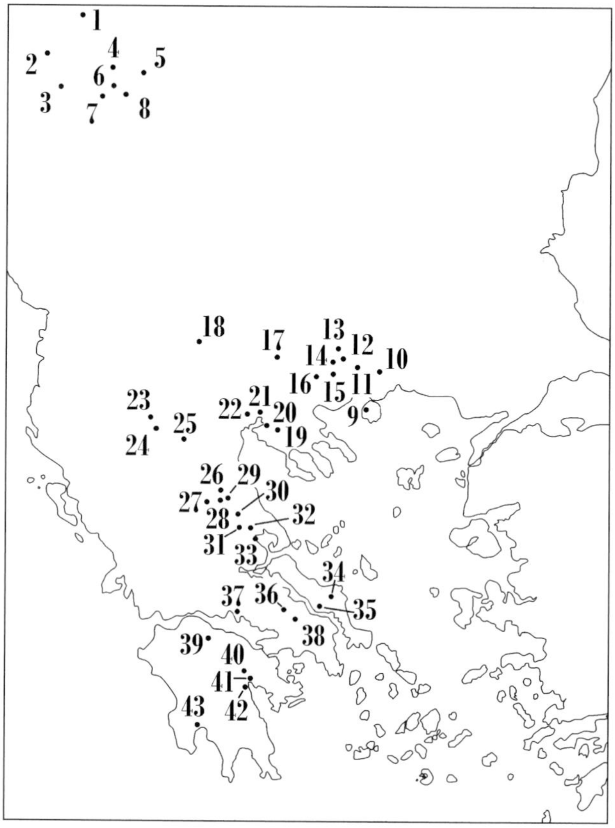

FIGURE 6.1. Map representing the 43 mainland of sites across the Balkan Peninsula from which zooarchaeological data were collated from.

Iconographic representation of red deer on pottery

In ancient Greek society, decorated vessels were much valued and produced in great numbers. Deer are one of the most frequently represented animals on these vessels and, for the preparation of this paper an extensive survey was conducted of the Beazley Archive (BA) and various corpora of vases. A quick review of deer depictions created on pottery, figurines, funeral monuments, architectural sculpture (Hofsten 2007), seals, gems and scaraboids revealed more than 1,500 images. However, it was necessary to leave aside the majority of these representations due to their ambiguous nature. For instance, many depictions appear unrealistic (e.g. Higgins 1967, 21, pl. 7D) and therefore

Relative chronology, phases	Site (map no.)	Total of vertebrated fauna	Wild fauna	Red deer		Fallow and/or Roe deer	Roe deer	References
		NISP	NISP	NISP	%	NISP %	NISP %	
EN, Starčevo culture (6100–5100 BC)	Bukovačka česma (8)	270	166	78	48.6	–	11.4	Greenfield 1992–93
I (6th–mid 5th millennium BC)	Divostin (6)	2401	203	45	22.2	–	0.5	Bökönyi 1988
II (4th M BC)		10785	1613	416	25.8	–	0.1	
LN (4100–3300 BC)	Opovo (1)	642	328	232	70.7	–	12.2	Greenfield 1986
MN, Vinča B (4250–4100 BC)	Petnica (3)	167	85	60	72.9	–	17.6	Greenfield 1986; Orton 2008, 217
LN, Vinča C (4100–3900 BC)		297	196	82	41.8	–	11.2	
LN, Vinča D (3900–3300 BC)		107	37	31	83.8	–	5.6	
Chalcolithic (3300–2500 BC)		250	139	120	86.3	–	3.6	
LBA–EIA (1300–800 BC)		194	66	42	63.6	–	3.6	
I–III, EBA–MBA (1950–1550 BC)	Ljuljac (7)	1719	712	257	38.5	–	2.5	Greenfield 1986
EB–EIA	Livade (5)	1033	221	95	43	–	4.5	Greenfield 1986
LN, Chalcolithic	Gomolava (2)	2592	919	467	53.6	–	13.9	Clason 1979; Orton 2008
EBA	Novačka Ćurpija (4)	532	12	1	8.3	–	8.3	Greenfield 1986
I (5500–5200 BC)	Sitagroi (12)	1848	159	59	37.1	7.5	–	Bökönyi 1986
II (5200–4600 BC)		6096	211	84	39.8	4.3	–	
III (4600–3500 BC)		13080	1062	480	45.1	5	–	
IV (3500–3100 BC)		4330	765	286	37.4	18	–	
V (3100–2200 BC)		9115	594	142	23.9	29.8	–	
3090–2210 BC	Springs of Anghitis cave (13)	493	144	40	31.2	38.2	–	Τρανταλίδου et al. 2007
3085–2775 BC	Orpheas cave, Anghitis gorge (11)	4055	460	10	2.2	1	–	Διώτη 2009, 30
Ia–b, MN (Sitagroi I)	Dimitra (15)	594	15	6	40	20	–	Yannouli 1997
II, LN (Sitagroi II)		553	17	4	23.5	–	–	
III, FN		1287	260	14	5.4	3.4	–	
EBA		235	23	2	8.7	8.7	–	

TABLE 6.1. Relative frequencies of cervids collected from Holocene sites (see Figure 6.1) across the Balkan Peninsula and Aegean region. Bone quantification is presented in terms of identifiable bones from each site (NISP), the identified red deer bones as well as the percentage of cervids within the total for wild fauna. For each table chronological horizons and terminology are used as proposed by the excavators. Continues pp. 62–65.
EN: Early Neolithic; MN: Middle Neolithic; LN: Late Neolithic; FN: Final Neolithic; EBA: Early Bronze Age; MBA: Middle Bronze Age; LBA: Late Bronze Age; EH: Early Helladic; MH: Middle Helladic; LH: Late Helladic; EIA: Early Iron Age.

Relative chronology, phases	Site (map no.)	Total of vertebrated fauna	Wild fauna	Red deer		Fallow and/or Roe deer	Roe deer	References
II (5300–5070 BC)	Promachon-Topolniča (17)	187	19	4	21	42.1	–	Kazantzis 2009
III (5070–4700 BC)		186	16	2	12.5	43.8	–	
LN		2007	282	115	40.8	0.7	–	Iliev and Spassov 2007
LN	Kryoneri (16)	419	60	32	53.3	3.3	–	Μυλωνά 2000
I, EBA a (Sitagroi IV–Va)	Pentapolis (14)	257	20	7	35	55	–	Yannouli 1994
II, EBA b (Sitagroi Vb)		511	51	13	25.5	56.9	–	
5th M (Karanovo V–VI, Sitagroi III, Dikili Tash II)	Paradeissos (Klisi Tepe) (10)	1766	296	186	62.8	9.5	–	Larje 1987
I, EBA a (Sitagroi Va)	Skala Sotiros (9)	4786	625	16	3.5	9.3	–	Yannouli 1994
II, EBA b (Sitagroi Vb)		1092	116	7	6	13.9	–	
Historic		1794	53	2	3.8	17	–	
I, MN	Vassilika (19)	253	9	5	55.55	33.33	–	Yannouli 1994
II, Early LN		1313	37	16	43.24	43.24	–	
III–IV, Late LN (Sitagroi III)		1179	21	10	47.61	42.85	–	
		538	11	6	54.54	9.09	–	
I–III (Sitagroi III)	Thérmi (B) (20)	1576	21	2	9.52	14.28	–	Yannouli 1994
I and II, end of MN–LN (5890–5531 BC)	Stavroupolis (21)	15504	1113	251	22.63	30.81	–	Γιαννούλη 2002; 2004
I, EN (end of 7th M)	Anza, Ovče Polje (18)	1198	41	4	9.75	–	–	Bökönyi 1976
II–III (5900–5500 BC)		750	22	5	22.72	9.09	–	
		1302	75	3	4.00	14.66	–	
IV, MN (5300–5000 BC)		3068	118	21	19.04	–	–	
EBA I (2400–1750 BC)	Kastanas (22)	111	52	6	12.53	34.61	–	Becker 1986
EBA II (2000–1680 BC)		123	34	5	14.70	73.52	–	
MBA (2150–1730 BC)		520	94	20	21.27	57.44	–	
LBA (1600–1300 BC)		2883	674	162	24.58	55.04	–	
1100–1200 BC		8460	4264	1125	26.38	61.86	–	
EIA (1000–800 BC)		6627	1612	360	22.33	80.32	–	
800–200 BC		6180	912	155	17.62	63.15	–	
C, MN (5459–5082 BC)	Dispilio (24)	361	75	44	58.66	36.00	–	Phoca-Cosmetatou 2008
B3b, MN		647	10	6	60.00	40.00	–	

Relative chronology, phases	Site (map no.)	Total of vertebrated fauna	Wild fauna	Red deer		Fallow and/or Roe deer	Roe deer	References
Late MN–LN Ia	Springs (Piges) of Koromilia Cave (23)	439	34	15	44.11	2.94	–	Trantalidou *et al.* 2011
LN/FN (5200–4950 BCE)	Megalo Nissi Galanis (25)	372	48	5	10.41	–	–	Greenfield and Fowler 2005
FN (4700–4450 BC)		2902	309	66	21.35	37.87	–	
FN/EBA (3rd M)		953	233	57	24.46	3.50	–	
I (*c*.64th–63th BC)	Achilleion (30)	961	66	15	22.72	13.63	–	Bökönyi 1989
II (*c*.62th–61st BC)		1489	102	30	29.41	8.82	–	
IIIa (*c*.61st BC)		2775	135	40	29.62	14.07	–	
IIIb–IVa (*c*.61st–59th BC)		504	24	5	20.83	12.5	–	
IVb (58th–56th BC)		2070	76	15	19.73	9.21	–	
Preceramic	Argissa (28)	2179	36	3	8.33	8.33	–	Boessneck 1962; von den Driesch 1987
EBA		656	134	98	73.13	0.74	–	
MBA		2352	312	193	61.85	6.08	–	
MN (5600–5200 BC)	Arapi and Otzaki (29)	464	5	3	60.0	–	–	Boessneck 1956; von den Driesch 1987
LN		266	13	1	7.69	7.69	–	
LN	Ayia Sofia Magoula (26)	3514	77	24	32.16	–	–	von den Driesch and Enderle 1976
Antiquity		137	17	1	5.88	–	–	
MN	Platia Magoula Zarkou (27)	951	39	19	49.71	30.76	–	Becker 1991
LN		1152	56	14	26.0	32.14	–	
EBA		3964	405	255	63.96	21.72	–	
EN	Sesklo (31)	722	56	14	37.89	12.5	–	Schwartz 1981
LN, early 5th M BC	Dimini (32)	2092	1943	113	4.81	3.53	–	Halstead 1992
LN	Pefkakia Magoula (33)	537	32	11	34.37	–	–	Jordan 1975; Amberger 1979; Hinz 1979
Chalcolithic (3rd M BC)		5742	489	292	58.63	0.61	–	
Chalcolithic–EBA		899	61	38	62.29	1.63	–	
EBA 2600–2000 BC)		8206	1136	906	79.75	0.26	–	
EBA–MBA		4072	2485	625	25.15	4	–	
MBA (2000–1550 BC)		10731	2920	2475	84.76	0.64	–	
LBA (1550–1200 BC)		1775	222	187	84.23	0.45	–	

TABLE 6.1. *continued.*
EN: Early Neolithic; MN: Middle Neolithic; LN: Late Neolithic; FN: Final Neolithic; EBA: Early Bronze Age; MBA: Middle Bronze Age; LBA: Late Bronze Age; EH: Early Helladic; MH: Middle Helladic; LH: Late Helladic; EIA: Early Iron Age.

Relative chronology, phases	Site (map no.)	Total of vertebrated fauna	Wild fauna	Red deer		Fallow and/or Roe deer	Roe deer	References
MN	Sarakenos cave (38)	1388	28	18	64.20	–	–	Trantalidou in prep.
LN Ia (5300–4800 BC)		5220	218	189	87.50	–	–	
LN Ib (4800–4300 BC)		2114	88	84	95.45	–	–	
LN IIa (4300–3800 BC)		786	25	22	88.00	–	–	
LN IIb (3800–3300/3200 BC)		941	42	41	97.61	–	–	
EH (EBA)		725	16	14	87.50	–	–	
MH (MBA)		1042	29	24	82.75	–	–	
Classical to Imperial period (c.4th BC to 180 AD)	Sanctuary of the Kabeiroi (36)	4001	176	69	39.20	22.15	–	Boessneck 1973
FN (LN Ib, 2nd half of 5th M)–LH Ia (mid of 2nd M)	Ayia Triada (37)	1523	93	19	20.43	3.22 (roe deer present)	–	Jensen 2006
I, LN Ia (5300–4800 BC)	Skoteini cave (34)	449	13	4	30.76	15.38	–	Kotjabopoulou and Trantalidou 1993
II, LN Ib		2511	122	29	23.77	6.55	–	
III, LN IIa (4300–3800 BC)		1099	72	19	26.38	–	–	
EH (2800–2100 BC)	Kaloyerovrissi (35)	69	6	4	66.66	44.44	–	Τραντάλίδου 1993
MH (2100–1550 BC)		58	8	2	25.00	75.00	–	
III, LN Ib (4700–4200 BC)	Cave of Lakes, Kastria (39)	967	86	46	54.48	–	–	Τραντάλίδου 1997
MH (2100–1550 BC)		52	13	3	23.07	–	–	
I, EN	Lerna (42)	166	19	2	10.52	–	–	Gejvall 1969
II, MN		594	19	11	57.89	–	–	
III, EH II (2500–2300 BC)		868	60	20	33.33	–	–	
IV, EH III (2300–2100 BC)		4292	189	75	39.68	5.33	–	
V, MH		5948	359	264	73.53	3.34	–	
VI, LH III (1400–1060 BC)		1124	91	48	52.74	–	–	

cannot be identified to species, whereas others depicting fallow deer reflect eastern influence (Καρδαρά 1963; Cook and Dupont 1998; Monaco 2002). All representations of juvenile and female deer were excluded, as they lack the defining traits that allow red, fallow and roe deer to be separated with confidence. With these restrictions in place, 30 images were examined for this paper.

TABLE 6.1. *continued.*

Relative chronology, phases	Site (map no.)	Total of vertebrated fauna	Wild fauna	Red deer		Fallow and/or Roe deer	Roe deer	References
*c.*13th–11th BC	Mycenae (40)	943	51	45	88.23	–	–	Trantalidou 2009
EH II, *c.*2400 BC	Tiryns (41)	2951	32	8	25.00	3.12 d.	–	von den Driesch and Boessneck 1990
EH II, *c.*1900 BC		2311	30	14	46.66	–	–	
*c.*16th–14th		532	9	7	77.77	–	–	
LH III B1 (*c.*13–14th BC)		2384	30	10	33.33	6.66 d.	–	
LH III B2, 1240/30–1200 BC		19503	359	238	66.29	3.62 d. and r.	–	
LH III C, early *c.*1200 BC		7368	65	30	46.15	9.23 d. and r.	–	
LH IIIc, developed		17788	337	258	76.55	5.04 d. and r.	–	
LH IIIC, late, *c.*1050 BC		4933	186	151	81.18	4.83 d. and r.	–	
MH I–II (2100–1600 BC)	Nichoria (43)	666	–	52	–	–	–	Sloan and Duncan 1978
LH I–LH IIB (1550–1420 BC)		1151	–	21	–	–	–	
LH IIIA1–LH IIB2 (1421–1200 BC)		1226	–	35	–	–	–	
Dark Age I–III (*c.*1050–775 BC)		506	–	19	–	–	–	
Byzantine (*c.*330–1204 AD)		105	–	3	–	–	–	

TABLE 6.1. *continued.*
EN = Early Neolithic;
MN = Middle Neolithic;
LN = Late Neolithic;
FN = Final Neolithic;
EBA = Early Bronze Age;
MBA = Middle Bronze Age;
LBA = Late Bronze Age;
EH = Early Helladic;
MH = Middle Helladic;
LH = Late Helladic;
EIA = Early Iron Age.

Morphometrical analyses of red deer antlers

It is well known that deer antler size, shape and growth exhibit phenotypic plasticity which are affected by habitat and nutritional quality as well as hereditary characteristics (e.g. Geist 1999; Kruuk *et al.* 2002; Eggeman *et al.* 2009). Some researchers in the past have stated that there are variations in antler morphology that appear to typify sub-species (Dolan 1988; Banwell 2009; Banwell 2011). However, antler traits are only useful if fully grown adult antlers are grown under very good nutritional conditions (Geist 1999). Whilst such conditions and factors that affect antler growth and traits cannot be assumed, spatially or temporally, within modern extant red deer populations, it is equally if not more important to consider these caveats within studies of antiquity.

Antlers preserved in the zooarchaeological record were examined in terms of their morphology. Although visual determination of approximate age from antlers is rather limited and subjective, due to the high degree of phenotypic plasticity exhibited by antler morphology, antler growth is quite rapid during the antlerogenesis process post-casting (Fennessy *et al.* 1992). Within a single individual antler, the architectural morphology largely becomes more complex with age: generally, red deer antlers increase the number of tines annually between 1–5 years before stabilising from 6–12 years of age until senescence is reached (>12 years old) and antler size/shape changes (usually overall size decreases accompanied by a decrease in the number of tines) with increasing age until death (Mysterud *et al.* 2005). Antlers that were considered to derive from animals older than five years, based on the morphology and antler size, were included in the analysis.

Results and discussion

Representation of red deer remains in archaeological wild animal assemblages

The presented zooarchaeological data (Table 6.1) demonstrate that the frequency of red deer relative to the wider wild mammal assemblage tends to vary between 20 and 95 percent. Their representation is lower (10 percent or less) at a number of sites but the majority of these date to the very Early Neolithic of the seventh millennium BC, when wild animal exploitation declined in favour of agriculture (e.g. as is seen at Anza, Argissa and Lerna). Frequencies of red deer are also low in other areas, such as district of Thérmi at the end of fifth millennium, but this is because other cervid species (roe deer and/or fallow deer) were also hunted, reducing the relative frequency of red deer.

From the end of the fifth to the beginning of the third millennium BC, red deer appear to become the predominant quarry in sites influenced by the Vinča culture, with particularly high frequencies noted at the Late Neolithic – Chalcolithic highland sites of Opovo and Petnica. Frequencies are also high in the Kopaïs lake basin between the end of the sixth to the beginning of the second millennium (Sarakenos), on sites in the Peneios valley dating to the third millennium BC (Argissa, Platia Magoula), and in assemblages from coastal Thessaly dating to the third to late second millennium (e.g. at Pefkakia). Red deer are well represented on all the Peloponnesian sites, particularly those dating towards the end of the second millennium: at Lerna they account for 53–74 percent of the wild fauna; 88 percent at Mycenae and 78 at Tiryns.

In summary, the zooarchaeological data suggest that across much of the Balkan Peninsula the red deer was the quarry *par excellence*. However, in the south-eastern part of the Peninsula and the eastern Aegean islands, fallow deer were also present and account for a high percentage, if not all, of the deer that were hunted in these regions. Regardless of these geographical variations, it is clear that cervids were the preferred game, followed secondarily by boars (*Sus scrofa*) and hares (*Lepus* sp.).

Representation of red deer in the iconographic record

Although red deer are often the best represented cervid in the zooarchaeological assemblages, the same cannot be said within the iconographic record. For instance, during the Bronze Age in Cyclades (Akrotiri on Théra) and in cities of the Mycenaean world, fallow deer were the only cervid depicted on mural paintings (Yannouli and Trantalidou 1999; Trantalidou 2002). Similarly, spotted deer are the only cervid on Ionian vases from eastern Greece (Καρδαρά 1963;

Cook and Dupont 1998; Monaco 2002), those produced under Corinthianising influence, as well as on Athenian red-figured pottery (on Athenian red figure vases see: Beazley 1963; Boardman 1989; Robertson 1992; Lissarague 1987, 95). It is possible some of these spotted deer represent juvenile or adult (in summer coats) red deer, rather than fallow deer, but the comparative absence of male adult red deer both in black and red figured vases is noteworthy. Nevertheless, some adult red deer are depicted and the timeline for their representation will now be considered.

From the collapse of Mycenaean civilisation to the eighth century BC decoration on pottery was composed entirely of abstract geometric forms. The earliest attempts to depict animals in Greek Geometric art were simple figures, executed in silhouette, in the prevailing linear style. Although highly schematised, their forms were apparently composed from actual observation of living animals. These early depictions were normally single figures isolated in panels near the handles of vessels, such as that showing a single stag under the handle of an Athenian geometric crater (Figure 6.2a).

By the Late Geometric period (*c.*760–700 BC) large funerary vases began to show continuous files of deer within the multi-banded scheme of geometric ornaments. Each frieze consisted of the same animal repeated continuously in the same pose: often this was a grazing female deer (e.g. on the Analatos hydria Athens, NAM: 313; Καλτσάς 2007, 181). Markoe and Serwint (1985) have argued that these single animal figures and the repeated files of animals were purely decorative motifs, ornamentation fit to be reproduced as a design element.

During the last quarter of the eighth century BC a transformation occurred in the vase painting, with the emergence of a new ceramic tradition known as Protocorinthian (720–630 BC). The diversification of animals is impressive with processions of exotic birds and four legged animals drawn in pure outline or in silhouette. In friezes, stags are, normally, depicted with their heads lowered to the ground in a grazing posture (Figure 6.2b). It was the varied juxtaposition of the basic schemes: the simple animal file and the pairing of confronted animals (Figure 6.2c) that gave the style its structural identity.

The Protocorinthian period was succeeded by the Corinthian style which is typified by the dominance of the animal frieze. This style had come to an end by the middle of the sixth century BC but its impact on the emerging Attic-black-figure tradition (600 to 500 BC) can be seen, with the ongoing preference for the animal frieze. The basic components of the frieze were a continuous line, in occurrence, of identical grazing deer (e.g. on a dinos; Figure 2d) or a repeated drawn carnivore – ungulate going to the left (e.g. on a lekanis; Figure 6.2e).

From about 560 BC through the third quarter of the sixth century, painters on cups explored a range of similar subjects in a delicate, minute style, either isolated on the lip zone (Lip cups) or simple compositions in a band in the handle zone on the vessel (Band cups). Those miniaturist painters are known as the Little Masters (see Figures 6.3a–f, which have been extracted from compositions associated with panthers (*Panthera* sp.), sphinxes, sirens and floral motifs).

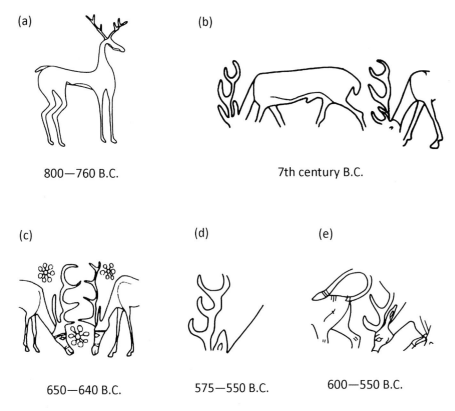

(a)

800—760 B.C.

(b)

7th century B.C.

(c)

650—640 B.C.

(d)

575—550 B.C.

(e)

600—550 B.C.

FIGURE 6.2. a. Stag under the handle of an Athenian Middle Geometric period crater. Collection Record: Paris, Louvre: A514. After Coldstream 1977, pl. 24e; b. Animal frieze. Row of stags on a conical oinochoe from Siphnos in the Cyclades, executed by the Grazing deer workshop during the Late Protocorinthian period. After Televantou 2008, pl. 55b; c. Animal frieze with two stags facing each other on a conical oinochoe from the ancient Agora of Syracuse. Produced by the Grazing deer workshop during the Late Protocorinthian period. After Benson 1969, pl. 23, 2; d. Stag grazing possibly from a dinos wall fragment animal frieze, found at the Athenian Agora (P 26758). After Moore and Pease Philippides 1986, pl. 57; e. Animal frieze on Athenian black-figure lekanis fragment, from Naucratis in Egypt. Collection Record: Oxford, Ashmolean Museum: G137.10. After the Beazley Archive, Vase Number: 300283.

The third constituent of the figure decorations in frieze was the animal attack (Markoe 1989). Among the compositions drawn we can refer to the lion attacking a stag on the *c.*570 BC krater signed by Ergotimos and Kleitias (Figure 6.3g). Another emblematic composition is the panther pouncing on a collapsing stag, as seen in the tondo of a Siana cup decorated by the Heidelberg Painter (Figure 6.3h). Here the panther bites the shoulder of its prey, while holding the neck with its forepaw, the stag's anguish expressed by the furrows painted on its brow (Brijder 1991, 366–367, 385, 459).

For nearly two centuries, then, animals played the dominant role on the painted vessels but by the middle of the sixth century BC, the *Animal Style* had come to an end and human figures progressively became the central characters. Hunting was an important theme throughout much of the period under consideration and there are many cups, amphorae, hydriai, and to a lesser extent lekythoi, that are painted with horsemen attacking an ungulate, sometimes demonstrably a red deer (Figures 6.4a and 6.4b). Often young men and adolescents are shown encircling an animal (Schnapp 1997, 212–236): this is seen on a hydria frieze attributed to Antimenes, *c.*550–500 BC, (Leiden, Rijksmuseum van Oudheden: PC63); on a Little Master Band cup attributed to the painter Elbows out, dated to *c.*550–500 (Naples, Museo Archeologico Nazionale: M943); and on the tondo of a Little Master Lip cup, *c.*575–525 (Altenburg, Staatliches Lindenau-Museum: 226). These designs were not arbitrary choices and it seems that the cup painters in particular

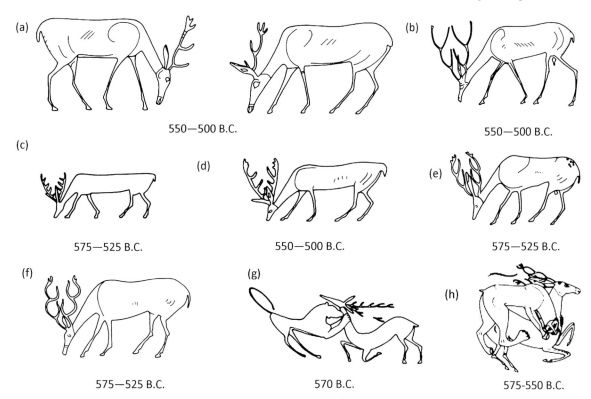

FIGURE 6.3. a. Stags from an animal frieze on an Athenian Little Master cup, from Vulci in Etruria. Collection Record: Munich, Antikensammlungen J972. After The Beazley Archive, Vase Number: 31911; b. Stag from an animal frieze on an Athenian black-figure Little Master Cup attributed to the mannerist painter Elbows out. Found at Akraiphia, Boeotia. Collection Record: Berkeley (CA), Phoebe Apperson Hearst Mus. of Anthropology: 8.61.65.45. After The Beazley Archive, Vase Number: 301429; c. Stag from an animal frieze deer on an Athenian black-figure Little Master cup, attributed to Tleson potter. Collection Record: Moscow, State Historical Museum: II1B367. After The Beazley Archive, Vase Number: 301199; d. Stag in a frieze on an Athenian black-figure Little Master band cup attributed to Tleson potter. Collection Record: Oxford, Ashmolean Museum, 1964.621. After The Beazley Archive, Vase Number: 350517; e. Stag from an animal frieze on an Athenian black-figure Little Master band cup, found at Vulci, Etruria. Note the five white dots between the rump and the beginning of the hind leg which is characteristic of an adult *Cervus elaphus maral*. Collection Record: Munich, Antikensammlungen, 2197. After the Beazley Archive, Vase Number: 31994; f. Stag grazing at the handle zone from an Athenian black-figure cup Little Master band cup animal frieze in the manner of Elbows Out, found at Vulci, Etruria. Collection Record: Munich, Antikensammlungen, J700. After the Beazley Archive, Vase Number: 31958; g. Lion attacking a stag in an animal frieze on a volute crater (the François vase, *c.*570 BC) signed by Kleitias painter and Ergotimos potter, found at Chiusi, Etruria. Florence, Museo Archeologico: 4209; h. Panther pouncing on a stag on the tondo of a Siana cup by the Heidelberg painter (*c.*550–late 540 BC). Collection Record: Brussels A 1578. After H. A. G. Brijder 1991, pl. 138f.

sought to recall the drinker's experiences: hunting was an act of initiation for adolescent men (Isler 1978). In this way the painters genuinely wanted to reflect not only the physical world but other values such as those that prevail in war (e.g. Lissarrague 1999, 95).

During the second quarter of the sixth century BC hunting scenes seem to draw increasingly upon myth and legend, with deer shown in association with gods and heroes (Figures 6.4c and 6.4d). Within this context the animal struggle

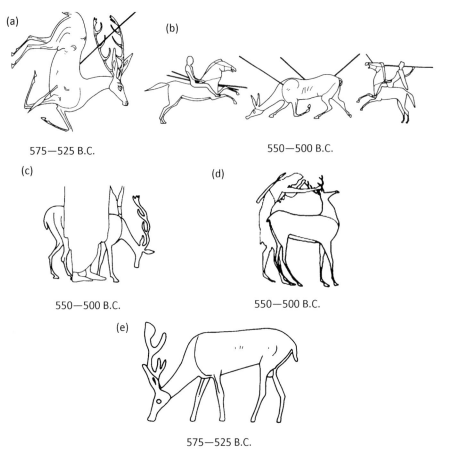

(a)

575—525 B.C.

(b)

550—500 B.C.

(c)

550—500 B.C.

(d)

550—500 B.C.

(e)

575—525 B.C.

FIGURE 6.4. a. Stag wounded by spear, in tondo, on Athenian black-figured lip cup signed by Tleson potter. Found at Vulci, Etruria. Collection Record: Boston, Museum of Fine Arts: 98.920. After The Beazley Archive, Vase Number: 301354; b. Hunting scene on an Athenian black-figured hydria attributed to Andokides. Paris, Louvre, F294. After The Beazley Archive, Vase Number: 302228; c. Stag and Apollo (from a scene depicting Muses, Dionysos and Hermes) on an Athenian black-figure neck amphora. Collection Record: Paris, Cabinet des Medailles: 231. After The Beazley Archive, Vase Number: 7840; d. Stag in depiction of Heracles capturing the Kerynitian Deer on an amphora by the Acheloos painter. Collection Record: Toledo, Museum of Art: 1958.69A. After Boardman 1989: pl. 209; e. Stag on the lip of an Athenian black-figure Little Master lip cup, found in Italy. Collection Record: Paris, Louvre, F94. Possibly a depiction of *Cervus elaphus hippelaphus*.

becomes more a metaphor of the wild strength and the bravery of the heroes illustrated on the vessel (Lissarrague 1999, 105, fig. 84; Vidal-Naquet 1981, 24). Based on the depictions outlined above, it seems that the hunting of game, small or large, was not simply an act of subsistence but rather an act that was central to men in society, a guarantee of human identity and of human supremacy over the beasts. Iconographic representations linked men with deities as well as tying and organising relations between men themselves (Durand and Schnapp 1984, 19, 55–57).

Clearly, fallow deer were the most commonly depicted species within art but the fact that it is possible to distinguish between fallow and red deer suggests that the iconographic record may be able to provide information about which subspecies of red deer are being depicted, particularly when the evidence from archaeological antlers are also considered.

Morphometrical analyses of red deer antlers

Inspection of archaeological antlers (data not shown) appeared to allow the possible identification of two groups on the basis of their traits. The first possible group consisted of antlers recovered from pre-ceramic levels of Argissa Magoula (Boessneck 1962, pl. 11.1, 6, 7), at Kitsos cave (Julien 1981, pl. LIII), at Skoteini and Sarakenos cave (Trantalidou under study). All these antlers are characterised by having only a brow tine over the coronet.

The second group consisted of those antlers found at Kastritsa cave in upper Palaeolithic Epirus (Ζάχος 2008, AMI 1061), Divostin (Lyneis 1988, pl. IIIf), Promachon, Kastanas (Becker 1986, pl. 43c,d), Argissa Magoula (Boessneck 1962, pl. 11.2, 4, 5) Pevkakia (Amberger 1979, pl. 6.17), Akrotiri on Thera (Trantalidou 2002, pl. 3d) and Kouveleiki cave. All are characterised by the occurrence of two tines over the coronet (burr).

There are several possible theories that may explain the differences between the antler phenotypes described in group 1 and group 2: (i) they come from sub-adult (3–4 years old) individuals, as proposed by Schmid (1972), or (ii) they may be from senescent adult red deer, or (iii) they were derived from an group of individual deer that had poor antler growth/development due to sub-optimum forage availability and quality, or (iv) are from a different deer species. Equally possible, these archaeological antlers may be derived from adult individuals of a particular population of red deer.

Synthesis of antler and iconographic: identifying subspecies

As we have already discussed earlier, antlers exhibit a high degree of phenotypic plasticity that is affected by numerous factors and therefore the use of specific antler traits in determining populations or subspecies of deer must be cautioned. For example, some studies have used the presence/absence of the second or bez tine as one of the primary taxonomic identifiers between *Cervus elaphus hippelaphus* (lacking bez) and *C. e. maral.* (e.g. Dolan 1988). But some modern red deer populations of *C. e. hippelaphus* found in Western/Central Europe do exhibit the bez tine either on one or both antlers, as well as within other populations where suitable high quality forage is available and supplemental feeding is conducted (e.g. Geist 1999; Mattioli *et al.* 2003). In reference to such incidences, the use of such specific characteristics as taxonomic identifiers within the archaeological record is highly problematic and uncertain.

If two subspecies of red deer were present in the ancient Balkans and Aegean region, it could be reasonable to assume that both, if distributed widely, might have been recorded by contemporary painters. In terms of antler morphology, nearly all those deer depicted on vases possess four to six tines above the pedicle and the bez tine on the basal segment emerges at a distance from the brow tine, which suggests this is the trez tine and the bez tine is absent. The only image that deviates from this pattern is the stag outlined on a Lip cup (Figure 6.4e). The portrayed antler exhibits four tines, the first (brow) and second (bez) tines are near the ear, on the basal segment, extremely close to each other. The distal antler extremity appears to exhibit a somewhat palmated or broaden beam structure that could be interpreted as a young adult fallow deer with growing antlers (velvet stage in the summer months), or equally an immature or mature red deer stag without the development of the crown tines.

Although the overwhelming majority of the deer depicted in Greek art appear to represent a single antler phenotype, here we tentatively suggest *C.e. maral*, may be portrayed, in the majority, within the iconographic record. However, it is important to recognise the limitations of the iconographic record and the highly limited use of specific antler traits within taxonomic studies (discussed earlier). Within the iconographic record, the inspiration of animal style motifs, such as the grazing animal and the animal struggle, are derived from the East – it was not only the ideas and vessels that were translocated but also the artists themselves that travelled to the west (Markoe and Serwint 1985, 7). With this in mind, and *if* the deer depicted in the corpus of Greek art is a certain red deer subspecies, two possibilities remain: it may represent either an iconographical borrowing from the East or it may be evidence that both subspecies of red deer lived in the Balkan Peninsula.

Conclusion

This is the first paper to bring together zooarchaeological and iconographic evidence for the representation of red deer in the Balkans and Aegean region and, as such, it provides an important foundation upon which others can build. Taken together, the zooarchaeological and iconographic evidence suggest that deer hunting was a rite-of-passage for adolescent males, their participation allowing boys to develop the physical and moral qualities (strength, quickness, courage) that enabled them to become men. The zooarchaeological data suggest that although cervid hunting was a socially important activity everywhere the deer species targeted varied geographically. Red deer were both common and widely hunted in the Balkan Peninsula but fallow deer were more abundant within the south-eastern part of the Peninsula and the eastern islands. This is largely consistent with the distribution of the iconographic evidence for the two species. For instance, in the zoomorphic tradition of Sicyon (Protocorithian vases) it is the red deer that is primarily depicted (Johansen 1923, 135) whereas the pottery produced in Miletos and spread to Chios or Rhodes focuses on representations of fallow deer. There is no clear explanation for the absence of red deer from Late Bronze paintings in the Aegean, since red deer bone are present in the Peloponnese and the Cyclades, although admittedly some may be imported. Nevertheless, as D. Palaiothodoros (*pers. comm.*) has highlighted, the provenance of the vases is not a safe criterion for the species geographical distribution – there are many cultural factors that influence the distribution of iconographic representation.

To some extent these findings challenge the applicability of iconography for reconstructing ancient patterns of deer biogeography. However, our study provides an initial basis that would benefit from interdisciplinary research. For example, since antler traits exhibit high phenotypic plasticity and therefore the use of specific antler traits as taxonomic identifiers is unreliable, advanced molecular techniques could be usefully employed. Extraction of ancient DNA could be attempted on the archaeological red deer antlers presented here to conclusively determine the subspecies present in this region.

Acknowledgements

This paper was presented in Paris during the 11th International Conference of Archaeozoology, in 2010. Dr E. Papaoikonomou, Dr F. Poplin and Dr J.-D. Vigne are thanked for their hospitality during this conference. Dr Papaoikonomou is also thanked for fruitful discussions on iconography. The authors would like to thank the following archaeological excavators that provided antlers for this study: Prof. C. Doumas, Prof. A. Sampson, Dr Koukouli-Chryssanthaki and C. Kontaxi. Thanks to Assistant Prof. D. Palaiothororos and Dr A. Athanassiou for reading and providing comments on early drafts of the text, and the editors of this volume for reviewing the text and making substantial suggestions.

Author contributions

M. Masseti provided the initial idea to search for the presence of the red deer subspecies (*C.e. hippelaphus* and *C. e. maral*) following his identification of antlers finds on the Aegean islands. M. Masseti also researched the recent non-archaeological literature. K. Trantalidou carried out osteological and iconographical research as well as prepared tables, written text and chose the figures, which were later, redrawn by L. Bouloti.

References

Amberger, K. P. (1979) *Neue Tierknochenfunde aus der Magula Pevkakia in Thessalien. Die Wiederkäuer,* Unpublished PhD dissertation. Institut fuhr Paläoanatomie, Domestikationsforschung und Geschichte der Tiermedizin der Universität, München.

Banwell, D. B. (2009) *The Red deer Part I. New Zealand Big Game Records Series Volume Five,* New Zealand Deerstalkers' Association, Incorporated, The Halcyon Press, Auckland, New Zealand.

Banwell, D. B. (2011) *The Red deer Part II. New Zealand Big Game Records Series Volume Six,* New Zealand Deerstalkers' Association, Incorporated, The Halcyon Press, Auckland, New Zealand.

Beazley, J. D. (1963) *Attic red figure vase painters,* OUP, 2nd edition, Oxford.

Becker, C. (1986) *Kastanas. Die Tierknochenfunde.* Prähistorische Archäologie in Sudosteuropa, 5, Wissenschaftsverlag Volker Spieß, Berlin.

Becker, C. (1991) 'Die Tierknochenfunde von der Platia Magoula Zarkou – neue Untersungen zu Haustierhaltung, Jagd und Rohstoffrverwendung im neolithischbronzezeitlich Thessalien', *Praehistorische Zeitschrift* 66, 14–64.

Benson, J. L. (1969) *Early Corinthian workshop,* Allard Pierson Museum, Amsterdam.

Boardman, J. (1989) *Athenian red figure vases. The Classical period,* Thames and Hudson, London.

Boessneck, J. A. (1956) 'Zu den Tierknochen aus Neolitischen Siedlung en Thessalien', *Berichte der Römisch Germanischen Kommission* 36, 1–51.

Boessneck, J. A. (1962) 'Die tierreste aus der Argissa Magula vom präkeramischen Neolithicum bis zur mittleren Bronzezeit', in , *Die Deutschen Ausgrabungen auf der Argissa Magula in Thessalien,* I., eds V. Milojčič, J. Boessneck and M. Hopf, Bonn, Habelt, 27–99.

Boessneck, J. (1973) *Die Tierknochenfunde aus dem Kabirenheiligtum bei Theben (Bootien),* Institut fuhr Palaeoanatomie, Domestikationsforschung und Geschichte der Tiermedizin der Universität München, Munich.

Bökönyi, S. (1986) 'Faunal remains', in *Excavations at Sitagroi. A prehistoric village in Northeastern Greece,* eds, C. Renfrew, M. Gimbutas and E. S. Elster, Monumenta Archaeologica 1, Los Angeles, 63–132.

Bökönyi, S. (1988) 'The Neolithic fauna of Divostin', in *Divostin and the Neolithic of Central Serbia,* eds A. Mcpherron and D. Srejovic, Ethnology Monographs 10, Department of Anthropology, Pittsburg, 419–445.

Bökönyi, S. (1989) 'Faunal remains', in *Achilleion. A Neolithic settlement in Thessaly, Greece, 6400–5600 BC,* eds M. Gimbutas, S. Winn and D. Shimabuku, Bookcrafters, Los Angeles, 315–332.

Brijder, H. A. G. (1991) *Siana Cups II. The Heidelberg Painter,* Allard Pierson Museum, Amsterdam.

Clason, A. T. (1979) 'The farmers of Gomolava in the Vinca and La Tène period', *Palaeohistoria* 21, 41–81.

Coldstream, J. N. (1977) *Geometric Greece,* Menthuen, London.

Cook, R. M. and Dupont, P. (1998) *East Greek Pottery,* Routledge, London.

Γιαννούλη, Ευτ. (2002) "Ήμερη και άγρια πανίδα από το Νεολιθικό οικισμό στη Σταυρούπολη Θεσσαλονίκης', in *Σωστικές Ανασκαφές στο Νεολιθικό οικισμό Σταυρούπολης Θεσσαλονίκης,* eds Δ. Β. Γραμμένος and Σ. Κώτσος, Archaeological Institute of Northern Greece 2, Thessaloniki, 683–742.

Γιαννούλη, Ευτ. (2004) 'Σταυρούπολη Θεσσαλονίκης, Νεότερα δεδομένα από την αρχαιοπανίδα του Νεολιθικού οικισμού', in *Σωστικές ανασκαφές στο Νεολιθικό οικισμό Σταυρούπολης Θεσσαλονίκης,* eds Δ. Β. Γραμμένος and Σ. Κώτσος, Archaeological Institute of Northern Greece 6, Thessaloniki, 728–729.

Διώτη, Α. (2009) *Σπήλαιο Ορφέα Αλιστράτης Σερρών. Μελέτη του οστεολογικού υλικού.* Unpublished MSc thesis. University of Thessaly, Volos.

Dolan, J. M. (1988) 'A deer of many lands. A guide to the subspecies of the red deer *Cervus elaphus* L.', *Zoonooz* LXII 10, 4–34.

von den Driesch, A. (1987) 'Haus und Jagdtiere im vorgeschichtlichen Thessalien', *Praehistorische Zeitschrift* 62, 121.

von den Driesch, A. and Enderle, K. (1976) 'Die Tierreste aus der Agia Sofia- Magula', in *Magulen um Larisa in Thessalien 1966,* eds Vl. Milojčič, A. von den Driesch, K. Enderle, J. Milojčič von Zumbusch and Kl. Kilian, Beiträge zur ur- und Frühgeschichtlichen Archäologie des Mittelmeer-Kulturraumes, R. Habelt Verlag, Bonn, 15–54.

von den Driesch, A. and Boessneck, J. (1990) 'Die Tierreste von der Mykenischen Burg Tiryns bei Nauplion/ Peloponnes', in *Tiryns,* eds H.-J. Weishaar, I. W. Hiden, A. von den Driesch, J. Boessneck, A. Rieger and W. Böser, Band XI, Deutsches Archealogisches Institut Athen, 87–129.

Durand, J. L and Schnapp, A. (1984) 'Boucherie sacrificielle et chasse initiatique', in *La Cité des images, religion et société en Grèce antique,* eds C. Bérard, C. Bron, J. L Durand., E. Frontisiducroux, F. Lissarague, A. Schnapp and J. P. Vernant, Université de Lausanne, Institut d'archéologie et d'histoire ancienne, Centre de recherches comparées sur les sociétés anciennes (France), F. Nathan, Paris, 49–66.

Eggeman, S. L., Hebblewhite, M., Cunningham, J. and Hamlin, K. (2009) 'Fluctuating asymmetry in elk Cervus elaphus antlers is unrelated to environmental conditions in the Greater Yellowstone Ecosystem', *Wildlife Biology* 15, 299–309.

Fennessy, P. F., Corson, I. D., Suttie, J. M. and Littlejohn, R. P. (1992) 'Antler growth patterns in young red deer stags', in *The Biology of Deer* ed. R. D. Brown, Springer, New York, 487–492.

Geist, V. (1999) *Deer of the world. Their evolution, behaviour and ecology*, Stackpole Books, Mechanicsburg PA.

Gejvall, N. G., (1969) *Lerna. A preclassical site in the Argolid. I. The fauna*, American School of Classical Studies at Athens, Princeton, New Jersey.

Greenfield, H. J. (1986) *The Paleoeconomy of the Central Balkans (Serbia), A Zooarchaeological Perspective on the Late Neolithic and Bronze Age (c.4500–1000 BC)*, BAR International Series 304 (i), Archaeopress, Oxford.

Greenfield, H. J. (1992–93) 'Faunal remains from Early Neolithic Starčevo settlement at Bukovačka Česma', *Starinar* 43–44, 103–113.

Greenfield, H. J. and Fowler, K. D. (2005) *The Secondary Products Revolution in Macedonia. The Zooarchaeological Remains from Megalo Nisi Galanis, a Late Neolithic – Early Bronze Age Site in Greek Macedonia*, BAR International Series 1414, Archaeopress, Oxford.

Halstead, P. (1992) 'Dhimini and the DMP faunal remains and animal exploitation in Late Neolithic Thessaly', *Annual of the British School at Athens* 87, 29–59.

Higgins, R. A. (1967) *Greek Terracotas*, Methuen, London.

Hinz, G. (1979) *Neue Tierknochenfunde aus der Magula Pevkakia in Thessalien I: Die Nichtwiederkäuer*. Dissertation, Institut für Paläoanatomie, Domestikationsforschung und Geschichte der Tiermedizin der Universität München, Munich.

Hofsten, S. V. (2007) *The feline prey theme in Archaic Greek art. Classification, Distribution, Origin, Iconographical Context*, Stockholm Studies in Classical Archaeology, 14, Stockholm University, Stockholm.

Iliev, N. and Spassov, N. (2007) 'Comparative study of the Domestic and Wild Animals from the sector Topolniča', in *The Struma/Strymon River Valley in Prehistory*, eds H. Todorova, M. Stefanovich, and G. Ivanov, Proceedings of the International Symposium Strymon Praehistoricus, 27/0901/10/2004, In the steps of James Harvey Gaul, 2, G. H. Stiftung, Sofia, 508–522.

Jensen, P. B. (2006) 'The faunal remains from the Bronze Age deposits at Haghia Triadha', in *Chalkis Aitolias I, The Prehistoric Periods*, eds S. Dietz and I. Moschos, Monographs of the Danish Institute at Athens, 7, Århus, 178–195.

Johansen, K. J. (1923) *Les vases Sicyoniens, étude archéologique*, Champion, Paris and Branner, Copenhagen.

Jordan, B. (1975) *Tierknochenfunde aus der Magula Peykakia in Thessalien*. Unpublished Dissertation, Institut für Paläoanatomie, Domestikationsforschung und Geschichte der Tiermedizin der Universität München.

Jullien, R. (1981) 'La faune des vertébrés, à l'exclusion de l'homme, des oiseaux, des rongeurs et des poissons', in *La grotte préhistorique de Kitsos (Attique)*, ed N. Lambert, Missions 1968–78. Ecole française d'Athènes, Paris.

Isler, H. P. (1978) 'The meaning of animal frieze in Archaic Greek Art, *Quaderni Ticinesi* 7, 728.

Καλτσάς, N. (2007) *Το Εθνικό Αρχαιολογικό Μουσείο*, Foundation I. Latsis, Olkos, Athens.

Καρδαρά, Χρ. (1963) Ροδιακή αγγειογραφία, Archaeological Society 49, Athens.

Kazantzis, G. (2009) *The faunal remains from the Greek sector of the Late Neolithic*

Settlement of Promachon – Topolniča. Unpublished MSc dissertation. University of Sheffield.

Kotjabopoulou, E. and Trantalidou, K. (1993) 'Faunal analysis at the Skoteini Cave', in *Skoteini at Tharrounia. The Cave, the Settlement and the Cemetery*, ed. A. Sampson, Ephoreia Palaioanthropologias, Spilaiologias, Athens, 392–434.

Kruuk, E. B., Slate, J., Pemberton, J. M., Brotherstone, S., Guinness, F. and Clutton–Brock, T. (2002) 'Antler size in red deer, Heritability and selection but no evolution', *Evolution*, 56, 1683–1695.

Larje, R. (1987) 'Animal bones' in *Paradeisos, A Late Neolithic settlement in Aegean Thrace*, ed. P. Hellström, The Museum of Mediterranean and Near Eastern Antiquities, memoir 7, Stockholm, 89–117.

Lissarrague, F. (1987) *Un flot d'images; une esthétique du banquet grec*, A. Biro, Paris.

Lissarrague, F. (1999) *Vases Grecs. Les Athéniens et leurs images*, Hazan, Paris.

Ludt, C. J.., Schroeder, W., Rottmann, O. and Kuehn, R. (2004) 'Mitochondrial DNA phylogeography of red deer (Cervus elaphus)', *Molecular Phylogenetics and Evolution* 31, 1064–1083.

Markoe, G. E. (1989) 'The "lion" attack in Archaic Greek art, heroic triumph', *Classical Antiquity* 8, 86–101.

Markoe, G. E. and Serwint, N. J. (1985) *Animal style on Greek and Etruscan vases*. R. H. Fleming Museum, University of Vermont, Queen City Press, Burlington.

Mattioli, S., Fico, R., Lorenzini, R. and Nobili, G. (2003) 'Mesola red deer: physical characteristics, population dynamics and conservation perspectives', *Hystrix, Italian Journal of Mammalogy* 14, 87–94.

Monaco, M. C. (2002) 'Representation of fallow deer in ancient Eastern Greek pottery', in *Island of deer. Natural history of the fallow deer of Rhodes and of the vertebrates of the Dodecanese (Greece)*, ed. M. Masseti, City of Rhodes, Environmental Organization, Rhodes, 133–138.

Moore, M. and Pease Phillipides, M. Z. (1986) *The Athenian Agora*, XXIII. *Attic black figured pottery*, ASCS, Princeton.

Mysterud, A., Meisingset, E., Langvatn, R., Yoccoz, N. G. and Stenseth, N. C. (2005) 'Climate dependent allocation of resources to secondary sexual traits in red deer', *Oikos* 111,245–252.

Μυλωνά, Δ. (2000) 'Οστά ζώων από τα ΥΝ στρώματα του Κρυονερίου Σερρών, Προκαταρκτική παρουσίαση', in Ανασκαφή στον προϊστορικό οικισμό Κρυονερίου, Νέα Κερδύλια, ed. Δ. Μαλαμίδου, *Αρχαιολογικό Έργο Μακεδονίας και Θράκης*, 11 (1997), 523–538.

Orton, D. (2008) *Beyond hunting and herding, humans, animals, and the political economy of the Vinča period*, Unpublished PhD dissertation. University of Cambridge.

Phoca-Cosmetatou, E. (2008) 'The terrestrial Economy of a lake settlement. The faunal assemblage from the first phase of occupation of Middle Neolithic Dispilio (Kastoria, Greece)', *Anaskamma* 2, 47–68.

Robertson, C. M. (1992) *The art of vasepainting in classical Athens*, CUP, Cambridge.

Schwartz, C. A. (1981) 'The fauna from the early Neolithic I phase at Sesklo', in *The Early Neolithic I settlement at Sesklo, an early farming community in Thessaly, Greece*, ed. M. H. J. M. N. Wijnen, Leiden University Press, Leiden, 135–136.

Schnapp, A. (1997) *Le Chasseur et la cité, chasse et érotique en Grèce ancienne*, Albin Michel, Paris.

Sloan, R. E. and Duncan, M. A. (1978) 'Zooarchaeology of Nichoria', in *Excavations at Nichoria in Southwest Greece*, I, eds G. Rapp and S. E. Aschenbrenner, University Press, Minneapolis, 60–77.

Skog, A., Zachos, F. E., Rueness, E. K., Feulner, P. G. D., Mysterud, A., Langvatn, R., Lorenzini, R., Hmwe, S. S., Lehoczky, I., Hartl, G. B., Stenseth, N. C. and Jakobsen, K. S. (2009) 'Phylogeography of red deer (*Cervus elaphus*) in Europe', *Journal of Biogeography 36*, 66–77.

Sommer, R. S. and Nadachowski, A. (2006) 'Glacial refugia of mammals in Europe: evidence from fossil records', *Mammal Review 36*, 251–265.

Televantou, C. (2008) *Siphnos. Acropolis at Aghios Andreas*. Ministry of Culture, XXI Ephorate of Prehistoric and Classical Antiquities, Athens.

Trantalidou, K. (1990) 'Animal and Human Diet in the Prehistoric Aegean', in *Thera and the Aegean World, 2*. Earth Sciences, ed. D. A. Hardy, Proceedings of the Third International Congress, Santorini, Greece, 03–09/09/1989, Thera Foundation, London, 392–405.

Τρανταλίδου, Κ. (1993) 'Οστεολογικό υλικό Καλογερόβρυσης. Παρατηρήσεις σε μικρά ανασκαφικά σύνολα', in *Kaloyerovrissi, A Bronze Age settlement at Phylla, Euboea*, ed A. Sampson, Ephoreia Palaioanthropologias – Speleologias, Athens, 163–168.

Τρανταλίδου, Κ. (1997) 'Θηράματα και οικόσιτα ζώα από το σπήλαιο των Λιμνών', in *The Cave of Lakes at Kastria of Kalavryta, A prehistoric site in the highlands of Peloponnese*, ed A. Sampson, Etaireia Peloponnisiakon Spoudon, 7, Athens, 415–455.

Trantalidou, K. (2002) 'The Rhodian fallow deer. Game and trophy since prehistoric times', in *Island of deer. Natural history of the fallow deer of Rhodes and the vertebrates of the Dodecanese (Greece)*, ed. M. Masseti, Prefecture of Rhodes, Rhodes, 159–164.

Τρανταλίδου, Κ. (2009) 'Διατροφικές επιλογές, κατανάλωση ζώων και στοιχεία οικονομίας στις Μυκήνες προς το τέλος της υστεροελλαδικής περιόδου', *AAA* 40–41 (2007–8), 115–146.

Trantalidou, K., Belegrinou, El. and Andreasen, N. (2011) 'Pastoral societies in the southern Balkan Peninsula, The evidence from caves occupied during the Neolithic and the Chalcolithic era', in *The phenomena of cultural borders and border cultures across the passage of time*, Conference 22–24/10/10, Trnava University, Anodos 10 (2010), 295–320.

Τρανταλίδου, Κ., Σκαράκη, Β., Καρά, Ε. and Ντίνου, Μ. (2007) 'Στρατηγικές επιβίωσης μετακινούμενων κυνηγών κτηνοτρόφων κατά τη διάρκεια της 4ης χιλιετίας. Στοιχεία από την εγκατάσταση στην ανατολική όχθη των Πηγών του Αγγίτη', Αρχαιολογικό Έργο Μακεδονίας και Θράκης, 18 (2005), 45–80.

Vermeule, D. T. (1979) *Aspects of death in early Greek art and poetry*, University of California Press, Berkeley.

Vidal-Naquet, P. (1981) *Le Chasseur noir. Formes de pensée et formes de société dans le monde grec*, La Découverte, Paris.

Yannouli, E. (1994) *Aspects of Animal Use in Prehistoric Macedonia, Northern Greece. Examples from the Neolithic and Early Bronze Age*, Unpublished PhD dissertation. University of Cambridge.

Yannouli, E. (1997) 'Dimitra, A Neolithic and Late Bronze Age village in northern Greece, the faunal remains', in *Νεολιθική Μακεδονία*, ed. Δ. Γραμμένος, The Archaeological Receipts Fund, Athens, 101–127.

Yannouli, E. and Trantalidou, K. (1999) 'The Fallow deer (*Dama dama* L. 1758) in Greece. Archaeological Presence and Representation', in *The Holocene history of the European vertebrate fauna. Modern aspects of research*, ed. N. Benecke, Berlin, 05–10/04/1998, D.A.I., Berlin Institut für Paläoanatomie, München, Institut für Haustriefkunde, Kiel, 247–281.

Ζάχος, Κ. (2008) Το Αρχαιολογικό Μουσείο Ιωαννίνων, ΙΒ' ΕΡΚΑ, Ioannina.

The Italian Neolithic Red Deer: Molino Casarotto

Katherine Boyle

Introduction

Hunting of red deer (*Cervus elaphus*) is a phenomenon commonly associated with Palaeolithic and Mesolithic communities in Western Europe (Boyle 1990; Boyle 2006). Its importance during the Neolithic is rarely considered in detail as domestic species become the archaeozoologist's focus of attention. However, in some parts of Europe it remains an important resource which is readily available for exploitation by early farming communities (Boyle 2006). Northern Italy is just one region where red deer frequently dominate faunal assemblages. This paper looks at the red deer material from the site of Molino Casarotto and what it may tell us about the economy in the region approximately 5600 years ago.

Molino Casarotto

Molino Casarotto lies in the Berici Hills region of Northern Italy (Figure 7.1). It is a former lakeside settlement not far from Lake Fimon. The area is one of Cretaceous-Miocene carbonates (limestone) and volcanic bedrock, Lake Fimon being blocked by alluvial sediments deposited by the Bacchiglione River. The area today is one of woodland environments, with lake vegetation which includes water-lilies (*Nymphaea* sp.), water chestnuts (*Trapa natans*) and buttercups (*Ranunculus* sp.). When the site was occupied during the Neolithic, the hills in the area were also wooded (Valsecchi *et al.* 2008), a fact which goes some way towards explaining the taxonomic structure of the faunal assemblage. In addition the lake provided a rich source of plant food (e.g. water chestnut), fish (e.g. carp (*Cyprinus carpio*), perch (*Perca fluviatilis*) pike (*Esox* sp.)) and freshwater shellfish.

The pottery and stone tool industry recovered during excavation at Molino Casarotto place the site (dated to between 5960±50 uncal. BP (R-754) and 5140±50 uncal. BP (R-750A) Bagolini and Biagi 1990, 14) in the earliest phase of the local Middle Neolithic or Square Mouthed Pottery Culture (VBQ: *Vasi*

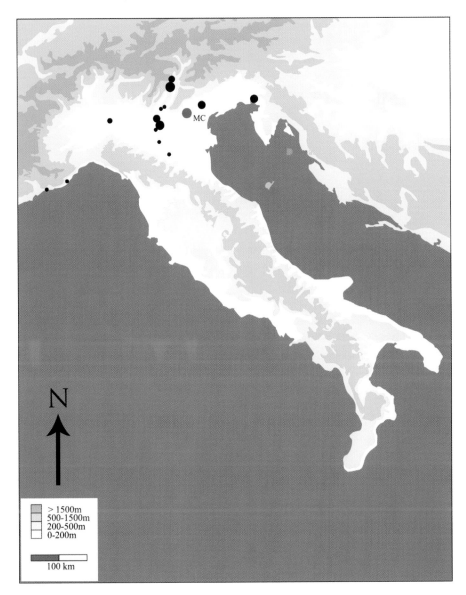

FIGURE 7.1. Map of the location of Molino Casarotto (MC) and other red deer assemblages in Northern Italy. Symbol size represents an approximate indication of red deer frequency. Large >50% NISP, Medium 20–50% NISP, Small >1–20% NISP.

a Bocca Quadrata, see Bagolini 1980), while the faunal material is a mixture of domestic and wild species. Several phases of occupation can be seen at the site, although the faunal material is treated here as a single palimpsest assemblage as there are no stratigraphic relationships observable between and within units (L. Barfield† *pers. comm.*) and the current analysis has revealed that a small proportion of the faunal material (eight bone fragments) from different 'layers' and sections of the site (squares) can be refitted exactly. At least three distinct settlement units or houses have been recorded. These structures include large central hearths of compacted layers of fresh water mollusc shells alternating with layers of water chestnut, and wooden platforms and posts of maple (*Acer platanoides*), European alder (*Alnus* glutinosa), hazel (*Corylus avellana*), ash (*Fraxinus ornus*) and beech (*Fagus sylvatica*) (Valsecchi *et al.* 2008). The units are about 20 m apart (Fontana 1998, 1068).

The faunal collection

The faunal material from Molino Casarotto considered here consists of a collection of material which was brought to Cambridge during the early 1970s after the site was excavated by Lawrence Barfield and colleagues (Bagolini *et al.* 1973). This material has remained in store and has thus far not been analysed and published; it supplements the material presented by Jarman in 1972. The material is representative of the complete assemblage (Barfield[†] *pers. comm.*): it includes fish bones and some fragmented microfauna (mainly watervole (*Arvicola amphibius*)), which affirms the undocumented assertion of the archaeologist working there at the time (Graeme Barker *pers. comm.*) that wet and dry sieving was carried out with a 13 mm and 3 mm mesh. It consists of 3330 complete bones and fragments of which at least one dimension exceeds 10 mm. Step-by-step analysis of this material included identification and recording (in text and photographic form, as appropriate) of (1) bone type, (2) bone break type as described by Shipman (1981, 105, fig. 5.2 A–G), (3) evidence of burning and other surface modification, (4) species or larger taxonomic group where known, (5) cut-marks coded (*sensu* Binford 1981 and Pérez Ripoll 1992), (6) bone fusion, tooth wear and eruption stage and an estimation of age (Carter 2006), and (7) roe deer (*Capreolus capreolus*) and red deer antler fragments. Only unshed antler bases, still attached to cranial bone were included in NISP values. No antler is included in MNE values. All bones were recorded to the square metre in which they were found at the site. Unidentifiable fragments in the assemblage, mostly under 20 mm in length, were also recorded in the original database. Where possible these were recorded as cranial, axial, long bone or foot. Fish bones were sorted according to body part and species; scales were not counted but, with the help of a fish scale atlas (Steinmetz and Müller 1991), several species of lake fish were identified.

Four methods were used in quantifying the faunal material from Molino Casarotto: the number of identifiable specimens of each species (NISP) was noted, an estimation of the minimum number of complete elements (MNE), a figure which takes into account conjoined fragments, and potential matches between long bone shafts and unfused proximal/distal epiphyses, Minimum Number of Animal Units *sensu* Binford 1978 (MAU and %MAU), and Minimum Number of Individuals per taxon (MNI). As MNE calculations involved refitting fragments and value judgements on the likelihood of specimens deriving from the same element, the disparity between NISP and MNE counts is much less than would be expected and MNE is the favoured quantification method in this research.

Taphonomy

Despite the fact that 86 percent of the material from Molino Casarotto is fragmented, and some of it (approximately five percent) burned, there is very little evidence of surface weathering and erosion, such as trampling and root etching. Surface marks reminiscent of those described by Behrensmeyer (1978) are absent. The near absence of gnawing damage caused by carnivores (only six red deer bones display carnivore tooth marks) and the complete absence of rodent gnawing are also notable and suggest that material was not exposed to the elements for long. Both the burning and most of the fragmentation observed are prehistoric: they do not result from excavation, retrieval and storage conditions. Correlation analysis of relationships between bone frequency (%MAU) and both Binford and Bertram's (1977, 109) and Lyman's (1984, 250) density figures yield Pearson product-moment correlation coefficient values of $r = 0.00015$ (p <0.001) and $r = 0.4$ (p <0.001) respectively. The major factor determining bone preservation therefore appears to be something other than natural bone density, here suggested to be processing-related fragmentation caused by human activity. The only other factor which might have played some part in assemblage formation is the dog (*Canis familiaris*). The dog is present in the assemblage (MNE = 28, MNI = 4) so the six red deer bones showing clear evidence of gnawing and chewing might represent at least one incident of dog feeding, intentional or otherwise.

Red deer exploitation at Molino Casarotto

Species representation

Almost 90 percent of the animal bones identifiable to species from Molino Casarotto stored in Cambridge (MNE = 2176) derive from wild taxa (Table 7.1). The most abundant species at Molino Casarotto is red deer (MNE = 1024), with boar (*Sus scrofa*) and roe deer forming a significant secondary component (MNE = 462 and 285 respectively). The major domestic species represent a third group (MNE = 187). Other wild taxa (e.g. badger *Meles meles*, fox *Vulpes* sp., otter *Lutra lutra*, bear *Ursus arctos* and horse *Equus caballus*) make up a substantially smaller group, the most important of which is badger, a boreal and temperate lowland forest or woodland-indicating species (Kowalczyk *et al.* 2000) which fits well into the faunal spectrum here. It is only the horse which some would consider to be out of place, as it has traditionally been thought of as being absent

TABLE 7.1. Species frequencies (NISP, MNE and MNI) at Molino Casarotto.

Species	NISP	MNE	MNI
Badger	25	25	2
Bear	17	17	1
Beaver	1	1	1
Boar	467	462	18
Cat	3	3	1
Cattle	68	68	5
Chamois (?)	2	2	1
Dog	28	28	4
Duck	1	1	1
Fish: carp	2	2	1
Fish: perch	3	3	1
Fish: pike	20	20	1
Fish: trout	2	2	1
Fish (undetermined)	19	19	-
Fox	21	21	2
Hare	10	9	1
Hawk	1	1	1
Horse	6	6	1
Lagomorph	6	6	-
Marten	1	1	1
Mustelid	2	2	1
Otter	19	19	4
Pine marten	3	3	1
Polecat	2	2	1
Rabbit	10	10	2
Red deer	1036	1024	29
Roe deer	288	285	11
Sheep/goat	119	119	13
Stoat	4	4	1
Water vole	8	8	1
Wolf	3	3	1

in Italy during the Neolithic. However it is recorded at a few other Italian Neolithic sites such as Maddelena di Chiomonte (Liguria) (Bertone and Fozzati 1998) and San Maria d'Agnano (Puglia) (Wilkens 2003), so its occurrence is not altogether surprising.

Age profile and season of occupation

The faunal assemblage from Molino Casarotto comprises material from a minimum of 71 individuals attributable to the major species which can be assigned to broadly defined age classes (*sensu* Lowe 1967). Of these, 29 are red deer individuals including three juvenile, five sub-adults and eighteen adults. There are also at least one neonate and one 'senile' individual. It is a mixed age profile which might suggest primarily targeting of individuals for meat and possibly antler.[1]

Unfortunately, no specific seasonality analysis has yet been carried out at Molino Casarotto. However, the existence of at least one neonate might indicate that the site was occupied at least once during summer, if we assume that, as is normally the case, red deer give birth between May and early July (see Carter 2006). The suggestion is supported by the nut and fruit collection which would suggest late summer to autumn occupation. Furthermore, until the lake was drained in the nineteenth and twentieth centuries winter waterlogging and flooding of much of the Lake Fimon basin (Jarman *et al.* 1982, 200) would surely have prevented winter occupation of sites close to the lake. Spring may have seen crops being sown elsewhere in the wider region. A late spring, summer to early autumn occupation (perhaps from shortly after the sowing season elsewhere to the period just before harvest) is much more probable than late autumn or winter. This is further supported by the presence of both young and sub-adult badger at the site, for badgers are usually born in the spring, frequently March (Bjarvall and Ullstrom 1986, 156). The presence of young, but not neonatal badger, suggests that when they died at or near Molino Casarotto, individuals were several months old, as they would be during the summer months. This suggests that the site of Molino Casarotto was not a permanent, year-round occupation but rather a short-term of temporary, perhaps seasonal site. It may well have been just one component of an annual subsistence cycle distributed over a much wider area.

Carcass butchery and processing

In order to determine the importance of red deer in the subsistence strategy of inhabitants of Molino Casarotto during the period of site occupation, both element frequency and frequency/distribution of cut-marks on bones are considered. These cut-marks all result from the use of flint tools in processing mainly large game (see Barfield and Broglio 1966; Bagolini *et al.* 1973; Alessio *et al.* 1974, 358).

The element frequency data recorded as %MAU values in Table 7.2 indicate that all red deer body segments are present and relatively well represented at

Bone type	%MAU	Bone type	%MAU
skull	30	carpal	22.1
mandible	50	sacrum	10
atlas	30	pelvis	10
axis	50	femur	30
cervical vertebra	14	tibia	100
thoracic vertebra	40.8	tarsal	0
lumbar vertebra	21.4	astragalus	65
caudal vertebra	1.7	calcaneum	75
rib	4.2	metatarsal	50
scapula	70	phalanx I	55
humerus	75	phalanx II	63.8
radius	50	phalanx III	40
metacarpal	100		

TABLE 7.2. Red deer element relative frequencies.

Molino Casarotto. Only bones of either particularly small size (complete or heavily fragmented) are present in much lower than average frequencies (mean %MAU = 45.2). Among the other body parts, it is the higher meat-yielding parts which are least abundant, including ribs, pelvis and femur. The ribs in particular appear to have been reduced to small fragments. These make up a substantial proportion of the bone component which is unidentifiable to species.

The fact that many parts are represented by values well in excess of 40 percent suggests that carcasses were not processed very intensively – instead red deer were taken at or near Molino Casarotto in sufficient quantity to warrant some discard. The practice of not processing carcasses of dominant species to a point of destruction is a feature frequently observed among prehistoric hunter-gatherer groups (Boyle 1997; 2000). It may well be that the hunters at Molino Casarotto were exploiting red deer not at a time of scarcity but as one of a range of alternative resources, perhaps supplementing domestic livestock on which their economy focused elsewhere. This is further demonstrated by the fact that evidence of marrow processing is rare. Marrow-rich bones are relatively abundant at the site and are frequently complete or nearly complete. Very few bear cut-marks or fracture patterns associated with marrow processing. Patterning among long bone fragments of the sort described by Enloe (1993) and Binford (1981, 154–161) is only rarely observed and only eleven (7 percent) of the 163 red deer bones which display cut marks have incisions which might result from marrow processing. Instead most cut marks can be attributed to dismemberment (60 percent), skinning (12 percent) and filleting (14 percent) for either consumption or storage of meat. Furthermore since most of the bones which do bear cuts come from sub-adult/adult individuals (79 percent) it is possible that hunters were aiming for high yield carcasses, which were then subjected to processing which followed a sequence or chain of specific activities.

Three major stages in post-kill red deer carcass processing have been identified on the basis of types and frequencies of cut-marks observed (Figure 7.6), with a fourth marrow-related stage of only minor importance.

Stage I: Skinning. Skinning marks, observed on 20 bones, generally occur on the lower limb and head bones (metapodia, phalanges, skull and mandibles). In the case of the head, marks are seen at the base of the antler, and possibly towards the nose or snout. Binford (1981, 107) attributes these marks to skinning (removal of pelt), when skins are sought for a specific purpose as opposed to being simply the initial stage in carcass processing. On the phalanges, marks are frequently deep and clear cut. They occur on the proximal epiphyses, suggesting, in the case of the first phalanx, disarticulation from the metapodia

– i.e. separation of the foot from the rest of the leg – and on the anterior aspect of the bone. These marks may well result from skinning activity – attempts to loosen the skin from the foot, above the hoof.

Stage II: Primary dismemberment and disarticulation. Marks resulting from primary carcass dismemberment and disarticulation occur on 60 bones, close to joints or articulations. They are caused by (a) cutting the tendons holding joints together (n = 27), and (b) inserting a tool to help leverage (n = 33). Dismemberment marks abound and include those seen on (i) crania – resulting from disarticulation of the head from the neck, (ii) proximal femora (Binford 1981, Fp-2 and Fp-5) – due to separation of the back leg from the pelvis (Binford 1981, PS-9 and PS-10), and (iii) deep notches on distal metapodia resulting from disarticulation of the foot from the leg (MTd-3; MCd-1 – transverse along the anterior articular margin: Binford 1981, 142). Evidence for separation of the head from the neck is also seen on the proximal face of the atlas, Binford attributes marks CV-2 and CV-3, which are similar to those seen at Molino Casarotto, to dismemberment of stiff carcasses. Evidence of separation of the forelimb from the axial skeleton and frequent subsequent disarticulation between carpals and the distal radius are also observed, as we see marks on the articular surface of the distal radius (RCd-2: Binford 1981, 125).

Stage III: Secondary dismemberment and filleting. These marks, although observed on only 39 bones, make up the largest group of individually-defined cut-marks and include those which relate to the removal of the tongue, while division of the vertebral column into segments can also be assigned to this category. At Molino Casarotto we see such evidence in the form of numerous cuts on the thoracic and cervical vertebrae (Binford 1981, TV-2, TV-3, TV-5; CV-4, CV-6). These are marks which relate to filleting rather than dismemberment, particularly marks TV-5 and TV-2 which result, according to Binford, from the removal of the tenderloin (TV-2) and ribs (TV-5) – one part of the carcass from which we have relatively few securely identified pieces at Molino Casarotto. A small number of marks on red deer lumbar vertebrae may also relate to filleting, although filleting marks at the site are usually long and relatively thin. They run along the longest length of the bone concerned (e.g. S-4 on the scapula). This observation is based on the regularity with which we see the centra of the thoracic and lumbar vertebrae sliced through completely – as cleanly as if the task were completed by a specialist butcher today. Abundant cuts also occur at the base of the vertebral dorsal spines, especially those which are sliced through in the manner attributed by Binford (1981, 112) to preparation for cooking. We see cut-marks on the proximal and distal articular faces of the lumbar vertebrae, where the bones have been separated one from another, while the ribs display cut-marks along much of their length (due to filleting?). Most of the large or relatively large, if not complete, ribs are missing distal ends (Binford 1981, RS-2), a fact which Binford (1981) attributes to dismemberment of the ribs and brisket. Cuts on the side of the rib head (Binford 1981, RS-3) are attributed to

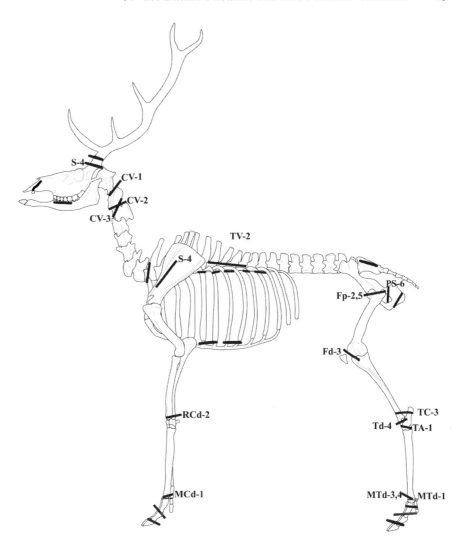

FIGURE 7.2. Sketch showing the distribution of cut-marks on red deer.

dismemberment. Short cuts attributable to dismemberment are also observed on the pelves (PS-3, PS-6: Binford 1981, 130), proximal and distal tibiae, metapodia (short chevron MTd-4: Binford 1981, 140) and distal humeri.

Stage IV: Marrow Processing. As already stated, evidence of marrow processing is rare at Molino Casarotto, only 0.5 percent of all complete bones and larger fragments (greater than one fifth of the whole), irrespective of species, showing relevant cut-marks. In all species pieces which most frequently bear marks attributable to marrow processing are the distal metapodia (cf. Binford 1981, 155, fig. 4.48). Indeed, six of the eleven red deer bones which display clear evidence of marrow extraction are metapodial fragments closely similar to the distal fragments described by Binford (1981, 155, fig. 4.48); another is a tibia and the remaining four are phalanges (phalanx I). These low values might suggest that this activity was only of minor interest at the site and that it was perhaps not

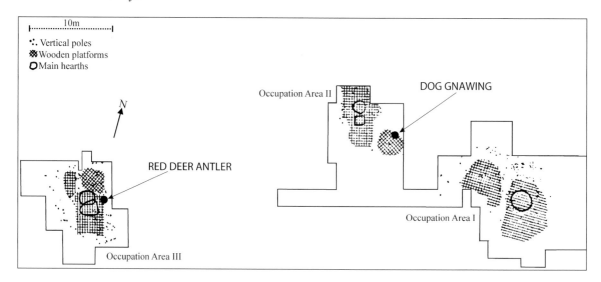

FIGURE 7.3. Sketch map of Molino Casarotto showing the location of three occupation areas, antler concentration and dog gnawing.

a long-term residential location where such activity regularly took place. Are we perhaps looking at a short-term hunting camp occupied only seasonally – a location used by a group of individuals who spent much of the rest of the year as 'Neolithic farmers' elsewhere?

One final group of cut-marks consists of a small number which might be associated with the use of red deer bone and antler as raw material. Although rare, a number of antler fragments show groups of cuts circling the base of the antlers (Binford 1981, S-4), already mentioned as a possible indication of skinning. These are sometimes accompanied by marks which result from the removal of the unshed antler from the skull. Two such cases of red deer antler-removal cut-marks are seen, which represent just 2.4 percent of the 64 larger antler fragments positively identified as red deer. In one case the edge is cut straight and then broken using some force, using some form of leverage to aid fracture.

Of the 64 red deer larger antler fragments recorded almost 70 percent were recovered on the western side of the site. Well over one third (37 percent) was recovered in one metre square of occupation area III (Figure 7.3, square O38, code: MC 72 3–38 O), suggesting that there was more intensive antler working activity here (L. Barfield[†] *pers. comm.*) or butchery of at least one complete red deer with antlers. In this square several of the pieces refit to form relatively large antler beams. There is also evidence of possible burning of the material in this square. It may be of some interest that the small quantity of roe deer antler is also clustered in this part of the site, although the frequencies are much smaller.

Finally, and perhaps most interestingly, 16 percent of the red deer bones show cut marks of some sort. Given the degree of fragmentation (86 percent of bones are fragmentary) this is a relatively small proportion. However it is typical of the assemblage as a whole, of which 13.5 percent are cut. For example, 14 percent of the wild boar bones from Molino Casarotto display cuts, while the equivalent figure for the roe deer is 15 percent. Interestingly values for the domestic species are similar: 14 percent for ovicaprids and 12 percent for cattle. Furthermore the proportion of each category of cut-marks is also relatively consistent between species. Skinning marks vary from 15 to 30 percent across common species (red deer, roe deer, boar, cattle (*Bos Taurus*), sheep (*Ovis aries*) and bear), primary disarticulation from 40 to 67 percent, secondary disarticulation and filleting from nine to 40 percent, with higher values associated with the larger species (red deer and cattle) and marrow processing zero to eight percent except for boar (27 percent). This patterning lends further credence to the suggestion that the individuals responsible for carcass processing were a consistent workforce, irrespective of the degree of skill amongst the butchers at the site. The apparent lack of difference between species might also suggest that there is little inter-species variation in processing, that the large game and domestic fauna were processed in a similar fashion. This hypothesis remains to be tested.

Conclusion

In conclusion, although some evidence for both domestic animal and crop husbandry is present at Molino Casarotto, much of the subsistence economy appears to have been based on exploitation of wild animals and plants throughout the period of occupation. The excavators strongly believed that there is no genuine stratigraphy to be seen at the site (L. Barfield[†] *pers. comm.*), although the range of radiocarbon dates obtained (from 5140±50 BP, 4000–3951 cal. BC R-750A to 5800±50 uncal. BP, 4760–4604 cal. BC, Alessio *et al.* 1974, 359–361; Bagolini and Biagi 1990, 14) suggest that there may be some chronological variation worthy of investigation. The preponderance of red deer at the site appears however, to be a feature throughout the stratigraphy, i.e. it is abundant in all the layers/spits labelled during excavation. There are no stratigraphic drawings available of the site. In the past the heavy predominance of red deer, roe deer and boar, plus the collection of mussels (*Mytilus* sp.), pike, turtle (Order: Chelonii) and water chestnuts, has meant that the site has been assigned a 'Mesolithic' character (Jarman *et al.* 1982, 129; Moeri 1990, 27), despite the development of farming elsewhere in the region and the fact that there are several other sites, such as Cornuda (Treviso, Veneto), where red deer and boar form over 50 percent of a VBQ ceramic assemblage (Moeri 1990). The VBQ occupants of Molino Casarotto have previously been described as 'outsiders' or '*allochtone*' (Moeri 1990, 27) who adapted to an environment in an area where the advantages of farming (or perhaps agropastoralism) were outweighed by those of a hunting economy. In an area of rich woodland but low arable potential where the soil was not suitable for crop-agriculture (Jarman and Bay-Petersen 1976; Moeri 1990) abundant evidence of a farming economy (arable and stock-breeding) is probably not to be expected. However, exploitation of a rich woodland environment, especially surrounding a lake where wild game was readily available does not necessarily mean that the region's population remained largely 'Mesolithic'. Instead, evidence of domestic animals and plants at the site, although rare, suggests an element of subsistence complexity which is often overlooked in studies of the period. What is suggested here is that the site was one at which red deer hunting formed a valuable component of a regional subsistence strategy which included available domestic stock, wherever it was bred.

Molino Casarotto was in all probability a short-term location, perhaps one where game could be obtained in order to supplement the staples of a domestic subsistence economy practised elsewhere: a specialised location, used for hunting purposes but to which a small quantity of domestic stock (or even meat, since the animals may have been in the form of partially processed carcasses before arrival at the site) may have been brought. Perhaps the increasingly successful farmers of the Middle Neolithic Bericci Hills made the most of an opportunity to broaden their subsistence base to include resources which required little or no effort to husband. Specialised hunting camps to which some domestic stock may have been brought are just one component of the record of a thriving farming community. It simply shows how complex the Neolithic economy had become and warns us against making simplistic assumptions regarding human behaviour in the past.

This is reflected by faunal assemblages from other sites in the north of Italy, where the wild faunas frequently exceed the domestic. For example, at Razza di Campegine (Emilia-Romagna) domestic stock (sheep, cattle and pig) makes up one third of the assemblage, while red deer exceeds 37 percent of the fauna (Cazzela *et al.* 1976). At Cornuda (Veneto), the domesticates represent only 21 percent, red deer, roe deer and boar making up the bulk of the remaining material (Riedel 1988), and at Mezzocorona-Borgonuovo, in the Adige valley north of Trento, red deer also dominates (Bazzanella *et al.* 1998). Meanwhile, further to the west, in Val Pennavaira region of Liguria, Arma di Nasino (level IX), is just one of several VBQ sites yielding a mainly wild faunal assemblage (Leale Anfossi 1958–61; Chiarenza 2007).

Just to the south of our region, the Lake Varese site of Isola Virginia, in Lombardy, yields VBQ deposits with clear evidence of a domestic economy which is not unlike that encountered at Molino Casarotto (Phillips 1975, 76–7); it is a lakeside village of similar age where both wild and domestic faunal material have been recovered. Further south again, at Maddalena di Muccia (Marche), analysis of faunal assemblage composition and the occurrence of pressure-flaked projectile points of the sort discovered at Molino Casarotto (Barfield 1971), led Barker (1975, 134) to conclude that hunting continued to play a significant part in the economy of Neolithic communities.

Molino Casarotto is therefore far from unique, and the complexity of the assemblage with both wild and domestic plant and animal components (mammalian, bird, fish and shellfish) firmly dated and cultural assigned to the *Vasi a Bocca Quadrata* Neolithic, shows that generalisations can be dangerous.

In summary, the red deer remained an important resource for hunters, foragers and farmers. The arrival of the Neolithic does not mean that it ceases to feature as food and raw material. Red deer may no longer have occupied the primary role in the economy, but it did continue to be an important resource embedded into the economic strategy of the regional population. Hunting groups of the sort seen at Molino Casarotto were not outsiders; nor were they Mesolithic remnants. Instead they were part of a vibrant Neolithic mixed economic group: cereal farming, livestock breeding, hunting, fishing and harvesting of wild plants all played an important role in the lifeways of the Middle to Late Neolithic of Northern Italy.

Acknowledgements

I should like to thank Professor Graeme Barker and Dr John Robb, both of Cambridge University, for their support and discussions while I was completing the analysis of and report on the material from Molino Casarotto – and of course the late Dr Lawrence Barfield, without whom this work would have been impossible. He was unfailingly supportive when I first contacted him to ask him what he knew of the material stored in Cambridge, and remained in contact throughout.

Note

1. Unfortunately the author has not yet seen bone/antler artefact material from the site.

References

Alessio, M., Bella, F. and Improta, S. (1974) 'University of Rome carbon-14 dates XII', *Radiocarbon* 16, 358–367.

Bagolini, B. (1980) *Introduzione al Neolitico dell'Italia Settentionale*, Società Naturalisti 'Silva Zenari', Podenone.

Bagolini, B., Barfield, L. H. and Broglio, A. (1973) 'Notizie preliminary delle ricerche sull'insediamento neolitico di Fimon Molino Casarotto (Vicenza) (1969–72)', *Rivista di Scienze Preistoriche* 28, 217–234.

Bagolini, B. and Biago, P. (1990) 'The radiocarbon chronology of the Neolithic and Copper Age of Northern Italy', *Oxford Journal of Archaeology* 9, 1–23.

Barfield, L. (1971) *Northern Italy before Rome*, Thames and Hudson, London.

Barfield, L. H. and Broglio, A. (1966) 'Materiali per lo studio del Neolitico del territorio Vicentino', *Bulletino di Paletnologia Italiana* 75, 307–344.

Barker, G. 1975. *The Archaeology of Early Man in Molise, Southern Italy*, Superintendency of Antiquities of Molise, Rome.

Bazzanella, M., Moser, L., Mottes, E. and Nicolis, F. (1998) 'The Neolithic levels of the Mezzocorona-Bordonuovo site (Trento): Preliminary data', *Preistoria Alpina* 34, 213–226.

Behrensmeyer, A. K. (1978) 'Taphonomic and ecologic information from bone weathering', *Paleobiology* 4, 150–162.

Bertone, A. and Fozzati, L. (1998) 'La preistoria del bacino della Dora Riparia oggi', *SEGUIUM – Ricerche e Studi Valsusni* 36, 11–82.

Bjärvall, A. and Ullström, S. (1986) *The Mammals of Britain and Europe*, Croom Helm, London–Sydney.

Binford, L. R. (1978) *Nunamiut Ethnoarchaeology*, Academic Press, New York.

Binford, L. R. (1981) *Bones; Ancient Men and Modern Myths*, Academic Press, New York.

Binford, L. R. and Bertram, J. (1977) 'Bone frequencies and attritional processes', in *For Theory Building in Archaeology*, ed. L. R. Binford, Academic Press, New York, 77–156.

Boyle, K. V. (1990) *Upper Palaeolithic Faunas from South West France, A Zoogeographic Perspective*, BAR International Series 557, Archaeopress, Oxford.

Boyle, K. V. (1997) 'Late Magdalenian carcase management strategies, The Périgord data', *Anthropozoologica* 25–6, 287–294.

Boyle, K. V. (2000) 'Intra-regional similarities in resource exploitation strategies: The late Magdalenian in the Vézère Valley', in *Regional Approaches to Adaptation in Late Pleistocene Western Europe*, eds G. L. Peterkin and H. Price, BAR International Series 896, Archaeopress, Oxford, 47–59.

Boyle, K. V. (2006) 'Neolithic wild game animals in Western Europe: The question of hunting', in *Animals in the Neolithic of Britain and Europe*, eds D. Serjeantson and D. Field, Neolithic Studies Groups Seminar Papers 7, Oxbow Books, Oxford, 10–23.

Carter, R. J. (2006) 'A method to estimate the ages of death of red deer (*Cervus elaphus*) and roe deer (*Capreolus capreolus*) from developing mandibular dentition and its application to Mesolithic NW Europe', in *Recent Advances in Ageing and Sexing Animal Bones*, ed. D. Ruscillo, Oxbow Books, Oxford, 40–61.

Cazella, A., Cremaschi, M. Moscoloni, M. and Sala, B. (1976) 'Siti neolitici in località di Razza di Campegine (Reggio Emilia)', *Preistoria Alpina* 12, 79–126.

Chiarenza, N. (2007) L'Eneolitico nell'Occidente Ligure – Revisione dei Materiali e Confronti. Tesi di dottorato di ricerca, Università di Pisa.

Enloe, J. (1993) 'Ethnoarchaeology of marrow cracking: Implications for the recognition of prehistoric subsistence organization', in *From Bones to Behavior: Ethnoarchaeological and Experimental Contributions to the Interpretation of Faunal Remains*, ed. J. Hudson, Southern Illinois University at Carbondale Center for Archaeological Investigations occasional Paper 21, Carbondale, 82–97.

Fontana, V. (1998) 'Procedures to analyse intra-site pottery distribution, applied to the Neolithic site of Fimon, Molino Casarotto (Italy), house site No. 3', *Journal of Archaeological Science* 25, 1067–1072.

Jarman, M. R. (1972) 'European deer economics and the advent of the Neolithic', in *Papers of Economic Prehistory*, ed. E. S. Higgs, CUP, Cambridge, 125–147.

Jarman, H. N. and Bay-Petersen, J. L. (1976) 'Agriculture in prehistoric Europe – the Lowlands', *Philosophical Transactions of the Royal Society of London series B* 275, 175–186.

Jarman, M. R., Bailey, G. N. and Jarman, H. N. (1982) *Early European Agriculture. Its Foundation and Development*, CUP, Cambridge.

Kowalczyk, R., Bunevich, A. N. and Jedrzejewska, B. (2000) 'Badger density and distribution of setts in Biaowieża Primeval Forest (Poland and Belarus) compared to other Eurasian populations', *Acta Theriologica* 45, 395–408.

Leale Anfossi M. (1958–61) 'Revisione dei materiali fittili e faunistici provenienti dagli scavi nella Grotta del Pertusello (Val Pennavaira-Albenga)', *Quaternaria* 5, 318–320.

Lowe, V. P. W. (1967) 'Teeth as indicators of age with special reference to red deer (*Cervus elaphus*) of known age from Rhum', *Journal of Zoology, London* 152, 137–153.

Lyman, R. L. (1994) *Vertebrate Taphonomy*, Cambridge Manuals in Archaeology, CUP, Cambridge.

Moeri, L. (1990) *Chasse et Elevage en Italie du Nord à l'Aube de la Domestication (Néolithique)*, Travail de diplôme Archéologie Préhistorique, Université de Genève, Fac. Des Sci., Section de Biologie, Département d'Anthropologie.

Riedel, A. (1988) 'The Neolithic animal bones deposit of Cornuda (Treviso)', *Annali dell'Università di Ferrara, sezione Scienze della Terra* 1, 71–90.

Pérez Ripoll, P. (1992) *Marcas de Carniceria, Fracturas Intencionada y Mordeduras de Carnivoros en huesos prehistoricos del mediterráneo español*, Instituto de Cultura Juan Gil Albert, Colección Patrimonio 15, Alicante.

Shipman, P. (1981) *Life History of a Fossil. An Introduction to Taphonomy and Paleoecology*, Harvard University Press, Cambridge, Massachusetts.

Steinmetz, B. and Müller, R. (1991) *An Atlas of Fish Scales: Non-Salmonid Species found in European Fresh Waters*, Samara Publishing, Cardigan.

Valsecchi, V., Finsinge, W., Tinner, W. and Ammann, B. (2008) 'Testing the influence of climate, human impact and fire on the Holocene population expansion of *Fagus sylvatica* in the southern Prealps (Italy)', *The Holocene* 18, 603–614.

Wilkens B. (2003) *Archeozoologia. Manuale per lo studio dei resti faunistici dell'area mediterranea.* CD-Rom, Schio.

Yvinec, J. H., Coutureau, M. and Tomé, C. (2007) 'Corpus of digitalized mammals skeletons', http://www.archeozoo.org/en-article134.html, accessed 6.11.2012.

8

Evidence for the Variable Exploitation of Cervids at the Early Bronze Age Site of Kaposújlak–Várdomb (South Transdanubia, Hungary)

Erika Gál

Introduction

The fortified site of Kaposújlak–Várdomb is located on a hillock orientated on a north-south axis on the western side of Hetesi drainage ditch in County Somogy (Figure 8.1). An area of 29,000 m2 was excavated during 2002 under the direction of Zsolt Gallina and Krisztina Somogyi. Remains of the Neolithic Period (Lengyel culture), the Copper Age (Pécel-Baden culture) and the Bronze Age (Somogyvár-Vinkovci culture and Urnfield culture) were recovered from over 1,400 features (Somogyi 2004). Archaeological finds indicate that the majority of these features were storage and refuse pits as well as fire places. The settlement was intensively inhabited by the beginning of the Early Bronze Age, *c.*2500–2300 BC (Gál and Kulcsár 2012).

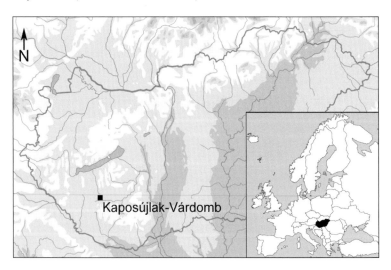

FIGURE 8.1. Location of Kaposújlak–Várdomb.

Species	NISP
Cattle (*Bos taurus*)	1,150
Sheep (*Ovis aries*)	213
Goat (*Capra hircus*)	1
Sheep and goat (*Caprinae*)	445
Pig (*Sus domesticus*)	626
Horse (*Equus caballus*)	31
Dog (*Canis familiaris*)	130
Domestic animals	2,596
Aurochs (*Bos primigenius*)	70
Red deer (*Cervus elaphus*)	208*
Roe deer (*Capreolus capreolus*)	109*
Wild boar (*Sus scrofa*)	245
Red fox (*Vulpes vulpes*)	5
Wild cat (*Felis silvestris*)	5
European polecat (*Mustela putorius*)	1
Badger (*Meles meles*)	12
Hare (*Lepus europaeus*)	84
European beaver (*Castor fiber*)	1
European hamster (*Cricetus cricetus*)	3
Rodent (Rodentia sp. indet.)	16
Goose (Anser sp. indet.)	2
Rook or Crow (*Corvus frugilegus*/*C. corone*)	1
Unidentifiable bird (Aves sp. indet.)	2
European pond turtle (*Emys orbicularis*)	4
Frog (Anura sp. indet.)	9
Fish (Pisces sp. indet.)	1
Wild animals	778
NISP total	3,374
Small ruminant	89
Large ruminant	113
Unidentifiable animal	13
Grand total	3,589

* Shed antler also counted

NISP = Number of identifiable specimens

TABLE 8.1. Abundance of remains by species according to the Number of Identified Specimens (NISP).

The site of Kaposújlak–Várdomb provided the second largest Early Bronze Age faunal assemblage in Transdanubia (the western part of Hungary), after the Makó/Somogyvár-Vinkovci culture assemblage from Paks–Gyapa that yielded 7,572 identifiable animal bones (Gál and Kulcsár 2012, 209). To date 3,374 animal bones have been identified from Kaposújlak–Várdomb, comprising six domestic and thirteen wild species (Table 8.1).

According to the dominance of domestic animals (76.9 percent), meat provisioning was based on animal husbandry. The frequency of remains indicated the importance of cattle (*Bos taurus*) (34.1 percent) above all, but sheep/goat (*Ovis*/*Capra*) (19.5 percent) and pig (*Sus domesticus*) (18.5 percent) were also well represented in the assemblage. The distribution of remains by age suggested, that cattle, sheep and goat were exploited for secondary products, while pig was mostly slaughtered at a young age. In addition, dogs (*Canis familiaris*) of different sizes and horse (*Equus caballus*) were raised at the settlement.

The quantity of identifiable specimens originating from game, totaling almost one quarter of the assemblage, is still noteworthy. The majority of the wild species' skeletal remains were from wild boar, *Sus scrofa* (7.3 percent), red deer, *Cervus elaphus* (6.2 percent) and roe deer, *Capreolus capreolus* (3.2 percent). The even representation of skeletal parts suggests entire skeletons of the hunted animals were transported to the site.

Following previous work at Kaposújlak–Várdomb on the relationships between people and animals (Gál 2009), as well as bone and antler manufacture (Gál 2011), this paper will focus on the exploitation of red and roe deer at the settlement.

Results

In addition to the bones representing food remains, a substantial number of worked bones and antlers (135 pieces, four percent of the assemblage), tool blanks as well as workshop debitage were found at Kaposújlak–Várdomb. Among the implements, points of various shapes and sizes, and hafted antler tools were the most frequent artefacts (Gál 2011).

The representation of species in the total refuse bone assemblage and tool

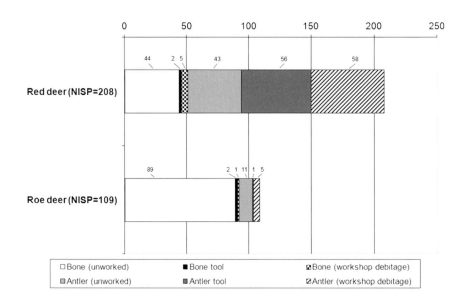

FIGURE 8.2. Distribution of worked and unworked bone and antler remains from red and roe deer.

assemblage showed a number of differences. For example, cattle and aurochs (*Bos primigenius*), that yielded more than one third of the remains from the main assemblage, barely account for three percent among the worked specimens. Similarly, only four implements were made from pig and wild boar skeletal elements, while these two species yielded 26 percent of the total bone assemblage. By contrast, only a few remains could be assigned to slaughtered red deer, while (gathered) red deer antler and antler artefacts were extremely frequent (37 percent). The representation of small domestic ruminants was quite balanced in the different assemblages. Sheep/goat remains were quite common both in the refuse bone (19.7 percent) and tool (28.1 percent) assemblage, while roe deer contributed few remains to these assemblages.

Further disparities were noted in the distribution of worked and unworked bone and antler remains from red and roe deer (Figure 8.2). The majority of the roe deer remains (88 percent: 92 bones and four antlers) and 51 bones and two antlers (25.5 percent) of the red deer assemblage were derived from slaughtered deer. The rest of red deer antler (155 pieces) could have been gathered as shed antler. In the case of roe deer, only 13 pieces of shed antler could be identified. The abundant faunal material from domestic taxa probably accounts for the small number of cervid bones that were utilised for tool manufacture. The only two tools comprise points made from red deer metapodia (one with articular end), a polished tool from a red deer mandible, and a small point with articular end carved from a roe deer metapodium (Gál 2011, table 3).

The antler of roe deer was also seldom used for making implements: only a polished antler tool and five antler pieces representing workshop debitage could be identified from this species. By contrast, red deer antler seems to have been regularly gathered and transported to the site, as demonstrated by the great number of un-worked antler fragments. Also, the frequency and degree of preservation of tools, unfinished implements and workshop debitage indicate the continuous processing of antler at the settlement.

Antler rose and beam tools that were drilled through in a medio-lateral direction, creating a tool used like an adze, represented the most abundant tool-type with 29 pieces. One specimen was drilled in a cranio-caudal direction, axe-like, exactly through the axis of the brow tine. The size of these implements varied between 72 and 235 mm (Figure 8.3).

The other group of hafted antler tools, comprising nine implements, was made from cut-off sections of beam and tine. Their greatest length varied between 83 and 266 mm. Usually the obliquely cut or bevelled end of the antler represented the working surface. Sometimes the end of the straight cut and drilled antler represented the haft for a blade. In contrast to the aforementioned adze-like tools, these objects have been identified as axe-like implements, since the working edge of the tool is usually set in line with the shaft (Figure 8.4). Other artefacts made from red deer antler included three large chisels, three curved scrapers and four polished tools (Gál 2011, tables 3 and 4).

Discussion

Red deer bone and antler have been important raw materials since the Upper Palaeolithic, and antler tools or unworked antler pieces have been found on sites, even where red deer skeletal elements were scarce in the bone assemblage. The preference of the Hungarian Bronze Age hunters for red deer (and wild boar) has already been discussed (Choyke 1987), but archaeozoological evidence for this phenomenon mainly concerned the Middle Bronze Age, and the eastern part of the Hungary (Choyke 1998; Choyke *et al.* 2004). This preference has been partly linked to the expansion of the use of red deer antler in tool and ornament industries (Choyke 1998, 173). Based on the level of craftsmanship, the existence of at least part-time specialisation creating new needs for steady sources of the raw material has been suggested (*ibid.*). The possibility of the trade of antler and goods made from antler was not excluded either. It was stated, that deer bone was rarely used in tool making, and cattle and Caprinae yielded most of the raw material used in producing the various bone implements.

Due to the cooling temperature and concomitant increasing rainfall between the Atlantic and Sub-boreal phases, a maximum of forest cover developed in the territory of present day Hungary by the beginning of the Bronze Age (2700–2500 BC). Even the Great Hungarian Plain was covered by a mixture of beech (*Fagus* sp.) and oak (*Quercus* sp.) forests, and hornbeam (*Carpinus* sp.) parkland forests in the floodplains of rivers. Transdanubia in the west was covered by mixed beech forests with occasional intrusions by oak, elm (*Ulmus* sp.) and lime (*Tilia* sp.) (Gyulai 1993, 16–17). The region, in which Kaposújlak–Várdomb is located, was assigned to the 'Area of Sub-Mediterranean Oak Forest' within the Carpathian Basin, being adjoined by the 'Area of Central European and Sub-

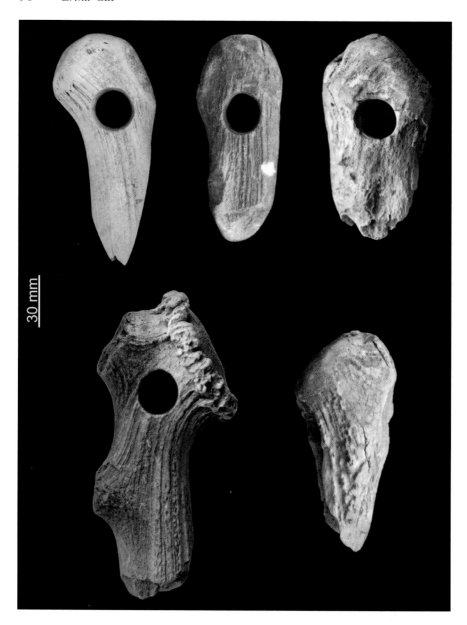

FIGURE 8.3. Hafted rose
and beam tools made
from red deer antler.

Mediterranean Mixed Forest' to the north, and the 'Area of Pannonian Forest Steppe' to the east (Kertész and Sümegi 1999, 69, fig. 4). According to the study on subfossil red deer bones from Hungary, the size of red deer gradually decreased from the Neolithic, and reached a minimum during the Bronze Age due to the cool and humid climate. In addition to the decline in size, it has been suggested that size differences between the two sexes would have decreased (Vörös 1979, 640).

Until the recently published results from Kaposújlak–Várdomb (Gál 2009) and Paks–Gyapa (Gál and Kulcsár 2012), both red and roe deer were rather

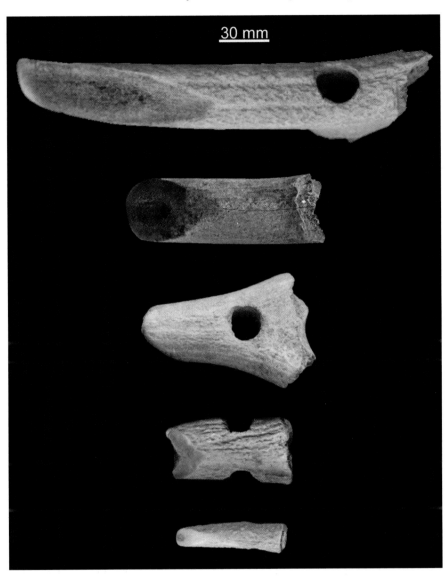

FIGURE 8.4. Hafted beam tools made from red deer antler.

underrepresented in the Early Bronze Age assemblages from Transdanubia. Red deer yielded 60 remains from a total of 12 sites, while roe deer were represented by only 11 remains (Bartosiewicz 1996, 33, table 1; Choyke and Bartosiewicz 1999, 242, table 1; Patay 2003, 45; Choyke *et al.* 2004, 180, fig. 2). Data regarding both species are summarised in Table 8.2.

The increased representation of red deer has been noted at a single Early Bronze Age site in Eastern Hungary. The tell settlement (accumulated remains of ancient settlements) of Gáborján–Csapszékpart, belonging to the Nyírség group of the Zók culture yielded 86 red deer (10.9 percent of the assemblage) and nine (1.1 percent) roe deer remains. Red deer (157 remains) were even more abundant at the Early and Middle Bronze Age (Otomani and Gyulavarsánd

Region	Site	Red deer		Roe deer	
		NISP	%	NISP	%
Western part of Hungary	**Kaposújlak–Várdomb (NISP = 3,374)**	208	6.2	109	3.2
	Paks–Gyapa (NISP = 7,573)	216	2.8	98	1.3
	Mezőkomárom–Alsóhegy (NISP = 1,381)	21	1.5	3	0.2
	Ravazd–Villibad-domb (NISP = 709)	12	1.7	5	0.7
	Százhalombatta–Földvár (NISP = 853)	14	1.6	0	0
Eastern part of Hungary	Gáborján–Csapszékpart (NISP = 787)	86	10.9	9	1.1
	Bakonszeg–Kádárdomb (NISP = 1,069)	157	14.7	7	0.6
	Gyulavarsánd–Laposhalom (NISP = 6,464)	808	12.5	39	0.6
	Tiszaug–Kéménytető (NISP = 3,194)	80	2.5	29	0.9
	Csongrád–Vidre-sziget (NISP = 1,010)	12	1.2	0	0
	Berettyóújfalú–Szilhalom (NISP = 851)	77	9.0	11	1.29

TABLE 8.2. Representation of red deer and roe deer in the most abundant Early (marked by shading) and Middle Bronze Age assemblages from Hungary.

cultures, respectively) tell site of Bakonszeg–Kádárdomb in the same region (Bökönyi 1988, 124, table 2). Details on the proportion of refuse remains and antler (tools) have not been published.

The larger proportions of wild animals, and of red deer in particular, were noted at the Middle Bronze Age tell sites of Gyulavarsánd–Laposhalom and Jászdózsa–Kápolnahalom in Eastern Hungary, and north of the Great Hungarian Plain, respectively. The number of worked pieces of antler was worthy of note at both sites (Choyke 1998, 163, table 1), while artefacts made from various osseous materials were prevalent (437 pieces) at Jászdózsa–Kápolnahalom. The importance of red deer antler seems to have increased through time at the latter settlement. As the majority of refuse and half-finished pieces came from the central mound of the tell, it has been suggested that differential access in the gathering and manufacture of this raw material could have existed. The special status of red deer has also been evidenced by three red deer skulls with intact antler racks found in the inner fortification ditch surrounding the tell at Jászdózsa–Kápolnahalom, which could have been linked to display or may have represented some kind of a foundation sacrifice. Ten to twelve skulls of other wild animals such as aurochs, wild boar and brown bear (*Ursus arctos*) were also found in the fortification ditch of this site (Choyke and Bartosiewicz 2009, 361–362, figs 4–5).

The majority of bones from Kaposújlak–Várdomb represented food and butchery remains. Nevertheless, a cattle skull found nearby human remains suggested that this may have originated from a sacrificed specimen. Similarly to the aforementioned tell site Jászdózsa–Kápolnahalom, skulls of various species such as cattle, horse and wild boar were found in a pit along with a skeleton of a dog. From an archaeological point of view, however, this feature seems to merely be a refuse pit (Gál and Kulcsár 2012, 210).

The outstanding number of red deer antler remains in contrast to the few bones of red deer suggests that the species was rarely hunted for meat, in spite of its probable frequency in the area. The meat quantity provided by red deer is also noteworthy: eviscerated stags weigh on average 160 kg, hinds 85 kg (Faragó 2002, 398). It is likely that domestic animals, especially cattle represented a continuous source of meat for people at the settlement. Domestic animal meat supplies seem to have been complemented by hunting wild boar and roe deer and, to a lesser degree, aurochs (Table 8.1). The bones of red deer as raw material could have been replaced by the skeletal parts of other large size species such as cattle and aurochs.

Our metrical data demonstrated that red deer were slightly larger than the average Bronze Age size for this species in Hungary, but were very close to data from the other recently identified assemblage (Paks–Gyapa) in Transdanubia. Complete antlers were not found at Kaposújlak–Várdomb, but the base/rose of the antlers could be measured only in ten cases. The greatest diameter of antler roses varied between 43.4 and 73.0 mm (mean = 56.1 mm; standard deviation = 9.0), while the smallest diameter varied between 33.4 and 68.6 mm (mean: 50.7 mm; standard deviation 9.4).

The abundance and variety of antler, antler tools and manufacturing debris provides evidence for the permanent collecting, deposition and working of red deer antler. As the archaeological study and interpretation of the site is ongoing, the spatial distribution of antler deposits and possible workshops is yet to be constructed. Similarly, any social differentiation within the settlement will be the subject of further studies. However, both the number and types of antler tools, especially the hafted adze and axe-like artefacts provide evidence for a developed antler working industry.

The preservation of hafted burr and beam tools suggested the long term use and curation of these implements. The functional significance of this type of tool consisted in its double use: beating and smoothing with the flat (burr) end, and cutting and piercing with the sharp bevel end or the inserted stone blade. Since only a single burr and beam tool can be made from an antler, its value must have been greater than that of other antler tools.

The number of roe deer remains is only half of those in the red deer assemblage found at Kaposújlak–Várdomb and most of these came from hunted animals. This would suggest that people were interested in the meat and skin rather than the antler of this species. The two-point antler and unfused cranial suture of a young buck further supports the suggestion that sub-adult roe deer were killed for meat (Gál 2009, 60, fig. 10a). Since sheep and goat were the second most abundant species in the assemblage accounting for almost 20 percent of specimens, their bones may have supplied enough raw material in this size category directly available at the settlement, and roe deer antlers were clearly rarely targeted. The small number of roe deer remains is characteristic of both Transdanubian sites and those located in the eastern part of Hungary (Table 8.2), although the environmental conditions would have been favourable for this species (Choyke and Bartosiewicz 1999–2000). The scarcity of roe deer remains was also emphasised in the case of other Hungarian Bronze Age sites (Choyke 1987, 112).

Evidence suggests that roe deer were being hunted more frequently than red deer, but the shed antler of red deer represented a more valued raw material than the antler of roe deer. This is understandable if one takes into account the size and quantity of raw material represented by the two species in question. A pair of red deer antlers may weigh from 1.3 kg (average of antler from two years old stags) to 9.2 kg (13 year old stags; Faragó 2002, 399); whereas a pair of roe deer antlers weigh on average only 0.3–0.4 kg (Bán and Fatalin, 1986, 203, table 5).

Not only are there size differences, but the antler morphology, thickness and density also differ between the two species. Roe deer develop short and erect antlers that usually have two, three, or occasionally four points. The maximum length that an antler reaches is 250 mm. In contrast, the length of the red deer antler may be up to 1,090 mm. The antler of a well-developed red deer stag consists of the beam, the brow tine, the bez tine, the trez tine and the terminal tines that form the crown. The circumference of the antler rose varies between 243–303 mm (Faragó 2002, 399–413). It is also worth mentioning, that antlers, in contrast with the boiled bones, retain both mineral and organic components making it strong, flexible and durable.

Conclusions

Hunting and gathering seem to have been regular activities in the life of people settled at Kaposújlak–Várdomb, as evidenced by a variety of game species and the outstanding number of antler remains. Disparities between the skeletal elements represented and abundance of red and roe deer remains, however, indicated different forms of exploitation of these two species. Roe deer was mostly killed for meat and skin, while red deer yielded raw material for antler manufacturing, mostly in the form of shed antler.

The abundance and preservation of hafted antler tools would suggest the importance of these implements in earth and wood workings at the fortified settlement of Kaposújlak–Várdomb. Nevertheless, based on the outstanding number of various antler implements, blanks and fragments representing workshop debitage and raw material, especially in comparison with other contemporaneous sites, it cannot be ruled out that due to its geographical position in Transdanubia, Kaposújlak–Várdomb represented a centre for antler gathering, manufacturing and trade during the Early Bronze Age.

Metal working pursued at the site as well as the fortified character of the settlement also distinguish Kaposújlak–Várdomb from other coeval sites in Transdanubia, and make its special status in the region likely. Further archaeological studies shall clarify whether its social structure resembles that of tell sites characteristic to the central and eastern part of Hungary.

Acknowledgements

This paper is the written version of my lecture presented in the session 'Cervids and society – Deer in Time and Space' organised by Jacqui Mulville and Naomi Sykes at the General Conference of ICAZ 2010 held in Paris. Archaeologists Zsolt Gallina, Gabriella Kulcsár and Krisztina Somogyi are thanked for having invited me to study the animal remains from Kaposújlak–Várdomb. Suggestions and linguistic corrections made by the two anonymous reviewers greatly improved the quality of this paper. My research is being supported by the Hungarian Scientific Research Fund (OTKA Project NF 104792).

References

Bán, I. and Fatalin, Gy. (1986) *Élőhely és trófeavizsgálat számítógéppel* [Computerized habitat and trophy evaluation], Akadémiai Kiadó, Budapest.

Bartosiewicz, L. (1996) 'Bronze Age animal keeping in Northwestern Transdanubia, Hungary', *Acta Musei Papaensis* 6, 31–41.

Bökönyi, S. (1988) 'Animal remains from the Bronze Age tells of the Berettyó Valley', in *Bronze Age tell settlements of the Great Hungarian Plain*, eds T. Kovács and I. Stanczik. Magyar Nemzeti Múzeum, Budapest, 123–135.

Choyke, A. M. (1987) 'The exploitation of red deer in the Hungarian Bronze Age', *Archaeozoologia* 1, 109–116.

Choyke, A. M. (1998) 'Bronze Age red deer: case studies from the Great Hungarian Plain', in *Man and the Animal World*, eds P. Anreiter, L. Bartosiewicz, E. Jerem and W. Meid, Archaeolingua, Budapest, 157–178.

Choyke, A. M. and Bartosiewicz, L. (1999) 'Bronze Age animal exploitation in Western Hungary', in *Archaeology of the Bronze Age and Iron Age: Experimental archaeology, environmental archaeology, archaeological parks*, eds E. Jerem and W. Meid, Archaeolingua, Budapest, 239–249.

Choyke, A. M. and Bartosiewicz, L. (1999–2000) 'Bronze Age animal exploitation on the Central Great Hungarian Plain', *Acta Archaeologica Academiae Scientiarum Hungaricae* 51, 43–70.

Choyke, A. M. and Bartosiewicz, L. (2009) 'Telltale tools from a tell: Bone and antler manufacturing at Bronze Age Jászdózsa–Kápolnahalom, Hungary', *Tisicum* 19, 357–376.

Choyke, A. M., Vretemark, M. and Sten, S. (2004) 'Levels of social identity expressed in the refuse and worked bone from the Middle Bronze Age Százhalombatta–Földvár, Vatya culture, Hungary', in *Behaviour Behind Bones: The Zooarchaeology of Ritual, Religion, Status and Identity*, eds S. Jones and A. Ervynck, Oxbow, Oxford, 177–189.

Faragó, S. (2002) *Vadászati állattan* [Zoology for hunters], Mezőgazda Kiadó, Budapest.

Gál, E. (2009) 'Relationships between people and animals during the Early Bronze Age: Preliminary results on the animal bone remains from Kaposújlak–Várdomb (South Transdanubia, Hungary)', in *Őskoros Kutatók VI. Összejövetelének konferenciakötete. Kőszeg, 2009. március 19–21. Nyersanyagok és kereskedelem (Proceedings of the 6th Meeting for the Researchers of Prehistory. Kőszeg, 19–21 March, 2009 Raw materials and trades)*, ed. G. Ilon, *ΜΩΜΟΣ* 6, Kulturális Örökségvédelmi Szakszolgálat – Vas megyei Múzeumok Igazgatósága, Szombathely – Budapest, 47–63.

Gál, E. (2011) 'Prehistoric antler- and bone tools from Kaposújlak–Várdomb (South-Western Hungary) with special regard to the Early Bronze Age implements', in *Written in Bones. Studies on Technological and Social Contexts of Past Faunal Skeletal Remains*, eds J. Baron and B. Kufel-Diakowska, Uniwersytet Wrocławski, Wrocław, 137–164.

Gál, E. and Kulcsár, G. (2012) 'Változások a bronzkor kezdetén-A dél-dunántúli gazdálkodás jellege az állatcsont leletek alapján (Changes at the beginning of the Bronze Age – Characterizing subsistence on the basis of animal remains in southern Transdanubia, Hungary)', in *Környezet – Ember – Kultúra. A természettudományok és a régészet párbeszéde (Environment – Human – Culture. Dialogue between applied sciences and archaeology)*, eds A. Kreiter, Á. Pető and B. Tugya, Nemzeti Örökségvédelmi Központ, Budapest, 207–214.

Gyulai, F. (1993) *Environment and Agriculture in Bronze Age Hungary*, Archaeolingua (Series Minor), Budapest.

Kertész, R. and Sümegi, P. (1999) 'Az Északi-középhegység negyedidőszak végi őstörténete [Palaeohistory of the Northern Mountain of Medium Height at the end of the Quaternary]', *A Nógrád Megyei Múzeumok Évkönyve* 23, 66–93.

Patay, R. (2003) 'Kora bronzkori leletek Balatonkeneséről (Early Bronze Age finds from Balatonkenese)', *Veszprém Megyei Múzeumok Közleményei* 22, 43–55.

Somogyi, K. (2004) 'Előzetes jelentés a Kaposvár-61-es elkerülő út 29. számú lelőhelyén, Kaposújlak–Várdomb-dűlőben 2002-ben végzett megelőző feltárásról (Preliminary report on the preceding excavation of site number 29. of the Route 61. encircling Kaposvár)', *Somogyi Múzeumok Közleményei* 16, 165–178.

Vörös, I. (1979) 'Archaeozoological investigations of subfossil red deer populations in Hungary', *Archaeozoology* 1, 637–642.

Red Deer Hunting and Exploitation in the Early Neolithic Settlement of Rottenburg-Fröbelweg, South Germany

Elisabeth Stephan

Introduction

The Linear Pottery Culture, abbreviated as LBK (from German: Linear-bandkeramik), has been regarded as the first Neolithic culture over much of Europe. The origin of the Linear Pottery Culture is thought to date between 5700 and 5600 BC, in the course of culture contact between the Starčevo-Kőrős-Criş groups, later Vinça Culture and an, as yet, unknown population of hunter-gatherers in present-day western Hungary (e.g. Lüning 1988; Gronenborn 1999). It expanded over hundreds of kilometres from the middle Danube in Hungary to the Rhine in the west, to central Poland to the north and the northern Ukraine to the east and ended approximately 5000 BC Since the basic work of Quitta (1960) there has been a controversial debate over the sudden appearance of the earliest LBK sites and the relatively rapid dispersal of similar kinds of domesticated plants and animals, architecture, pottery, burials, and other remains (e.g. Lüning 1988; Whittle 1996). The focus of the discussions has been on the early part of the LBK and on the nature of contacts, interactions and transformations between incoming farmers, indigenous Mesolithic hunter and gatherers and the La Hoguette pottery group which is thought to have derived from very early Neolithic groups in the south of France (Tillmann 1993; Kind 1997; Kind 1998; Gronenborn 1999; Frirdrich 2005; Gehlen 2006; Gronenborn 2007). South-west Germany is an interesting region for tackling these problems as it forms the south-western extent of the earliest phase of the LBK, the La Hoguette pottery is present in a number of earliest LBK sites and the Late Mesolithic is well investigated (Lüning *et al.* 1989; Kind 1997, 1998; Gronenborn 1999; Kieselbach *et al.* 2000; Strien 2000; Kind 2003; Gronenborn 2007*)*.

Faunal studies have long indicated a remarkable uniformity in the LBK subsistence economy. According to Müller (1964), cattle were the most important component of the animal economy and the contribution of hunting was negligible.

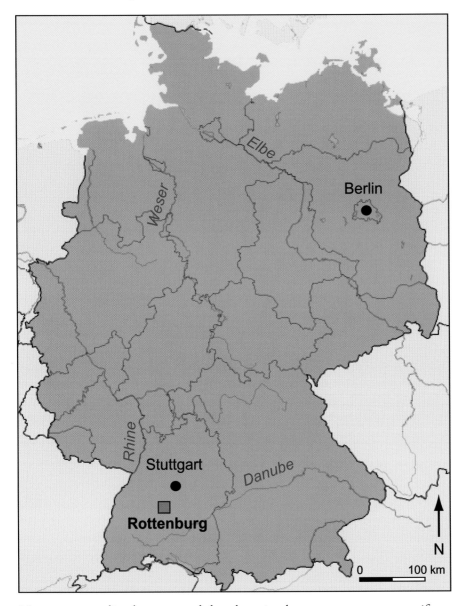

FIGURE 9.1. Location of Rottenburg-Fröbelweg at the river Neckar between the Swabian Alb and the Black Forest in southwest Germany.

More recent studies demonstrated that the animal economy was not as uniform as previously thought (Döhle 1993; Döhle 1994a; Döhle 1997). This trend was confirmed by studies made during recent years on faunal assemblages from several sites of the earliest phase of the LBK indicating no clear pattern of similarities in the animal economy for the earliest farming communities (Pucher 1987; Döhle 1994b; Uerpmann 2001). In this paper the site of Rottenburg-Fröbelweg in southwest Germany, dated to the earliest LBK, will be discussed in terms of wild and domestic faunal exploitation strategies (Figure 9.1). The site is exceptional in the quantity of faunal remains compared to other earliest LBK sites in the immediate vicinity of Rottenburg such as Ammerbuch-Reusten, -Poltringen, and -Pfäffingen

FIGURE 9.2. Extension of the earliest LBK (yellow) and the La Hoguette Culture (red) with the locations of LBK 1 sites in Middle Europe (modified after Gronenborn 1994, fig. 1; Gronenborn 2007, fig. 2).

GERMANY:
BADEN-WÜRTTEMBERG:
1. Rottenburg-Fröbelweg
2, 3. Ammerbuch-Reusten and -Poltringen
4. Ammerbuch-Pfäffingen
5. Vaihingen a.d. Enz
BAVARIA:
6. Kleinsorheim
7. Enkingen
8. Aich
9. Mintraching
10. Wang
11. Schwanfeld
HESSE:
12. Bruchenbrücken
13. Goddelau
14. Niedereschbach
LOWER SAXONY:
15. Eitzum
SAXONY-ANHALT:
16. Eilsleben
AUSTRIA:
17. Neckenmarkt
18 Strögen.

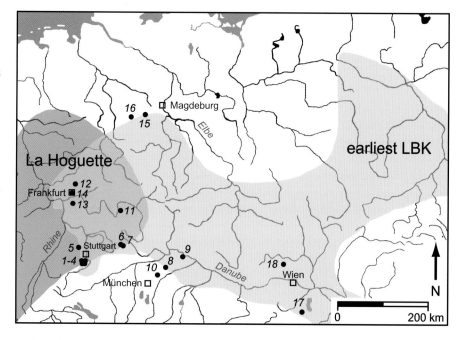

and in the German states Bavaria and Hesse such as Kleinsorheim, Enkingen, Aich, Mintraching, Wang, Schwanfeld, Niedereschbach, and Bruchenbrücken (Stork 1993; Uerpmann 2001), and in its location at the south-western border of the LBK 1 extension and at the intersection of the LBK and the La Hoguette Culture (Figure 9.2).

Site, finds and dating

The site of Rottenburg-Fröbelweg is located at the river Neckar between the Swabian Alb and the Black Forest in south-west Germany (Figure 9.1). The site was partially excavated in an area of development from 1989 to 1995 (Bofinger 2005). Ditches, post-holes and pits proved the existence of a small village with four or five long houses. These houses are characteristic of the earliest phase of the LBK because contrary to the houses of the later LBK phases, their ground plans do not have central posts or strengthening of the north-western part (Bofinger 2005). The majority of the pottery assemblage belongs typologically to the earliest LBK phase and a minor amount of La Hoguette pottery found in several pits together with the LBK ware point to the existence of contact and exchange between the two cultures (Bofinger 2005, 72–101). Calibrated AMS ^{14}C measurements of charred emmer and einkorn grains and of mammal bones date the settlement to between 5300 and 4550 cal. BC (Bofinger 2005, 112–127). According to these dates the settlement lasted for several hundred years and extended into later LBK phases. This means that the radiocarbon dates are too late for the earliest LBK phase, which started in south-western Germany at approximately 5600 BC (Kind 1997, 118–144; Bofinger 2005, 112–127). Comparably late dates exist for other German and Austrian earliest LBK sites such as Bruchenbrücken, Goddelau, Eitzum, Ammerbuch-Pfäffingen, Neckenmarkt, and Strögen (Lenneis *et al.* 1996; Stäuble 2005). For the Rhine-Main Area, Lüning (2005, 67–72) described two phases of the earliest LBK based on calibrated radiocarbon dates: the earliest LBK 1 from 5500 to 5350 cal. BC and the earliest LBK 2 from 5300 to 5150 cal. BC.[1] The later phase is contemporaneous with the early LBK or so called Flomborn phase and Lüning (2005, 71–72) supposed that two linear pottery 'traditions' coexisted for approximately 150 years. Strien (2000; 2011) proposed an alternative relative chronology for the Neckar region based on the ceramic decorated motifs linked with dendrochronological dated sequences (see Bogaard *et al.* 2011; see Knipper 2011, 10, fig. 2.3). This chronology includes only one earliest LBK phase which is contemporaneous with the earliest LBK 1 of Lüning. At present both chronologies are not consistent and further investigations of the relationship between the earliest and the earlier LBK and the origin of Flomborn are needed.

The faunal assemblage

During the excavations 5,812 fragments of bones, teeth, and antler were recovered from LBK features at the site. A substantial proportion (62 percent) of the faunal remains was found in pits and ditches along the sides of the long houses. The rest originated from two overlying layers. All bones were poorly preserved, highly fragmented and fragile. The bone surfaces were often abraded and showed extensive root etching. This poor preservation and the very precise recovery techniques employed resulted in a substantial amount of very small fragments and only 25 percent of the bone finds could be determined to species level (Table 9.1). In weight this identifiable percentage is equal to 80 percent of the assemblage. A total of 30 percent of the unidentified fragments belong to medium-sized animals (sheep *Ovis aries*, goat *Capra hircus*, roe deer *Capreolus capreolus*, domestic pig *Sus domesticus*), and 29 percent derived from large animals (wild boar *Sus scrofa*, red deer *Cervus elaphus*, cattle *Bos taurus*). The rest consists of tiny fragments with an average weight of 0.6 g which provides no information about animal size.

Taxonomic composition

The taxonomic composition demonstrates that the earliest farmers in Rottenburg kept few domestic animals. Only ten percent of the identified bones are from cattle, sheep, goat, and pig (Table 9.1). Approximately two thirds of identified specimens can be confidently attributed as wild animals. Red deer predominate with approximately 70 percent of the identified specimens (80 percent by weight); roe deer (9.3 percent by NISP/4.3 percent by weight) and wild boar (5.1 percent by NISP/10.8 percent by weight) were also common. In contrast to other earliest Neolithic sites in Germany such as Kleinsorheim, Schwanfeld and Eitzum (Uerpmann 2001) and Austria such as Neckenmarkt and Strögen (Pucher 1987)

	NISP	%	Weight (g)	%
4: Without size estimation	1820	41.3	338.2	11.9
3: Sheep/Goat/Roe deer – pig	1315	29.8	666.3	23.5
2: Wild boar – Red deer	1011	22.9	1203.1	42.4
1: Cattle – Aurochs	263	6.0	629.0	22.2
Unidentified specimens	4409	100.0	2836.6	100.0
Cattle, *Bos taurus*	37	2.6	661.9	6.1
Sheep, *Ovis aries*	11	0.8	51.4	0.5
Goat, *Capra hircus*	1	0.1	4.3	0.0
Sheep/Goat	77	5.5	209.1	1.9
Pig, *Sus domesticus*	17	1.2	158.0	1.4
Dog, *Canis familiaris*	4	0.3	1.9	0.0
Domestic animals	147	10.5	1086.6	10.0
Sheep/Goat/Roe deer	144	10.3	231.5	2.1
Pig/Wild boar	160	11.4	275.7	2.5
Domestic or wild animals	304	21.7	507.2	4.6
Aurochs, *Bos primigenius*	1	0.1	39.0	0.4
Red deer, *Cervus elaphus*	708	50.5	7552.0	69.2
Roe deer, *Capreolus capreolus*	131	9.3	473.2	4.3
Wild boar, *Sus scrofa*	72	5.1	1183.6	10.8
Wild cat, *Felis silvestris*	1	0.1	1.6	0.0
Pine marten, *Martes martes*	2	0.1	0.5	0.0
Marten/Polecat, *Martes/Mustela* spec.	2	0.1	0.4	0.0
Beaver, *Castor fiber*	22	1.6	64.8	0.6
Rodentia indet.	4	0.3	0.2	0.0
Northern goshawk, *Accipiter gentilis*	1	0.1	0.2	0.0
Grey heron, *Ardea cinerea*	1	0.1	0.7	0.0
Fish, Pisces indet.	4	0.3	1.1	0.0
Snails	3	0.2	0.9	0.0
Wild animals	952	67.8	9317.3	85.4
Total NISP	1403	100.0	10912.0	100.0
Total NISP	1403	24.1	10912.0	79.4
Total unidentified specimens	4409	75.9	2836.6	20.6
Total	5812	100.0	13748.6	100.0

TABLE 9.1. Rottenburg-Fröbelweg. Taxonomic composition.

where aurochs finds are most numerous, in Fröbelweg aurochs are represented by only one phalanx II. A few bones originate from small mammals (wild cat, pine marten, rodents), birds (northern goshawk, grey heron), and fish which could not be further identified. A total of 22 beaver bones were found exclusively in three pits. The bones were not articulated and represent one juvenile and two adult individuals. No differences could be observed in the taxonomic composition of the faunal remains from pits and ditches with exclusively LBK pottery and from features with additional finds of La Hoguette ceramics. In all features wild animals, especially red deer, predominate over domestic animals. Due to the extremely small number of faunal assemblages from La Hoguette contexts the economy of this group is not well known (Bofinger 2005, 129–135; Gehlen 2006, 50–51; Gronenborn 2007, 80) and it is difficult to track down possible influences on the earliest LBK economy.

Animal economy and exploitation of red and roe deer

Cattle were generally slaughtered at an advanced age and meat may not have been the only exploited resource, but specific indications of milk and milk products or their use as draught animals were not observed. The exploitation strategy of the small domestic ruminants could not be determined based on the few finds of sheep and goat. Domestic pig and wild boar were bred or hunted for meat, but their contribution to the food supply appears to have been small.

Meat requirements were mainly satisfied by hunting and red deer was the most important game. All parts of the red deer skeleton were represented (Stephan 2005, 345, fig. 17),[2] which implies that the complete animals were brought to the settlement and dismembered on the site. Although predominantly subadult-adult animals were hunted, all red deer finds originate from remarkably small animals in the range of female deer (hinds) from other Neolithic sites in Germany and Switzerland such as Ammerbuch (Stork 1993), Eilsleben (Döhle 1994a; Döhle 1994b), Burgäschisee-Süd (Boessneck *et al.* 1963), Twann (Becker 1981; Becker and Johanssen 1981) and modern collections (Stephan 2005, 346). Concerning the environmental conditions in the area of Rottenburg it is unlikely that only small red deer lived here and a preference for hind hunting must be considered. The almost complete absence of antler, which is represented only by two small fragments and one artefact,[3] would support this suggestion, although no convincing reasons for such a targeted hunting strategy are known.

As with red deer, all parts of the roe deer skeleton are represented, and it appears that complete animals were brought to the settlement (Stephan 2005, 348, fig. 20). The large amount of fragments of metatarsals used as raw material for tool production is remarkable (Bofinger 2005, 101–102; Stephan 2005, 352–356). Similarly to red deer, only one unworked antler fragment was found and predominantly subadult-adult roe deer were hunted. But unlike the red deer the measurements indicated roe deer of the similar sizes as those recovered at other early LBK sites in Germany and Switzerland (e.g. Eilsleben and Middle Germany: Müller 1964; Döhle 1994a, b; Pfäffingen 'Lüsse': Stork 1993; Twann, Switzerland: Becker 1981; Becker and Johanssen 1981).

Red deer: worked incisors and canines

An extraordinary find was made in a pit west of house No. 3. Right and left incisors and canines of a red deer mandible were found together with no trace of the jawbone. There is no evidence of manipulation on the third incisors or canines, but the roots of the first and second incisor each show a semi-circular drill-hole in the lingual-buccal direction (I2 left mesial surface, I1 left distal surface, I1 right distal surface, I2 right mesial surface) resulting in two complete holes (Figure 9.3). These drill-holes could have been useful only if the teeth were held together, either by the mandible/gingiva or by being bound in wood or another material.

Throughout the Neolithic single teeth were used as jewellery. However, in contrast to the find from Rottenburg-Fröbelweg, their roots are drilled from the mesial to the distal surface generating a complete hole in a single tooth. These pendants are frequently made from the maxillary canines of stags as well as incisors and canines of pigs, wild boars and dogs (*Canis* sp.) (e.g. Schibler 1981; Schibler 1997, 173–175; Haack 2002, 100–103). Drilled teeth of large ruminants are rarely found. Examples include cattle teeth pendants from the LBK site of Herxheim in Germany (Haack 2002, 101), and a few pendants made of cattle and red deer incisors found at the lake dwelling sites in Switzerland (Schibler 1997, 174).

Teeth resembling the Fröbelweg finds were described for the Magdalenian sites Gönnersdorf und Petersfels in Germany (Poplin 1972; Poplin 1983). In Petersfels more than 200 and in Gönnersdorf about 45 single incisors and canines of reindeer (*Rangifer* sp.) mandibles were excavated, most of them

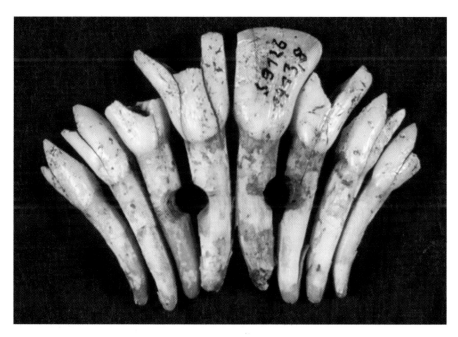

FIGURE 9.3. Buccal face of the right and left incisors and canines of a red deer mandible found in pit No. 128 at Rottenburg-Fröbelweg with semi-circular perforations in the roots of the first and seconds incisors.

with broken or sawn off roots. Poplin (1972, fig. 3; 1983, 136–144, figs 6 and 7) explained these finds as follows: only 50 percent of the long and thin roots of the teeth are covered by the jawbone, and the gums would have had to have been present to attach them to the bone, if they were not attached to another material. These anatomical features allow the removal of the teeth all together from the jaw and enable them to be mounted on clothes as described for the manufacturing of belts by Alaskan Eskimo women (Poplin 1994).

Jewellery made from red deer and wild boar teeth were found in Mesolithic graves in Denmark (e.g. Dragsholm and Vedbæk, Brinch Petersen 1974; Alberthsen and Brinch Petersen 1976). Predominantly single teeth drilled from the mesial to the distal surface are utilised, but some rows of incisors and canines were also found. In these cases, the roots were complete but show small notches on the buccal surface. They were also interpreted as jewellery attached to clothing or belts and are very similar to the Rottenburg teeth. The perforations of the Rottenburg incisors are located in the middle of the root and were therefore situated outside the jawbone. It is supposed that they were carefully removed from the jawbone together with the gums and served as jewellery as described above. As no comparable pieces are known from the Neolithic and the most similar examples are from the Mesolithic or Upper Palaeolithic this special find could be a reference to the survival of old traditions and points to the significance of hunting.

Livestock management and hunting during the earliest LBK in Central Europe

Earliest LBK sites in the immediate vicinity of Rottenburg-Fröbelweg are also characterized by high percentages of wild animals (e.g. Ammerbuch-Reusten, Poltringen, Pfäffingen 'Lüsse': Stork 1993). This indicates that hunting was an essential part of subsistence during the earliest phase of the LBK in this region. Hunting activities focussed principally on red deer. Roe deer (*Capreolus capreolus*), aurochs (*Bos primigenius*), and wild boar were exploited to a lesser extent (Figure 9.4; for the location of the sites see Figure 9.2).[4] The picture becomes more complex looking at the taxonomic composition of the faunal assemblages of the earliest LBK sites in Baden-Württemberg (Vaihingen a. d. Enz: Schäfer 2010[4]), Bavaria, Hesse, Lower Saxony, Saxony-Anhalt, and Austria (Figures 9.2 and 9.4). There is no uniform pattern of livestock breeding and hunting and the subsistence strategies of the first settlers differed between different regions and between single settlements within the regions (for detailed discussions see Uerpmann and Uerpmann 1997; Stephan 2005, 357). The exploitation of large herbivores: aurochs, cattle and red deer was a common feature at most of the sites. Bruchenbrücken and Strögen are exceptions, where sheep, goats, and pigs were much more frequently exploited.

Importance of hunting

It is highly likely that the inhabitants of Rottenburg-Fröbelweg and other earliest LBK settlements in this region were the first farmers in this region. They adapted animal husbandry strategies to a new environment and different climatic conditions compared to the regions they came from. It has to be considered that breeding, especially of cattle, needs constant care and competes with crop cultivation, which is characterised by highly seasonal workloads. Restrictions on fodder supply could occur during wintertime and could also have caused problems for feeding the animals properly (Ebersbach 2002, 137–166). Therefore, it could have been economical for the first settlers to restrict breeding and to reduce livestock management by sustaining diet through hunting the common wild mammals especially red and roe deer. Hunting for meat could have related to the necessity to protect the fields from being damaged and the crops from being eaten by wild animals (Boessneck *et al.* 1963; Uerpmann 1977; Uerpmann and Uerpmann 1997). This protective hunting might have been important if the Early Neolithic fields were small and surrounded by forest, as suggested by Gregg (1988, 132) based on palaeoethnobotanical data, and/or if the early farmers practiced an intensive garden cultivation, as proposed by Bogaard (2004) based on archaeobotanical investigations. Additionally, the opening of forested areas to create fields for crop production could cause extended hunting activities through reduced forest shelter (cf. Lüning 2000, 193).

Due to the very low proportion of young domestic animals in Rottenburg-Fröbelweg it should be questioned whether the inhabitants bred the livestock on site at all. Alternatively, they could have imported the animals from other settlements, and kept them for only a short time before slaughter or they could have imported joints of meat from animals slaughtered elsewhere. Due to the small numbers of cattle, sheep, goat and pig the element representations do not provide (strong) evidence for either of these possibilities. The skull fragments of cattle, sheep/goat and pig found at the site point to the on-site butchering of these animals.

It is also possible that the site was used as a hunting station. This suggestion is supported by the fact that the site is situated on the flood plain of the Neckar above clayey soil, less suited to agriculture and not above fertile and easily tilled loess soils as it is the case in most of the other LBK settlements. Furthermore, the radiocarbon dates indicate a long settlement period, and it is conceivable that the settlement was used not only during the earliest phase of the LBK but served as hunting station during later LBK phases coexisting with other farming villages. This possibility has been discussed before for settlements in south Germany and Switzerland: Burgäschisee-Süd/Switzerland (Boessneck *et al.* 1963), Polling/Bavaria (Blome 1968), Hornstaad/southwest Germany (Kokabi 1991; Lüning 2000, 127–128) and for comparable sites in north Germany (Lüning 2000, 193–196). Seasonal hunting activities were suggested for these settlements by Lüning (2000, 194–195), but data

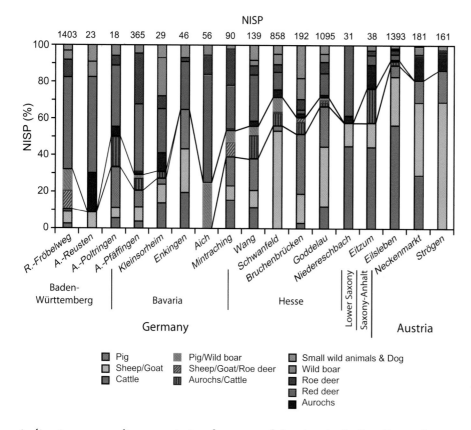

FIGURE 9.4. Taxonomic composition of faunal assemblages from Rottenburg-Fröbelweg and earliest LBK sites in Germany and Austria (Rottenburg-Fröbelweg: this study; Ammerbuch-Reusten and -Poltringen: Uerpmann 2001; Ammerbuch-Pfäffingen: Stork 1993; Bavaria: Kleinsorheim, Enkingen, Aich, Mintraching, Wang, Schwanfeld: Uerpmann 2001; Hesse: Bruchenbrücken, Goddelau, Niedereschbach: Uerpmann 2001; Lower Saxony: Eitzum: Uerpmann 2001; Saxony-Anhalt: Eilsleben: Döhle 1994b; Austria: Neckenmarkt, Strögen: Pucher 1987).

indicating seasonality are missing for most of the sites including Rottenburg-Fröbelweg.

Pollen analyses provide evidence for highly forested landscapes in south-west Germany during the Mesolithic and the earliest Neolithic, sufficient for feeding large numbers of red deer, roe deer, and wild boar (*c.*7000–6000 BC; Liese-Kleiber 1991; Kind 1997; Smettan 2000). Due to the low proportions of cattle a degradation of the natural forests in the immediate environment of Rottenburg-Fröbelweg is not likely and the environs should have been well suited as deer habitat. Most of the wild species found in the Fröbelweg settlement are consistent with deciduous mixed forest and less dense coniferous forest with clearings and riverside forests along the Neckar and other waters (e.g. roe deer: Lehmann and Sägesser 1986; Stubbe 1990, 26–29; red deer: Bützler 1977; Wagenknecht 1981, 196–200, 244–245; Beninde 1988). Due to their dietary requirements red deer need large areas for feeding, and herds migrate over large distances during spring and autumn. They tend to follow rivers because of reliable food supply and cover (Bützler 1977; Wagenknecht 1981, 196–200, 244–245; Beninde 1988). Therefore, during the Early Neolithic deer might have migrated along the Neckar valley in the Rottenburg area. In the south-west of the site the Neckar flows in a narrow valley, while to the east, the valley opens to a small basin offering excellent opportunities for hunting (Bofinger 2005, 14–21, fig. 17).

Relationships to Mesolithic hunter-gatherers

Based on several arguments it is conceivable the first farmers and the hunter-gatherers of the Late Mesolithic were in contact in south Germany. The earliest LBK site Rottenburg-Fröbelweg is located in the close proximity (600 m) to the Mesolithic site of Rottenburg-Siebenlinden. The taxonomic composition of the faunal assemblage from the Late Mesolithic horizon at Rottenburg-Siebenlinden, dominated by red deer, aurochs, roe deer and a few finds of wild boar is, with the exception of the domestic animals, comparable to the composition of the Fröbelweg assemblage (Kind *et al.* 2012, 128–129). The radiocarbon dates for the Late Mesolithic at Rottenburg-Siebenlinden show a bimodal distribution and although there is no strong evidence from the stratigraphy or from the lithic assemblage, the existence of two settlement phases, the main phase between 6500 and 6100 cal. BC and a minor phase between 5700 and 5500 cal. BC is plausible (Kind *et al.* 2012, 47–58, 69–72). A continuous transition from the Mesolithic to the Neolithic cannot be proven because of a temporal gap of centuries between the settlements of the last Mesolithic people and the first farmers at Rottenburg. Considering single ^{14}C dates the difference between the latest Mesolithic and the earliest LBK is reduced to about 250 years (Bofinger 2005, 112–127, table 16; Kind *et al.* 2012, 51, table 9). The consideration of all dated sites in Germany demonstrate that Mesolithic and earliest LBK radiocarbon dating overlap and the latest Mesolithic in the Danube and the Oberschwaben region is contemporary with the earliest LBK sites in the Neckar region (Kind 1997, 118–144). Geographically no overlaps are observable, and there is no site yielding layers of the earliest LBK phase above Mesolithic layers. Late Mesolithic sites are known from the Danube, Oberschwaben, and Basle in Switzerland. Sites dated to the earliest LBK phase are located only in the Neckar region and in Alsace (Bofinger 2005, 131, fig. 78; Gehlen 2006, fig. 3). Although, the gap between the radiocarbon dates from the Late Mesolithic and the earliest LBK at Rottenburg is short and could be narrowed by new investigations, at present evidence suggests that the Mesolithic hunter-gatherers had vanished in the Neckar region at the time the first farmers arrived. Hunter-gatherers might have survived in different regions like the Schwabian Alb–Danube region and Oberschwaben as these regions were more slowly colonised as they were less suited to the LBK mode of subsistence (Kind 1997).

Regardless of these facts, the remarkable resemblance of the production techniques of lithic artefacts and of the tool types used at earliest LBK sites (e.g. Rottenburg-Fröbelweg and Bruchenbrücken) with lithic inventories from Mesolithic sites provide a strong argument for relationships between the two groups (Gronenborn 1994; Kind 1998; Kind 2003; Kind 2005; Strien 2000; Gehlen 2006). During the Neolithic tool sizes increased but otherwise no innovations were observed. Kind (1998) hypothesised that the first farmers established their techniques on the basis of Mesolithic traditions. Additionally, he suspected a coexistence and manifold relationship of hunter-gatherers and farmers in the same geographical region between 5500 and 5000 BC, as it is observed for modern farmers and hunter-gatherer communities in South Africa, India, South America and the Philippines (cf. Gregg 1988, chapter 3). The interactions could have been of a different nature. Besides the exchange of animal and plant dietary resources, non-economic relationships can be developed such as cross-border marriages as Bentley *et al.* (2002) supposed for the inhabitants of the Stuttgart-Mühlhausen site of Viesenhäuser Hof based on the strontium isotope ratios in tooth enamel of young women (see Bentley 2007, 125; Knipper 2011, 350–353). These ratios differ from the isotope signatures of the other inhabitants of the settlement, but are consistent with the signatures of the highlands at both sides of the Rhine. Therefore the young women could have originally belonged to hunter-gatherer communities and migrated to the Stuttgart site for marriages.

Conclusions

The taxonomic composition provides evidence that the earliest farmers in Rottenburg kept few domestic animals and influences on the animal economy by the La Hoguette culture could not be identified. About two thirds of identified specimens are from wild animals. Red deer predominate with approximately 70 percent of the identified specimens; roe deer and wild boar were also common. The needs of meat were satisfied mainly by hunting red and roe deer, with red deer being the most important game. The small size of the red deer and the lack of antler finds could indicate a specialisation in hind hunting.

The importance of hunting could be explained as follows: The first farmers in Rottenburg adapted animal husbandry to a new environment. The opening of forested areas to create fields for crop planting and the hunting of herbivores for the protection of plant crops could cause the extended hunting activities and high proportions of wild animals in the archaeological record. During the Atlantic period (*c.*7000 to 4000 BC) the environment of the site offered good conditions for hunting deer and other wild animals. Its location at the river Neckar with surrounding vegetation of less dense forests and riverside forests was optimal for this purpose, too. Therefore hunting could have been a fitting alternative to satisfy the needs of meat especially in consideration of likely initial problems with livestock breeding.

Additionally, it seems possible that the site was used as a hunting station as discussed for other settlements in north and south Germany and Switzerland. Due to the temporal overlaps, the similarities of the silex inventories and the faunal assemblages it is also suspected that the first farmers and the hunter-gatherers of the Late Mesolithic in south Germany and especially in Rottenburg were in contact and the first farmers established some technological methods on the basis of Mesolithic traditions. A set of drilled incisors and canines of a red deer mandible with comparable finds only known from Mesolithic graves in Denmark and the Upper Palaeolithic sites in Germany could be a reference to the survival of old traditions and points to the continued significance of hunting.

Acknowledgements

I would like to express thanks to Jörg Bofinger und Claus-Joachim Kind, Landesamt für Denkmalpflege im Regierungspäsidium Stuttgart, Esslingen, for advice concerning the archaeology of the site Rottenburg-Fröbelweg, and the discussions about the distribution of the faunal remains within the settlement.

Notes

1. The radiocarbon dates could be affected by wiggles in the calibration curve for the mid-later sixth millennium cal BC, although there is no real plateau between 5500 and 5000 BC. All dates used by Lüning were calculated using the wiggle matching method of Bronk Ramsey *et al.* (2001).
2. Deer ribs and vertebrae are underrepresented due to poor preservation and the problems of determining them to species level. The overrepresentation of tibiae and metapodials can be explained by easy identification and the good preservation of these dense skeletal elements.
3. The artefact was made of a tine of antler (Stephan 2005, 352, fig. 22) and resembles the holders for axe blades from the Neolithic site of Twann in Switzerland (Suter 1981, 38–43).
4. The site Vaihingen a. d. Enz was not included in Figure 9.9 as detailed data are yet to be published.

References

Alberthsen, S. E. and Brinch Petersen, E. (1976) 'Excavation of a Mesolithic cemetery at Vedbæk, Denmark', *Acta Archaeologica* 47, 1–28.

Becker, C. and Johanssen, F. (1981) *Tierknochenfunde. 2. Bericht. Mittleres und oberes Schichtenpaket (MS und OS) der Cortaillod-Kultur. Die neolithischen Ufersiedlungen von Twann 11*, Staatlicher Lehrmittelverlag, Bern.

Becker, C. (1981) *Tierknochenfunde. 3. Bericht. Unteres Schichtenpaket der Cortaillod-Kultur sowie eine zusammenfassende Betrachtung über das gesamte Knochenmaterial aus Twann (Cortaillod- und Horgener Kultur). Die neolithischen Ufersiedlungen von Twann 16.* Staatlicher Lehrmittelverlag, Bern.

Beninde, J. (1988) *Zur Naturgeschichte des Rothirsches*. Reprint of the Original Edition of 1937, Parey, Hamburg/Berlin.

Bentley, A., Price, T. D., Lüning, J., Gronenborn, D., Wahl, J. and Fullager, P. (2002) 'Prehistoric migration in Europe: Strontium isotope analysis of early Neolithic skeletons', *Current Anthropology* 43, 799–804.

Bentley, A. (2007) 'Mobility, specialisation and community diversity in the Linearbandkeramik: Isotopic evidence from the skeletons' in *Going Over. The Mesolithic-Neolithic Transition in North-west Europe*, eds A. Whittle and V. Cummings, OUP, Oxford, 117–140.

Blome, W. (1968) *Tierknochenfunde aus der spätneolithischen Station Polling*. Unpublished PhD dissertation. University of Munich, Munich.

Boessneck, J., Jéquier, J.-P. and Stampfli, H. R. (1963) *Seeberg Burgäschisee-Süd. Teil 3. Die Tierreste*, Acta Bernensia II, Stämpfli and Cie, Bern.

Bofinger, J. (2005) *Untersuchungen zur neolithischen Besiedlungsgeschichte des oberen Gäus*, Materialhefte zur Archäologie in Baden-Württemberg 68, Theiss, Stuttgart.

Bogaard, A. (2004) *Neolithic Farming in Central Europe. An Archaeobotanical Study of Crop Husbandry Practices.* Routledge, London/New York.

Bogaard, A., Krause, R. and Strien, H.-C. (2011) 'Towards a social geography of cultivation and plant use in an early farming community', *Antiquity* 85, 395–416.

Brinch Petersen E. (1974) 'Gravene ved Dragsholm. Fra jægere til bønder for 6000 år siden', *Nationalmuseets Arbejdsmark* 1974, 112–120.

Bronk Ramsey, C., van der Plicht, J. and Weninger, B. (2001). '"Wiggle matching" radiocarbon dates', *Radiocarbon* 43/2A, 381–9.

Bützler, W. (1977) *Rotwild*, 2nd Edition, BLV Verlag, Munich.

Döhle, H. -J. (1993) 'Haustierhaltung und Jagd in der Linienbandkeramik, ein Überblick', *Zeitschrift für Archäologie* 27, 105–124.

Döhle, H.-J. (1994a) 'Betrachtungen zum Haustier-Wildtier-Verhältnis in neolithischen Tierknochenkomplexen', *Forschungen und Berichte zur Vor- und Frühgeschichte in Baden-Württemberg* 53, Theiss, Stuttgart, 223–230.

Döhle, H.-J. (1994b) *Die linienbandkeramischen Tierknochen von Eilsleben, Bördekreis. Ein Beitrag zur neolithischen Haustierhaltung und Jagd in Mitteleuropa*, Veröffentlichungen des Landesamtes für archäologische Denkmalpflege Sachsen-Anhalt 47, Halle, Saale.

Döhle, H.-J. (1997) 'Husbandry and hunting in the Neolithic of central Germany', *Anthropozoologica* 25–26, 441–448.

Ebersbach, R. (2002) *Von Bauern und Rindern. Eine Ökosystemanalyse zur Bedeutung der Rinderhaltung in bäuerlichen Gesellschaften als Grundlage zur Modellbildung im Neolithikum*, Basler Beiträge zur Archäologie 15, Schwabe and Co. AG, Basel.

Frirdrich, C. (2005) 'Struktur und Dynamik der bandkeramischen Landnahme', in *Die Bandkeramik im 21. Jahrhundert. Symposium in der Abtei Brauweiler bei Köln vom 16.09.19.9.2002*, Internationale Archäologie 7, eds J. Lüning, C. Frirdrich and A. Zimmermann, Rahden/Westf., 81–124.

Gehlen, B. (2006) 'Late Mesolithic – Proto-Neolithic – Initial Neolithic? Cultural and economic complexity in south-western central Europe between 7000 and 5300 cal BC', in *After the Ice Age. Settlements, Subsistence and Social Development in the Mesolithic of Central Europe*, ed. C.-J. Kind, Materialhefte zur Archäologie in Baden-Württemberg 78, 241–257, Theiss, Stuttgart.

Gregg, S. A. (1988) *Foragers and Farmers: Population Interaction and Agricultural Expansions in Prehistoric Europe*, University of Chicago Press, Chicago.

Gronenborn, D. (1994) 'Überlegungen zur Ausbreitung der bäuerlichen Wirtschaft in Mitteleuropa – Versuch einer kulturhistorischen Interpretation ältestbandkeramischer Silexinventare', *Prähistorische Zeitschrift* 69, 135–151.

Gronenborn, D. (1999) 'A variation on a basic theme: the transition to farming in Southern Central Europe', *Journal of World Prehistory* 12, 123–210.

Gronenborn, D. (2007) 'Beyond the models: 'Neolithisation', in central Europe' in *Going Over. The Mesolithic-Neolithic Transition in North-west Europe*, eds A. Whittle and V. Cummings, OUP, Oxford, 73–98.

Haack, F. (2002) *Die bandkeramischen Knochen-, Geweih- und Zahnartefakte aus den Siedlungen Herxheim (Rheinland-Pfalz) und Rosheim (Alsace)*. Unpublished MA dissertation. University of Freiburg.

Kieselbach, P., Kind, C.-J., Miller, A. M. and Richter, D. (2000) *Siebenlinden 2. Ein mesolithischer Lagerplatz bei Rottenburg am Neckar, Kreis Tübingen*, Materialhefte zur Archäologie in Baden-Württemberg 51, Theiss, Stuttgart.

Kind, C. -J. (1997) *Die letzten Wildbeuter. Henauhof Nord II und das Endmesolithikum in Baden-Württemberg*, Materialhefte zur Archäologie in Baden-Württemberg 39, Theiss, Stuttgart.

Kind, C.-J. (1998) 'Komplexe Wildbeuter und frühe Ackerbauern. Bemerkungen zur Ausbreitung der Linearbandkeramik im südlichen Mitteleuropa', *Germania* 76, 1–23.

Kind, C.-J. (2003) *Das Mesolithikum in der Talaue des Neckars – die Fundstellen von Rottenburg Siebenlinden 1 und 3*, Forschungen und Berichte zur Vor- und Frühgeschichte in Baden-Württemberg 81, Theiss, Stuttgart.

Kind, C.-J. (2005) 'Stratigraphie der ältesten Bandkeramik von Rottenburg-Fröbelweg', in *Untersuchungen zur neolithischen Besiedlungsgeschichte des oberen Gäus*, ed J. Bofinger, Materialhefte zur Archäologie in Baden-Württemberg 68, Theiss, Stuttgart, 255–322.

Kind, C.-J., Beutelspacher, T., David, E. and Stephan E. (2012) *Mesolithikum in der Talaue des Neckars 2. Die Fundstreuungen von Siebenlinden 3, 4 und 5*, Forschungen und Berichte zur Vor- und Frühgeschichte in Baden-Württemberg 125, Theiss, Stuttgart.

Kokabi, M. (1991) 'Ergebnisse der osteologischen Untersuchungen an den Knochenfunden von Hornstaad im Vergleich zu anderen Feuchtbodenfundkomplexen Südwestdeutschlands', *Berichte der Römisch-Germanische Kommission* 71, 145–160.

Knipper, C. (2011) *Die räumliche Organisation der linearbandkeramischen Rinderhaltung: naturwissenschaftliche und archäologische Untersuchungen*, BAR International Series 2305, Archaeopress, Oxford.

Lehmann, E. and Sägesser, H. (1986) *Capreolus capreolus (Linnaeus, 1758) – Reh*, in *Handbuch der Säugetiere Europas Bd. 5/Raubsäuger II*, eds J. Niethammer and F. Krapp, Aula, Wiebelsheim, 233–268.

Lenneis, E., Stadler, P. and Windl, H. (1996) 'Neue ^{14}C-Daten zum Frühneolithikum in Österreich', *Préhistoire Européenne* 8, 97–116.

Liese-Kleiber, H. (1991) 'Züge der Landschafts- und Vegetationsentwicklung im Federseegebiet. Neolithikum und Bronzezeit in neuen Pollendiagrammen', Siedlungsarchäologische Untersuchungen im Alpenvorland, *Berichte der Römisch-Germanischen Kommission* 71, 58–83.

Lüning, J. (1988) 'Frühe Bauern in Mitteleuropa im 6. und 5. Jahrtausend v. Chr.', *Jahrbuch des Römisch-Germanischen Zentralmuseums Mainz* 35, 27–93.

Lüning, J. (2000) *Steinzeitliche Bauern in Deutschland. Die Landwirtschaft im Neolithikum*, Universitätsforschungen zur Prähistorischen Archäologie 58, Habelt, Bonn.

Lüning, J. (2005) 'Bandkeramische Hofplätze und die absolute Chronologie der Bandkeramik', in *Die Bandkeramik im 21. Jahrhundert. Symposium in der Abtei Brauweiler bei Köln vom 16.09.19.9.2002*, Internationale Archäologie 7, eds J. Lüning, C. Frirdrich and A. Zimmermann, Rahden/Westf., 49–74.

Lüning, J., Kloos U. and Albert, S. (1989) 'Westliche Nachbarn der bandkeramischen Kultur: La Hoguette und Limburg', *Germania* 67, 355–393.

Müller, H.-H. (1964) *Die Haustiere der mitteldeutschen Bandkeramiker*, Naturwissenschaftliche Beiträge zur Vor- und Frühgeschichte Teil 1, Akademie Verlag, Berlin.

Poplin, F. (1972) 'Abgeschnittene Rentier-Schneidezähne von Gönnersdorf', *Archäologisches Korrespondenzblatt* 2, 235–238.

Poplin, F. (1983) 'Die bearbeiteten Zähne vom Rentier und anderen Tieren vom Petersfels', in *Naturwissenschaftliche Untersuchungen an Magdalénien-Inventaren vom Petersfels, Grabungen 1974–1976*, eds G. Albrecht, H. Berke and F. Poplin, Tübinger Monographien zur Urgeschichte 8, Archaeologica Venatoria, Tübingen, 133–153.

Poplin, F. (1994) 'Menschen- und Pferdeknochen in Viereckschanzen am Beispiel von Gournay-sur-Aronde (Nordfrankreich)', *Beiträge zur Archäozoologie und*

Prähistorischen Anthropologie. Forschungen und Berichte zur Vor- und Frühgeschichte in Baden-Württemberg 53, 315–322.

Pucher, E. (1987) 'Viehwirtschaft und Jagd zur Zeit der ältesten Linearbandkeramik von Neckenmarkt (Burgenland) und Strögen (Niederösterreich)', *Mitteilungen der Anthropologischen Gesellschaft in Wien (MAGW)* 117, 141–155.

Quitta, H. (1960) 'Zur Frage der ältesten Bandkeramik in Mitteleuropa', *Prähistorische Zeitschrift* 38, 1–38.

Schäfer, M. (2010) 'Viehzucht- und Jagdstrategien der ersten Bauern in Süddeutschland', in *Familie – Verwandtschaft – Sozialstrukturen: Sozialarchäologische Forschungen zu neolithischen Befunden. Fokus Jungsteinzeit*, eds E. Claßen, T. Doppler and B. Ramminger, Berichte der AG Neolithikum 1, Kerpen-Loogh, 107–118.

Schibler, J. (1981) *Typologische Untersuchungen der cortaillodzeitlichen Knochenartefakte. Die neolithischen Ufersiedlungen von Twann* 17, Staatlicher Lehrmittelverlag, Bern.

Schibler, J. (1997) 'Knochen- und Geweihartefakte', in *Ökonomie und Ökologie neolithischer und bronzezeitlicher Ufersiedlungen am Zürichsee*, eds J. Schibler, H. Hüster-Plogmann, S. Jacomet, C. Brombacher, E. Gross-Klee and A. Rast-Eicher, Monographien der Kantonsarchäologie 20, Zürich/Egg, 122–219.

Smettan, H. W. (2000) *Vegetationsgeschichtliche Untersuchungen am oberen Neckar im Zusammenhang mit der vor- und frühgeschichtlichen Besiedlung*, Materialhefte zur Archäologie in Baden-Württemberg 49, Theiss, Stuttgart.

Stäuble, H. (2005) *Häuser und absolute Datierung der Ältesten Bandkeramik*, Universitätsforschungen zur prähistorischen Archäologie 117, Habelt, Bonn.

Stephan, E. (2005) 'Die Tierknochenfunde aus der ältestbandkeramischen Siedlung Rottenburg "Fröbelweg"', in *Untersuchungen zur neolithischen Besiedlungsgeschichte des oberen Gäus*, ed J. Bofinger, Materialhefte zur Archäologie in Baden-Württemberg 68, Theiss, Stuttgart, 323–383.

Stork, M. (1993) 'Tierknochenfunde aus neolithischen Gruben in der Gem. Ammerbuch, Kr. Tübingen', *Zeitschrift für Archäologie* 27, 91–104.

Strien, H.-C. (2000) *Untersuchungen zur Bandkeramik in Württemberg*, Universitätsforschungen zur prähistorischen Archäologie 69, Habelt, Bonn.

Strien, H.-C. (2011) 'Chronological and social interpretation of the artefactual assemblage', in *Plant Use and Crop Husbandry in an Early Neolithic Village: Vaihingen an der Enz, Baden-Württemberg*, ed. A. Bogaard, Frankfurter Archäologische Schriften 16, Habelt, Bonn.

Stubbe, C. (1990) *Rehwild*, 3rd Edition, Deutscher Landwirtschaftsverlag, Berlin.

Suter, P. J. (1981) *Die Hirschgeweihartefakte der Cortaillod-Schichten*. Die neolithischen Ufersiedlungen von Twann 15, Staatlicher Lehrmittelverlag, Bern.

Tillmann, A. (1993) 'Kontinuität oder Diskontinuität? Zur Frage einer bandkeramischen Landnahme im südlichen Mitteleuropa', *Archäologische Informationen* 16, 157–187.

Uerpmann, H.-P. (1977) 'Betrachtungen zur Wirtschaftsform neolithischer Gruppen in Südwestdeutschland', *Fundberichte Baden-Württemberg* 3, 144–161.

Uerpmann, M. and Uerpmann, H.-P. (1997) 'Remarks on the faunal remains of some early farming communities in Central Europe', *Anthropozoologica* 25/26, 571–578.

Uerpmann, M. (2001) 'Animaux sauvages et domestiques du Rubané "le plus ancien" (LBK 1) en Allemagne', *Actes de la première table-ronde. Rôle et statut de la chasse dans le Néolithique ancien danubien (5500–4900 av. J.-C.)*, Premières Rencontres Danubiennes Strasbourg 20 et 21 Novembre 1996, Leidorf, Rahden, Westfalen, 57–75.

Wagenknecht, E. (1981) *Rotwild*, 4th Edition, Neumann-Neudamm, Berlin.

Whittle, A. (1996) *Europe in the Neolithic: The Creation of New Worlds*, CUP, Cambridge.

Red Deer Antlers in Neolithic Britain and their Use in the Construction of Monuments

Fay Worley and Dale Serjeantson

Introduction

The Neolithic monuments of southern England include some of the most impressive and substantial earthworks in British archaeology. The flint mines and monuments represent massive feats of engineering; they have deep quarries and ditches dug through solid chalk, greensand or limestone and banks and mounds constructed of the quarried stone. They are found throughout most of southern Britain (Figure 10.1). Yet the monuments were built between six

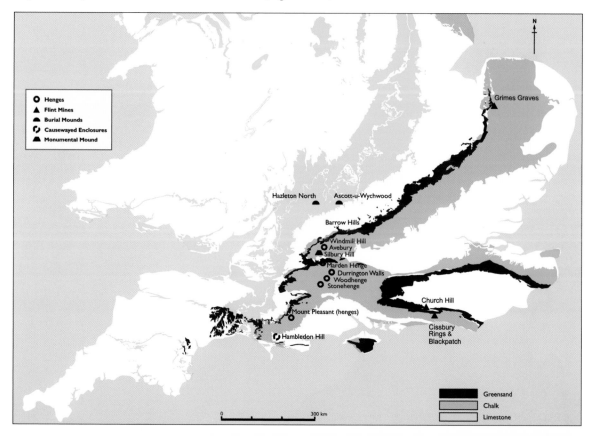

and four thousand years ago, without the use of metal tools, by illiterate farming communities. The primary tool used for the construction of the monuments was a pick made from a red deer (*Cervus elaphus*) antler (Figure 10.2). Here, we will show how the physical remains of these antler picks tell us about prehistoric tool technologies, deer populations and the activities of the communities responsible for the monuments. Excavators of Neolithic sites since the nineteenth century have mentioned antler picks. The account here builds on these early descriptions as well as on the analysis of the two largest assemblages, those from Durrington Walls henge enclosure and from the flint mines at Grime's Graves excavated in 1972–6 (Clutton-Brock 1984), on the authors' own research on the antlers from Stonehenge (Serjeantson and Gardiner 1995), Silbury Hill and Marden henge (Worley 2011a; Worley 2011b; Worley and Serjeantson 2011) and on the antlers from various other sites (Table 10.1).

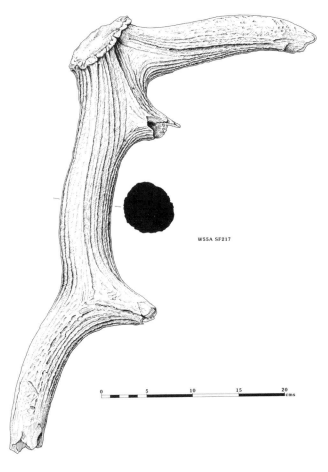

W55A SF217

FIGURE 10.2. *(above)* Antler pick from the Stonehenge Lesser Cursus (Richards 1990, fig. 57). The antler is shed, the bez and trez tine have been removed and the brow tine is only slightly worn. © English Heritage

Site	Date	Site type
Ascott-under-Wychwood	Early Neolithic	Long barrow
Avebury	Late Neolithic	Henge
Barrow Hills, Radley (611)	Late Neolithic	Ring ditch
Barrow Hills, Radley (4969)	Final Neolithic (Beaker)	Burial
Blackpatch	Early/Mid Neolithic	Flint mine
Church Hill	Early Neolithic	Flint mine
Cissbury	Early Neolithic	Flint mine
Durrington Walls	Late Neolithic	Henge
Grimes Graves	Late Neolithic	Flint mine
Hambledon Hill	Early Neolithic	Causewayed enclosure
Hazleton North	Early Neolithic	Long barrow
Marden henge	Late Neolithic	Henge
Mount Pleasant	Late Neolithic	Henge
Silbury Hill	Late Neolithic	Henge/mound
Stonehenge	Mid/Late Neolithic	Henge
Windmill Hill	Early/Mid Neolithic	Causewayed enclosure
Woodhenge	Late Neolithic	Henge

TABLE 10.1. *(left)* Sites with assemblages of antler picks, showing date and type of site. 'Late Neolithic' sites have Grooved Ware pottery.

Making antler picks

The antler picks were generally created in a standard fashion, although with variations. The beam of the antler was divided to remove the crown, usually on the superior side of the trez tine, but sometimes between the bez and trez tines. Superfluous tines were detached. This left a 'pick' with a handle formed from the beam and a blade tine or tines (Figure 10.2). Shorter handled picks, where the crown was removed between the bez and trez tines, are known but they are not common; they comprised ten percent of the assemblage from Stonehenge and were even more infrequent in the 1914 and 1972–6 assemblages from Grime's Graves. Occasionally the handle or any remains of a pedicle were smoothed; presumably this was done either to enhance the ease and comfort of use or for aesthetic effect. This mode of manufacture resulted in one or more surplus tines together with the antler crown. The latter is often referred to as a 'rake' (Sanders 1910, 101), the assumption being that the crowns were used to scrape up chalk rubble. The detached tines seem to have been used as wedges, as discussed below. Surplus tines could also be used as raw material for smaller tools and artefacts.

We do not understand what governed the retention of the brow, bez or both tines as the blade of the pick. At Grime's Graves, Durrington Walls and Avebury (Gray 1935), the picks overwhelmingly retained only the brow tine (Table 10.2) but those from Stonehenge are less standardised, with 22 percent of picks retaining both tines, 36 percent retaining only the bez and 42 percent only the brow tine. It is not clear why the bez was sometimes retained rather than

Site	N	Variable		
Division of beam		*Above trez*	*Below trez*	
Stonehenge	94	85	15	
Tine selected		*Brow*	*Bez*	*Brow and bez*
Stonehenge	59	42	36	22
Avebury	26	92	4	4
Durrington Walls	328	96	4	-
Grime's Graves (1972–6)	282	99	1	-
Scorch marks present		*Scorched*	*Not scorched*	
Durrington Walls	332	14	86	
Avebury	46	17	83	
Grime's Graves (1972–6)	283	52	48	
Stonehenge	84	56*	44	
Battering		*Battered*	*Not battered*	
Woodhenge	10	(1)	(9)	
Hambledon Hill	55	34	66	
Stonehenge	76	46	54	
Avebury	27	59	41	
Marden henge	21	67	33	
Grime's Graves (1972–6)	281	68	32	
Hazleton North	16	69	31	
Durrington Walls	332	91	9	
Grime's Graves (1971–2)	41	93	7	

TABLE 10.2. Modifications to antler picks; data are given as a percentage or N (in brackets) when ten or fewer antlers were found. * Percent of scorch marks on beam.

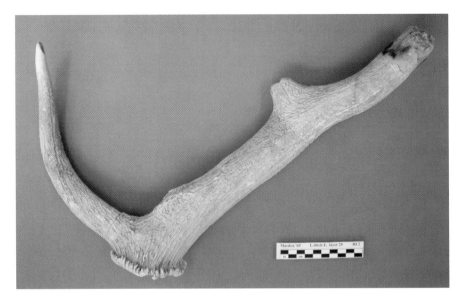

FIGURE 10.3. Antler pick from Marden henge: the bez and trez tines have been removed and the beam is charred where heat was applied to assist with chopping it from the crown. The brow tine is slightly chipped from use but otherwise hardly worn. Photograph: Fay Worley, English Heritage.

the brow tine but it has been suggested that retaining both tines increased the use-life of the pick, whereby when one tine was exhausted the second could be used. We tested for correlations in the Stonehenge antlers between the relative length and robustness of the tine, the curvature of the tine, and the angle of the tine to the beam and the choice of tine, but found none.

Several picks are scorched at the point where the antler was to be divided (Figure 10.3) and some of the skull fragments attached to unshed antlers are also scorched. Charring traces on antler picks have been recognised since at least 1935: at Avebury the picks '*bore marks of fire, especially in the position of the grip on the beam of the antler*' (Gray 1935, 149). Applying heat at the point of impact makes bones and antlers more brittle and easier to break with stone tools (Clutton-Brock 1984, 26). Heating also assisted with smoothing the broken surface of the antler or the skull. The prevalence of scorched antlers is variable: at Stonehenge and Grime's Graves half of the antler picks show some traces of charring, while for unknown reasons at the late henges fewer than twenty percent were affected by fire (Table 10.2). One antler from Silbury Hill shows ephemeral traces of scorching only in the form of a very slight colour change associated with a localised, finely fractured surface at points of division (Worley 2011a). It is possible that more picks at other sites were scorched, but that the traces were obliterated by erosion of the antler surface, or not recognised in analysis, or not commented on in the published report.

Excavation records from Stonehenge (e.g. Cleal *et al.* 1995, 127) and Windmill Hill (Smith 1965, 9) describe small temporary fires or fire debris on the base of some of the ditch segments. Small fires were also noted on the base of several flint mines (Barber *et al.* 1999). These may have been related to on site antler modification.

Other methods of modifying antler were occasionally used. Some were cut or sawn around the circumference of the beam or tine with a flint until they could be snapped (Legge 2008, 574). Yet another method of removing tines was to macerate the antler in water, which softened it prior to snapping it off (Penn *et al.* 1984; Casseyas 1997; Riedel *et al.* 2004). This has been found elsewhere in Britain and also in northern Europe, but there is no evidence for the use of this technique with the picks discussed here.

Antler selection

The structural properties of antler were no doubt central to its selection for quarrying tools. Antler has superior strength and robusticity to wood and it has greater elasticity than bone or stone (Clutton-Brock 1984, 16–7; MacGregor 1985, 25–9) making it more effective than either material alone. The builders of the prehistoric monuments did not use all the antlers available to them to make picks, but selected antlers on the basis of their development stage, regeneration cycle stage and, sometimes, body-side. The larger, stronger, more developed antlers, from mature, but not post-prime, stags of four years and older were almost always chosen. Some antlers had just a brow and a trez tine (Stage D as defined by Schmid 1972, fig. 23) but most of those chosen originally had the full complement of brow, bez and trez (Schmid Stages E and F). These would have been stronger, and would have provided more useable tines, than the antlers of younger or very old stags. For example, 90 percent of the antlers from Stonehenge were Schmid's stages E or F, and nearly 80 percent of those from Avebury. More than 70 percent of antlers from Marden henge (Worley 2011b) but only just over 60 percent of those from the Hazleton North long barrow (Levitan 1990) were at this stage (Table 10.3). Marden henge had some antlers that seem be only at Schmid's Stage B, the earliest stage at which the antler has a projecting tine to form the pick blade; this may reflect the relatively easier working of the greensand bedrock.

Side was a lesser consideration. Legge (1981) reports 43 left and 22 right antlers from the 1971–2 shaft at Grime's Graves, and judged that left side picks would have been more efficient for right-handed miners in the confines of the shaft and galleries, but at other sites there is little or no bias towards left side antlers (Table 10.2). This suggests that, out of the confines of a mine, the monument builders used all the available mature antlers.

The most notable feature of the picks is the preference for shed antlers rather than those cut from the head of deer. The proportion of shed antlers from all sites together is over 80 percent (Table 10.2). It varies from 65 to 87 percent in the Early to Middle Neolithic and from 65 to as much as 98 percent in the Late Neolithic henge of Mount Pleasant (Wainwright 1979). The choice of shed antlers was not simply because they had attained full development for that year and so were harder and larger than unshed antlers, since antlers reach their optimal size and are fully hardened for several months before they are shed (Chapman 1975, 137–9). While shed antlers can be collected year-round, it is more likely that that the majority were collected in the spring soon after

Site	N	Variable								
Percent (or N) using shed antler		*Shed*					*Not shed*			
Radley Barrow Hills (4969)	3						(3)			
Blackpatch	10	(9)					(1)			
Silbury Hill	5	(5)								
Radley Barrow Hills (611)	8	(8)								
Woodhenge	10	(10)								
Windmill Hill	23	65					35			
Stonehenge	79	65					35			
Marden henge	23	74					26			
Hambledon Hill	63	76					24			
Grime's Graves (1972–6)	282	82					18			
Grime's Graves (1914)	198	84					16			
Avebury	34	85					15			
Hazleton North	23	87					13			
Durrington Walls	332	87					13			
Mount Pleasant	61	98					2			
Percentage using left or right antler		*Left*					*Right*			
Marden henge	24	42					58			
Durrington Walls	330	51					49			
Stonehenge	91	53					47			
Grime's Graves (1972–6)	283	53					47			
Hazleton North	26	58					42			
Hambledon Hill	40	58					42			
Grime's Graves (1971–2)	65	66					34			
Development stage (%)	N	B?	B+	B–D	C+	C–D	D	D–E	E–F+	
Stonehenge	70	-	-	-	-	-	-	-	90	
Avebury	38	-	13	-	3	-	5	-	79	
Marden henge	20	5	-	20	-	5	-	-	70	
Hazleton North	34	-	-		-	3	21	15	62	

TABLE 10.3. Biological variables in antler pick manufacture, data are given as a percentage or N (in brackets) when ten or fewer antlers were found; development stages after Schmid (1972, fig. 23).

shedding, as shed antlers lying around in the landscape would be subject to gnawing by scavengers, something which has only rarely been recorded on picks.

Using antler picks

Evidence for how antler picks were used comes from three sources: the artefacts themselves; modern experiments with quarrying chalk with antler picks; and the marks left by picks in the chalk of the monuments. It has been suggested that the antlers were swung and used as levers like modern metal pick-axes to excavate chalk and limestone (Ashbee and Cornwall 1961) or that the tines were used as wedges to fissure the rock. The primary source of evidence for the efficacy of antler picks to excavate chalk and build earthworks is in experiments by Pitt Rivers in the 1870s (Lane-Fox 1876) and in the Overton Down Earthwork project of the 1960s (Ashbee and Cornwall 1961).

If tines were used as wedges or as the point of picks, one might expect use-wear including polishing, pitting bevelling or scratches on tine tips, and such marks are often recorded. However, polishing, scratches and gouges also occur naturally, caused by the behaviour of the stags themselves, through spreading scent, rutting, digging into the ground and sloughing antler velvet (Olsen 1989; Jin and Shipman 2010) so damage of this nature to tines is not in itself evidence for human use. There is one source of damage that we can be confident is of human origin: battering or bruising on the posterior side of the burr, coronet and beam. This is commonly found: it was was recorded on nearly all the picks from Durrington Walls and the Grime's Graves 1971–2 shaft (Table 10.2). It is present on 59 percent of picks with surviving burrs at Avebury. It was seen on approximately two thirds of those from from Marden henge, Grime's Graves and Hazleton North but on only one-third of those from the Hambledon Hill causewayed enclosure (Legge 2008). The location of the battering suggests that it was caused either by using the posterior burr and beam as a hammering tool, perhaps to wedge tines into the chalk, or by using another tool to strike the antler in this region. Battering has been recognised since at least 1915 when, in discussing the picks from his excavations at Grime's Graves, Clarke (1915b, 129–130) wrote '*The hammered condition of the heads of so many of the deer-horn picks suggests a method of splitting the chalk in its line of fissure by means of wedges, detached tines, or first year horns [antlers]… Possibly, too, one horn may have been held with its tine against the chalk, while another was used to hammer it, and some of the straight tined horns could have been used in this way. The majority, however, have the brow tine curved in one direction, and a blow on the head of such a pick would not be very effective*'. The Overton Down experiments made it clear that the most extreme examples of battering must have been caused by hitting the antler with a stone (Ashbee and Cornwall 1961). Battering is most frequent at large Late Neolithic sites (Table 10.2) and it is frequent on picks from open sites as well as those from the confined galleries of flint mines. Though there is some battering on the Marden henge antlers, it is always relatively slight, again perhaps because the greensand required less force to fragment than chalk and limestone.

The marks left in chalk by antler tools also attest to their use for quarrying. The most famous of these were discovered in the Blackpatch flint mine, where one wall of the mine was scarred by two rows of 5–10 cm deep holes. These were interpreted as pick strikes defining a block of chalk which was then to be removed (Goodman *et al.* 1924). Some large chalk blocks on the mine's floor also bore multiple pick marks and the excavators described '*several longitudinally split pick holes… seen on various parts of the walls where blocks of chalk have been wedged out*'. Similarly, Pull illustrated pick marks in the walls and on the chalk rubble of Church Hill flint mine (Pull 2001, 124; Russell 2001, 113). In the Cissbury flint mine: '*Many of the chalk blocks encountered bore the deep impressions of antler picks and punches while broken tines were observed within the walls of some galleries*' (Russell 2001, 186). Unfortunately, Russell and colleagues could not trace the picks themselves from these last two sites, nor any reports on them. At Stonehenge, too, the tip of an antler was found imbedded in a chalk block (Cleal *et al.* 1995, 194). Silbury Hill was constructed of rough chalk block revetments interfilled by finer chalk dumps (Leary 2010, 146), and one of those blocks bears three antler pick strikes (Worley 2011a). The first deep perpendicular strike corresponds to a plane of fracture of the block, while the second and third overlying strikes are seen in longitudinal section, curving across a second plane of fracture. At both sites, the strikes defined and created the block of chalk. The curvature and dimensions of the strike marks are compatible with an antler tine.

Use-life and efficiency of antler picks

The nineteenth and twentieth century experiments have provided insight into the effectiveness and use-life of the tools. In the 1870s, Pitt Rivers' workman excavated three cubic feet (0.085 m³) of chalk in 90 minutes (0.057 m³/hr), using picks and tine wedges made from two antlers; they noted that the wedges were particularly effective (Lane-Fox 1876, 382). Atkinson and Sorrell (1959, 59) suggested that approximately three m³ of solid chalk could be excavated in a day, although do not provide a basis for this estimate, and the Overton Down experimental earthwork project showed that it was possible to excavate five cubic feet (0.142 m³) of chalk in an hour (Ashbee and Cornwall 1961, 131). These calculations, combined with the volume of chalk quarried at each site, made it possible to estimate the number of man days work required to construct some of the monuments: 68,000 days work to excavate the ditches at Avebury (Ashbee and Cornwall 1961) and 78–80,000 days (Worley 2011a) to construct the mound at Silbury Hill. It is not clear how much work can be done with an antler pick before the tine is too worn for further use. The excavator of the Grime's Graves mine shaft opened in 1972–6 thought that each pick would have lasted a single day so estimated that there might have been an annual requirement of approximately 400 picks per shaft (Clutton-Brock 1984, 16). Legge (1981, 101) proposed a slightly lower demand for 100–150 picks a year, based on the number found in the 1971–2 shaft. Baysian radiocarbon analysis of the datable material from Silbury Hill suggests that the mound may have been constructed over the course of about a century, which implies an average annual requirement of 500–1,400 picks (Worley 2011a). These estimates are based on a number of assumptions, namely the use-life of a pick, the daily volume of chalk which can be excavated with each pick and a constant rate of construction as well as the calculations of radiocarbon dates. At each site they imply a massive demand for antler.

Deposition of picks

Given the effort expended obtaining antler picks in the Neolithic period, and the value of the antler as a useable pick and as a raw material for further tools, it is interesting that we recover picks in quantity at all. The reason seems to be that picks were often not casually discarded, but were deliberately left where they had been used, usually at the base of the ditch or pit. Some were broken or worn to the point where they were no longer useable but others were discarded when still only slightly worn (e.g. Figure 10.2). At the Early Neolithic long barrows of Hazleton North and Ascott-under-Wychwood picks were found in the base of the shallow quarries from which stone had been taken to build the mounds (Levitan 1990, 205, Mulville and Grigson 2007, 244, 246). Sometimes picks were found in groups as with the 57 picks found together at Durrington Walls, the eight picks together at the base of a shaft at Grime's Graves and those entangled at the base of Y hole 30 at Stonehenge (Cleal *et al.* 1995, fig. 152). Eight antlers were deposited in Ring Ditch 611 at Barrow Hills, Radley, evenly arranged around the base of the ditch and, uniquely, four antlers were found arranged around a burial of the Beaker period at the same site (Barclay and Halpin 2002, figs 4.1, 4.62). There is little doubt that some of the picks were deliberately placed in the mines or ditches features where they had used. The reason is clearly *not* merely functional; it is more likely that it was depositing the antler pick had some spiritual significance. Perhaps the re-use of a tool which had been used on a sacred site was taboo (Leary 2010, 146) or the antler which had been 'borrowed' for the monument's construction was symbolically returned into the ground.

Discussion

Antler technology

We have shown here that the basic pick illustrated in so many archaeological textbooks – which uses the beam and the brow tine of a shed antler with the crown and bez and trez tines removed – has many variations. The process of manufacture – known in French archaeology as the '*chaîne opératoire*' – normally involved the removal of the crown, though even this was occasionally left attached to the beam. Tines other than the brow were sometimes used, especially at the smaller sites. The picks were made with varying degrees of skill and finish both between and within sites. It might be expected that these differences in choice of antler form and technique would reflect the type of site on which it was used, but – as we have seen – up to now none of the variants in size, tine use and modification appear to correlate closely with the type of monument at which they were used. There was a greater degree of standardisation in the tools from flint mines than in those from other sites, but even there the use of fire to modify antler seems to be the choice of the individual worker. In this the antler pick was like any other craftsman-made artefact: it combined knowledge of the material, a visual idea of the end product and individual variations according to the skill and taste of the maker (Steele 1997).

Neolithic deer populations

In view of the fact that antlers were of such crucial importance for the creation of the monumental earthworks and flint mines, red deer in the Neolithic period must have been more valuable for their antler than for their meat (Legge 1981, 100). Some postcranial bones of red deer are present in half of all assemblages in Neolithic southern Britain but only in small numbers. They make up only two percent of bones from food remains by NISP from all sites, confirming that few red deer were killed and eaten (Serjeantson 2011, 44). The economic importance of antler is also emphasised by the selection of predominantly shed antlers. The use of shed antler means that there was no impact on the yield of the following years. Killing a mature stag provided two antlers, but it would have removed a pair of antlers from the annual local supply in subsequent years. It would then have taken several years before a newborn calf could provide suitable replacement antler.

The calculation of how many antlers would have been required per year raises some problems. For Grime's Graves, Legge (1981, 101) calculated that a population of 120 red deer could have supplied enough picks if one flint mine was opened each year. A herd or herds of this size could have been supported in the Breckland in the vicinity of the flint mines. However, the immense annual demand for antlers at Silbury Hill raises the question of whether they were acquired through trade or barter. Trade in antlers would not be unexpected in view of what we know of the circulation of ground stone axes in Neolithic Britain; indeed, trade in antler to the flint mines may well have been tied in with the export of flint axes to communities further afield. A current research project is evaluating the application of isotopic analysis to antlers to test whether they came from close to the excavated site or from a different environment further away (Montgomery *et al.* 2011).

People and deer

There must have been a sophisticated and probably standardised system by which antler was collected and passed on. Individuals and communities must have met the need for antler through organised seasonal collection of shed antler. Fletcher (2011, 32) found that deer could be tempted by the provision of fodder such as ivy to congregate in a certain place at the time of year when antlers are shed. Even if people in the past did not actively tempt deer to congregate, they must at least have been familiar with the places where their local herds regularly shed their antlers.

Conclusion

It is only when we are able to compare the antlers from both small and large assemblages that we will better understand the variations in the *chaîne opératoire* of Neolithic tool manufacture and use. At present it is still unclear to what degree these variations were determined by practicality and to what extent they reflect the individual skills and preferences of the toolmakers. As artefacts formed from vertebrate hard tissues, antler picks sit uncertainly between the remit of general archaeologists, zooarchaeologists and finds specialists. As a result, the full range of information that they can provide has sometimes been lost. Understanding how antler picks were made and used requires a familiarity with bone taphonomy, the technologies of bone working and quarrying as well as knowledge of deer biology and behaviour. Only then can we appreciate the full range of skill and knowledge of the pick makers and monument builders. The skills used in obtaining suitable antler and manufacturing picks in Neolithic Britain were based on two thousand years of experience of antler and the behaviour of red deer herds. They reveal a sophisticated degree of knowledge of the material properties of antler and in addition a close relationship with the deer amongst which the people lived.

Acknowledgements

The authors would like to thank the following people for access to assemblages, advice and inspiring discussion of Neolithic antler tools: David Field, Jim Leary, Tony Legge, Jacqui Mulville, Ian Riddler, Lisa Webb (Wiltshire Heritage Museum, Devizes) and Alasdair Whittle.

References

Ashbee, P. and Cornwall, I. W. (1961) 'An experiment in field archaeology', *Antiquity* 35, 129–134.

Atkinson, R. J. C. and Sorrell, A. (1959) *Stonehenge and Avebury*, Her Majesty's Stationary Office, London.

Barber, M., Field, D. and Topping, P. (1999) *Neolithic Flint Mines in England*, English Heritage, London.

Barclay, A. and Halpin, C. (2002) *Excavations at Barrow Hills, Radley, Oxfordshire. Volume 1: The Neolithic and Bronze Age Monument Complex*, Oxford Archaeological Unit, Oxford.

Casseyas, C. (1997) 'New light on old data: a Neolithic (?) antler workshop in Dendermonde (Belgium, O.VI.)', *Notae Praehistoricae* 17, 199–202.

Chapman, D. I. (1975) 'Antlers – bones of contention', *Mammal Review* 5, 121–172.

Clarke, W. G. (1915b) *Report on the Excavations at Grime's Graves, Weeting, Norfolk, March–May 1914*, Prehistoric Society of East Anglia, London.

Cleal, R. M. J., Walker, K. E. and Montague, R. eds (1995) *Stonehenge in its Landscape: Twentieth-Century Excavations*, English Heritage, London.

Clutton-Brock, J. (1984) *Excavations at Grimes Graves Norfolk 1972–1976. Fascicule 1: Neolithic Antler Picks from Grimes Graves, Norfolk, and Durrington Walls, Wiltshire: A Biometrical Analysis*, British Museum Publications, London.

Fletcher, J. (2011) *Gardens of Earthly Delight: A History of Deer Parks*, Windgather Press, Oxford.

Goodman, C. H., Frost, M., Curwen, E. and Curwen, E. C. (1924) 'Blackpatch Flint-mine excavation, 1922: Report prepared on behalf of the Worthing Archaeological Society', *Sussex Archaeological Collections* 65, 69–111.

Jin, J. J. H. and Shipman, P. (2010) 'Documenting natural wear on antlers: a first step in identifying use-wear on purported antler tools', *Quaternary International* 211, 91–102.

Lane-Fox, A. (1876) 'Excavations in Cissbury Camp, Sussex', *Journal of the Anthropological Institute of Great Britain and Ireland* 5, 357–390.

Leary, J. (2010) 'Silbury Hill: a Monument in Motion', in *Round Mounds and Monumentality in the British Neolithic and Beyond*, eds J. Leary, T. Darvill and D. Field, Oxbow, Oxford, 139–152.

Legge, A. J. (1981) 'The agricultural economy', in *Grimes Graves, Norfolk Excavations 1971–72: Volume I*, ed. R. Mercer, Her Majesty's Stationary Office, London, 79–103.

Legge, A. J. (2008) 'Livestock and Neolithic society at Hambledon Hill', in *Hambledon Hill, Dorset, England: Excavation and survey of a Neolithic Monument Complex and its Surrounding Landscape*, eds R. Mercer and F. Healy, English Heritage, London, 536–585.

Levitan, B. (1990) 'The non-human vertebrate remains', in *Hazleton North: The Excavation of a Neolithic Long Cairn of the Cotswold-Severn Group*, ed. A. Saville, English Heritage, London, 199–214.

MacGregor, A. (1985) *Bone, Antler, Ivory and Horn: The Technology of Skeletal Materials since the Roman Period*, Crook Helm, London.

Montgomery, J., Evans, J., Worley, F. and Warham, J. (2011) 'Building Silbury Hill – where did the antler picks come from?', British Academy Small Research Grant, Awarded 2011.

Mulville, J. and Grigson, C. (2007) 'The animal bones', in *Building Memories: The Neolithic Cotswold Long Barrow at Ascott-Under-Wychwood, Oxfordshire*, eds D. Benson and A. Whittle, Oxbow, Oxford, 237–253.

Olsen, S. L. (1989) 'On distinguishing natural from cultural damage on archaeological antler', *Journal of Archaeological Science* 16, 125–135.

Penn, J., Field, D. and Serjeantson, D. (1984) 'Evidence of Neolithic occupation in Kingston: excavations at Eden Walk, 1956: with notes on Medieval animal bone and ground axes from Kingston', *Surrey Archaeological Collections* 75, 207–224.

Pull, J. H. (2001) 'Pit A', in *Rough Quarries, Rocks and Hills: John Pull and the Neolithic Flint Mines of Sussex*, ed. M. Russell, Oxbow, Oxford, 122–125.

Richards, J. (1990) *The Stonehenge Environs Project*, London, English Heritage.

Riedel, K., Pohlmeyer, K. and von Rautenfeld, D. B. (2004) 'An examination of Stone Age/Bronze Age adzes and axes of red deer (*Cervus elaphus* L.) antler from the Leine Valley, near Hannover', *European Journal of Wildlife Research* 50, 197–206.

Russell, M. ed (2001) *Rough Quarries, Rocks and Hills: John Pull and the Neolithic Flint Mines of Sussex*, Oxbow Books, Oxford.

Sanders, H. W. (1910) 'On the use of the deer-horn pick in the mining operations of the ancients', *Archaeologia* 62, 101–124.

Schmid, E. (1972) *Atlas of Animal Bones for Prehistorians, Archaeologists and Quaternary Geologists*, Elsevier, London.

Serjeantson, D. (2011) *Review of Animal Remains from the Neolithic and Early Bronze Age of Southern England (4000 BC–1500 BC)*, Research Department Report Series 29/2011, English Heritage, Portsmouth.

Serjeantson, D. and Gardiner, J. (1995) 'Red deer antler implements and ox scapula shovels', in *Stonehenge in its Landscape: Twentieth-century Excavations*, eds R. M. J. Cleal, K. E. Walker and R. Montague, London, English Heritage, 414–430.

Smith, I. F. (1965) *Windmill Hill and Avebury: Excavations by Alexander Keiller 1925–1939*, OUP, Oxford.

Gray, H. St George (1935) 'The Avebury excavations, 1908–1922', *Archaeologia* 84, 99–162.

Steele, J. (1997) 'Traditional crafts: learning by doing', *Cambridge Archaeological Journal* 7, 314–316.

Wainwright, G. J. (1971) 'The excavation of a Late Neolithic enclosure at Marden, Wiltshire,' *Antiquaries Journal* 51, 177–239.

Worley, F. (2011a) *The Antler Assemblage Excavated from Silbury Hill in 2007–8*, Research Department Report Series Report 17/2011, English Heritage, Portsmouth.

Worley, F. (2011b) *A Reanalysis of the Red Deer Antlers Excavated from Marden Henge in 1969*, Research Department Report Series Report 20/2011, English Heritage, Portsmouth.

Worley, F. and Serjeantson, D. (2011) 'The importance of red deer antlers for the creation of Neolithic monuments', in *BoneCommons*, Item #1653, http://alexandriaarchive. org/bonecommons/items/show/1653 (accessed July 27, 2012).

11

Antler Industry in the Upper Magdalenian from Le Rond du Barry, Polignac, Haute-Loire, France

Delphine Remy and Roger de Bayle des Hermens[†]

Introduction

Located in the valley of the Borne in Velay on the high plateau of the southeast Massif Central (Figure 11.1), Le Rond du Barry is the largest cave site in the Auvergne region with stratified archaeological deposits. This vast cavity, the mouth of which faces west, is more than 40 m in length and fourteen m wide. It developed in a weakness formed by a radial crack in an ancient volcanic cone at about 850 m above sea-level. It is well known since the eighteenth century (Robert 1937), and was initially excavated at the end of the nineteenth century by J. Penide. From these excavations, Penide recovered human remains including a well-preserved skull. Nothing remains as testament to his activities; neither material objects nor documentary files (Bayle Des Hermens (De) 1967).

Following a survey carried out by A. Laborde in 1965, R. de Bayle des Hermens undertook methodical excavations each summer from 1966 until 1988. When he commenced his excavations, the whole floor of the cave was covered with broken rock which had protected the site from clandestine and unauthorised excavation but which hampered the excavation (Bayle Des Hermens (De) 1972).

At the top of a Mousterian layer and sealed in by ceiling collapse, R. de Bayle des Hermens found a succession of Magdalenian layers:

(i)   Layer F2, rich in bone and lithic material attributed to Ancient Magdalenian,
(ii) Layer E, also very rich in the same materials attributed to Upper Magdalenian,
(iii) Layer D, which provided only a few finds, attributed to Final Magdalenian.

The reappraisal of all the finds, namely the lithic raw material (V. Delvigne, thesis in preparation), the lithic industry (Lafarge 2007) and the bone industry, has allowed us to make a general reinterpretation of the settlement patterns evident in the cave particularly through the discovery of Badegoulian units which until now had remained unrecognised (Lafarge *et al.* 2012). Moreover, new AMS radiocarbon dates have highlighted the presence of Middle Magdalenian deposits which were first considered to be upper Magdalenian.

The preliminary results of these studies, as well as a new series of AMS radiocarbon dates, suggest that the previously excavated material from Le Rond du Barry should no longer be interpreted simply layer by layer. A different prehistoric situation may be revealed if each layer is examined as per the occupancy time period rather than taken as one unit of time. Five areas have been defined (see Figure 11.1), mainly on the basis of the lithic industry analysis and the AMS radiocarbon dates. But the validity of this new approach can only be demonstrated by the complete study of all objects within each area. This paper reviews and characterises the antler industry of Area 4 of the Le Rond du Barry cave site.

Area 4 is the most 'simple' area, with a homogeneous Magdalenian collection without a Badegoulian layer. Its industry has been attributed to Upper Magdalenian on the basis of one radiocarbon date (Beta 306186: 12,400 ± 50 years BP) and the occurrence of characteristic elements of this period; in particular two antler harpoons and a perforated antler baton decorated with a schematised female figure (Bayle Des Hermens (De) 1986). For practical reasons, the division into different areas was made using the excavation grid, which, as a spatial division made at the time of excavation does not represent the prehistoric use of space. Therefore, it is likely that some mixing of material may have occurred, particularly at area boundaries/interfaces. It is important to point out, therefore, that Area 4 is in contact with Area 2 where Badegoulian and middle Magdalenian have been recognised.

Material and methods

The antler industry from Area 4 consists of 61 pieces, which fall into the following technological categories: waste from debitage and shaping (23), blanks (7 fragments), roughout (1) and finished objects (30). This distribution follows the methodology proposed by Averbouh and Provenzano (1999) and Averbouh (2000).

The question of determination of age and sex from antler has already been largely discussed (Bouchud 1966; Spiess 1979; Averbouh 2000), and it appears that there are no reliable criteria for precise determination. So, the common classification (Averbouh 2000; Goutas 2004; Pétillon 2006) by size class: small, medium and large (Table 11.1) was employed.

Area 4 was fully excavated, and the sediment was then sieved. All of the bone material, especially faunal remains, was carefully sorted for pieces of industry

TABLE 11.1. Criteria of the determination of modules (after Averbouh 2000, Goutas 2004, Pétillon 2006).

Criteria Module	Large module	Medium module	Small module
Circumference of beam A	+ 120 mm	95/110 mm	70/85 mm
Diameter of pedicle	+ 33 mm	25/35 mm	15/25 mm
Thickness of compacta	7/10 mm	4/7 mm	2/4 mm

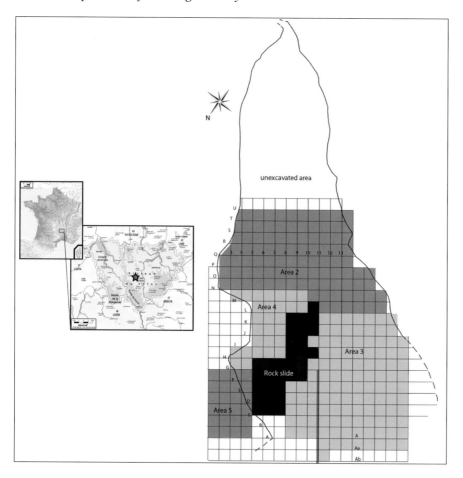

FIGURE 11.1. Location and map of Le Rond du Barry (modified from Delvigne, R. de Bayle and Lafarge). Area 1, in front of the cave is not illustrated.

that had not been previously identified. Therefore, it was assumed that we have the complete collection of the antler industry from Area 4.

The goal of the study was to identify the technical terms that have governed antler working in Le Rond du Barry for the upper Magdalenian period. The results obtained could then be compared with those from other areas with two mains objectives. Firstly, to define more precisely the different settlements in the cave and propose a chrono-cultural attribution for the material which is not clearly attributed to middle or upper Magdalenian (especially in Area 3). Then, secondly, based on typological and technological criteria, the continuity or the evolution of antler industry between middle Magdalenian (identified in Areas 1 and 2) and upper Magdalenian may be highlighted.

Antler industry in Area 4

Area 4 yielded a bone and antler industry as well as ivory items. The present study focuses on the 60 antler artefacts, of which 55 are made from reindeer (*Rangifer tarandus*) antler. The remaining five are of unidentified cervid antler. All the basal parts of the antler found in Area 4 indicate that they are from shed (or cast) antlers suggesting that these antlers were collected. No object made of red deer (*Cervus elaphus*) antler was clearly identified; however, a single unworked piece (a long tine) from a red deer was found. Therefore, even if the 'unidentified' pieces were made of red deer antler, the exploitation of this species appears marginal relative to reindeer.

Waste

The term waste refers to all items whose production, related to debitage, shaping or repair, and was not sought. We excluded from this category the fragments of finished objects when they only showed functional fracture.

Blocks

Three blocks were present in Area 4. The first was a basal part of a left large module reindeer antler (Figure 11.2:1). This piece was 201 mm long, with a circumference of 120 mm and a thickness of compact tissue greater than eight mm all around the beam. This antler was exploited by peripheral double grooving which resulted in the extraction of a large flat stick from the posterior side and at least two thin sticks from the anterior side.

The second block was the anterior side of a medium sized antler beam with an incomplete second tine (Figure 11.2:2). The edge of a longitudinal groove is visible on the anterior aspect, above the tine. The posterior aspect of the beam was entirely torn off. A removal scar resulting from direct percussion is visible on the mesial aspect. The point of impact which caused this triangular removal was directed from the spongiosa towards the more compacted outer surface. Antler tool production using direct percussion is a process characteristic of the Badegoulian and, until now, the technique of extraction by double grooving has not been identified in this culture (Allain *et al.* 1974; Pétillon and Ducasse 2012). Thus, the joint presence of one side of a groove along with signs of direct percussion on the same artefact makes its cultural attribution difficult. The third block was a section of antler beam which bears an unfinished groove on one side. The grooving activity remained incomplete and so the block remained more or less intact.

Finally, four fragments belonged to the so-called 'inter-stick' category of waste. These triangular fragments (width measurements range from 10.8 to 14.9 mm) came from a part located between two extractions of sticks on a block; hence they were derived from the fragmentation of the blocks.

Waste from shaping and repair of other pieces

In Area 4, there were two kinds of objects that resulted from the repair and shaping of final pieces. The first kind included sticks which showed traces of shaping and/or an end which showed traces of activity which resulted

FIGURE 11.2. Material from area 4. 1. reindeer antler base; 2:. reindeer antler beam A; 3. fragment of stick; 4. 'pointe à base raccourcie'.

from shortening of the stick. Eight items displayed these traces. They were all fragments of narrow sticks with rectangular cross-section. In all cases, the reduction in length was accomplished by bifacial or trifacial scraping.

The second kind of object was waste debris from the making of 'pointe à base raccourcie' (shortened base edges) (Figure 11.2:4) which were elongated objects, with a penetrating distal end and a proximal end, shortened by irregular removals (Mons 1988). The origin and function of these points has been discussed by several authors (Bertand 1999; Goutas 2004; Chauvière and Rigaud 2005) and it appears that this category comprises two types: 'real' spear-points and waste from shaping and re-shaping. Concerning the points from Le Rond du Barry, if they are typologically defined as 'pointes à base raccourcie', there is no doubt that they represent waste derived from shaping or re-shaping. These three points illustrate that peripheral scraping was the technique used for shortening the tool.

Finally, six fragments of antler, all showing at least one groove, were defined as waste, but without more precise determination. They all showed post-depositional fractures and they probably come from the fragmentation of blocks.

Blanks

The term blank designates every element from which a finished object can be obtained by shaping. It excludes items that were partially shaped, which were categorised as roughouts.

No unbroken blanks were found in Area 4, the seven pieces were fragments of narrow sticks, extracted by double grooving (Figure 11.2:3). Their lengths ranged between 29 and 74 mm, their widths between 10.8 and 14.9 mm, and they were generally rectangular in cross-section. They all exhibited post-depositional fractures on each extremity.

Roughouts

Roughouts represent an intermediary state between the blank and the finished object and include all the items that were partially shaped.

A single roughout was found in Area 4. The shaping was sufficiently advanced to indicate that the object was destined to be a short spear point with single bevelled base. The bevelling on the object appeared to be unfinished; the line where the grooves meet was still forming a peak. The last shaping stage to be undertaken on this object was probably going to be the scraping of this peak in order to regularise the bevelled surface. One discarded piece previously highlighted by R. de Bayle (1972) exhibits complementary morphology. It was a fragment of a stick extracted by double grooving and shortened at one end by slanted bifacial grooving.

Finished objects

Finished objects are the aim of all the technical transformation operations. They can be 'like new' (without use-wear), at different stages of wear, repaired, or broken.

Weapons

Spear or lance armatures were well represented in the collection, with thirteen artefacts, including two objects fitting the class of harpoons. One fragmentary example was a distal portion with an elongated point that had two barbs. Another example, which was complete, had a flattened section with double protuberances and a row of short barbs on the left edge. The form of its base suggests it may have been destined for attachment to a shaft using a detachable hafting method. However the striations made on either side of the base, presumably to improve adhesion (Julien 1999), were indicative of a permanently fixed hafting technique.

Most other objects in this class were different types of spear points among which there were several complete specimens. One spear had a point and a simple base, a large circular shaft and a long thin groove on each edge, probably intended as the position where lithic fragments could be mounted in order to make a composite point. Another specimen had a very fine spear point with a long thin groove on its lower side. One end was pointed; the other end was shaped into a very thin bevel, giving the appearance of a point. Both these spear point types were uncommon in the Upper Magdalenian.

The remaining points, including those with a single-bevelled base, were more characteristic of the Upper Magdalenian. The few specimens recovered from Area 4 were all short points, with a bevel on the lower side. In all cases, their entire surface had been worked by scraping. Within this group, there was a single fragment of a double-bevelled point. Its distal part was missing, however the morphology and size of this piece, as well as the presence of striations on the bevel, support its placement within the category of spear points.

Domestic tools

Bevelled tools were the only type of domestic tool present amongst the antler material from Area 4. The five bevelled tools have had one end modified by the formation of a double or single faced bevel. In the absence of specific traces of use-wear, it is difficult to ascertain their precise function. Four of them were worked over their entire surface by scraping. The remaining one, showed signs that scraping was less important. On it, the edges still showed traces of grooving and only the distal end was worked. This artefact also carries on both ends characteristic traces of percussion produced by its use in the extraction of sticks.

At least four fragments come from unidentified artefacts: two fragments of mesial parts and two with bevelled ends, which cannot be attributed to bevelled tools (distal end) or spear points (proximal end).

Objects on a complete segment of beam

Six objects were made from whole antler sections; however, given their fragmentary state, a typological attribution is sometimes difficult. The most spectacular of these was a perforated baton (Figure 11.3:1) which was not available to be examined and therefore a publication describing this artefact was relied upon (see Bayle Des Hermens (De) 1969; Bayle Des Hermens (De) 1972). The artefact was made on an almost entire small piece of antler, the departure of the palm was still visible in the proximal part and the fracture at this point appeared somewhat irregular. The base and the second tine have been cut off. A circular perforation was made through the area of the junction between the beam and the tine. Although several suggestions have been put forward for the use of the perforated batons, there is no general consensus regarding their use (Rigaud 2001). The variety in size and the shape of these types of objects suggests the possibility that they had not one, but probably several different uses. The perforated baton from Le Rond du Barry was engraved with a schematic female figure, a very rare decorative element for this type of object.

A second object was complete enough for its use to be determined (Figure 11.3:3). It was a bevelled tool, produced on the long section of a small module beam. The base and tines were completely removed and the surface was entirely worked by scraping. The bevel, which showed evidence of crushing, was made using a combination of scraping and abrasion. Additionally, the proximal part of this tool showed the characteristic traces of use found on a 'retoucher'; which is how we classify it.

Lastly, four portions of antler showed traces of shaping but their fragmentation makes any typological identification impossible. Among these was a piece to which unique shaping processes were applied (Figure 11.3:2). It was a curved portion of a beam approximately 15 cm long. In section, the object goes from circular in the proximal part to rectangular in the distal part. Near the base, the edges were modified to reduce the dimensions of the cross-section. Beyond this contraction, the upper side was levelled following which two long incisions were engraved along the artefact's edges. The nominal proximal end appeared to have been adapted for a gripping function however, as the tip of the distal end was missing it was impossible to determine what kind of object this is.

Extraction techniques and operational schemes

Grooving to obtain blanks was the only extractive technique observed on the waste material. Grooving was also used for severing sections of the blanks, as was surface scraping. For shaping finished objects, scraping was the technique usually employed, although one bevelled tool was partly worked using abrasion. Incision with a sharply pointed stone flake was used for the addition of the decoration. Incision was also the technique used to produce the bifacial grooves required to form the barbs of harpoons. As almost all of the excavated objects had varnish applied as a preservative measure following their excavation, it is now impossible to observe any traces of use-polish on them.

The only operational scheme evident in this assemblage was that of transformation by extraction, using the method of double grooving. This technique was seen both on finished objects and on fragments of sticks, which exhibited traces of grooves on one or more edges. However, considering the lack of waste on beam B and C, no precise operational scheme could be determined.

Economical approach

There was one basal section of antler in the assemblage. This object showed evidence of the extraction of at least three sticks.

If a large beam produces approximately 30 cm of material from the section between the second tine and branch B (see Averbouh 2000), then this antler could have provided 90 cm of stick, with a compacta thickness of around 8.5 mm. For this antler, there were no clues regarding how much exploitation beam

FIGURE 11.3. Objects on a complete segment of an antler beam. 1. perforated baton decorated with a schematized female figure (Photo J.-P. Raynal); 2. undetermined tool; 3. bevelled tool.

B/C underwent, but such activity cannot be excluded. Consequently, it was likely that this antler provided a substantial amount of useful material.

The compacta thickness retained for the large module class attribution of finished objects was seven mm (note: shaping reduces the thickness). The total overall length of finished objects recovered that were made on large sticks reached almost 40 cm. This length could have been provided with some excess by the exploitation of beam A of the large module antler even when the fragmentary pieces were considered. However, there was at least one spear point from the assemblage which cannot come from this antler as its compacta thickness reached 9 mm.

The second block provides little information about the intensity of its exploitation, which raised the possibility that a medium module antler could have been exploited.

The overall length of finished objects on sticks made from medium module antler was 1.30 metres. Further, no finished object was made on sticks from small modules antlers (compacta thickness below 4 mm). All the finished objects made on sticks that were recovered from the excavation could have derived from the peripheral exploitation of just one large antler (objects with greater compacta thickness extracted from the lower antler section) and the partial exploitation of one medium antler (objects with lower compacta thickness extracted from the upper section of an antler). Therefore, the two antlers exploited could have produced enough blanks to make the finished objects of the assemblage. But even with the apparent presence of at least two different antlers, the absence of the major part of these antlers would be noticeable. Indeed, even in the case of very high exploitation, some parts remained rarely exploited: in particular parts of beam B with the posterior tine, the pedicle and beam A between the first and the second tine and the palms. Even considering that antler palms could have been pruned when antlers were collected, some waste was still missing. These waste pieces could be either mixed with the material of another area, or in the unexcavated part of the cave. Only the ongoing analysis of the whole antler material of all areas will shed light on this situation.

The finished artefacts provided the only evidence of the production of objects on a complete segment of small modules of antlers (transverse exploitation of antler). There was a complete absence of waste corresponding to this production technique, which should consist of tines and basal parts. Given the complexity of settlement in Le Rond du Barry and the arbitrary character of the limits that define different areas of occupation, this absence could be either explained by the waste being in another area of the cave or the exogenous production of objects on portions of small antlers was being carried out. The latter explanation may be more likely, given that the bone industry from Le Rond du Barry affirms that this kind of waste was absent in the osseous material.

Origin of raw material

Area 4 revealed 40 fragments of antler of more than 5 cm long, which exhibited breaks on dry antler and no technical traces. Thus, these fragments were excluded from the technological study but they still provide information about raw material procurement. For reindeer antler, all portions of the antler were represented: beam A/B/C, basal parts, tines and palms.

The five basal portions (one among the industry and four among the unworked material) of antler found in Area 4 were derived from shed antler, thus their presence in the cave cannot be related to hunting activities. Moreover, the thickness of compact tissue measured on other antler parts indicated that most antlers were from mature individuals. Archeozoological analysis by Costamagno (1999) showed that reindeer was the third most commonly hunted species, after horse (*Equus* sp.) and ibex (*Capra* sp.), and that cranial remains were frequently abandoned on the killing site. These data may suggest that antlers were mostly collected. Nevertheless, even if the method of collection for these antlers cannot be determined (as equally the deer could have been hunted and antlers subsequently removed), it can still be proposed that the collection occurred near the antler shedding period.

Most of the reindeer antlers are from the medium and large module class, and only four tine fragments are from small module antlers (young male/female reindeer). So the majority of antlers are from adult mature males (large class) and young males (two to three years old) or adult females (medium class size) (Averbouh 2000; Goutas 2004). Adult males shed their antlers after the rutting season, in November, while young males shed them in winter (from December to February) and adult females in April (Bouchud 1966). These antlers showed no visible deterioration caused by rodents, carnivores or reindeer themselves; therefore they could have been collected soon after they were cast. This indicates that antlers from Area 4 were collected between the beginning of winter and the beginning of spring. This is particularly interesting as the presence of Magdalenian groups in the Massif Central during winter has not yet been attested (Fontana *et al.* 2009; Fontana and Chauvière 2009). However, there are no data available on the presence of reindeer in the Massif Central during the winter. Thus, these antlers could have been collected close to the site during a winter occupation. Alternatively, they could have been collected in another region during the winter and brought to the site during another season. These current data do not make it possible to decide in favour of one assumption over the other at present.

Conclusion

The antler industry of Area 4 of Le Rond du Barry showed characteristics indicative of the Upper Magdalenian; in particular, debitage produced by double longitudinal grooving, as well as formally identified objects typical of this period. However, Area 4 also contained objects not characteristic of this attribution. The presence of these objects may be the result either of mixing of the deposits or perhaps could result from a certain originality of osseous production techniques used during the Upper Magdalenian at Le Rond du Barry. Economic interpretation remains limited for Area 4, given the small quantity of material available for analysis.

Two types of productions were identified. Firstly, a production of finished objects on segments of small-sized antlers, as there was no waste in the whole cave corresponding to such production and therefore they may have originated away from this cave site. The second type of production concerned finished objects made from sticks from medium and large-sized antler. Some waste pieces testify to a local production of sticks, which could correspond to the finished objects from Area 4.

Finally, the study highlighted the lack of certain types of waste in contrast to the specific types of waste present in the assemblage. Given the complexity of the fill which represented many successive occupations and settlements, the preliminary results presented herein for Area 4 will need to be correlated with analyses from all other locations of Le Rond du Barry cave.

Acknowledgments

I wish to thank Peter Bindon and Amy Clark for their help in translating and correcting this manuscript. This paper was greatly improved by the comments of the anonymous reviewers. I also address my sincere thanks to Jean-Paul Raynal for his support and to Vincent Delvigne and Audrey Lafarge for their work on Le Rond du Barry stratigraphy. This work was supported by the association l'Archéo-logis/CDERAD.

References

Averbouh, A. (2000) *Technologie de la matière osseuse travaillée et implications palethnologiques*, Thèse de Doctorat de l'Université Paris I Panthéon-Sorbonne.

Allain, J., Fritsch, R., Rigaud, A. and Trotignon, F. (1974) 'Le débitage du bois de renne dans les niveaux du Badegoulien de l'Abri Fritsch et sa signification' in *Premier colloque international sur l'industrie de l'os dans la préhistoire*, ed. H. Camps-Fabrer, Sénanque, Editions de l'Université de Provence, Provence, France, 67–71.

Averbouh, A. and Provenzano, N. (1999) 'Propositions pour une terminologie du travail préhistorique des matières osseuses. 1 – Les techniques', *Préhistoire Anthropologie Méditerranéennes* 7–8, 5–25.

Bayle Des Hermens (De), R. (1967) 'La Grotte du Rond du Barry, à Sinzelles, commune de Polignac (Haute-Loire)', *Bulletin de la Société Préhistorique Française* 64, 155–174.

Bayle Des Hermens (De), R. (1969) L'industrie osseuse du Magdalénien final de la grotte du Rond du Barry, commune de Polignac, Haute-Loire: note préliminaire, *L'Anthropologie* **73**, 253–260.

Bayle Des Hermens (De), R. (1972) 'Le Magdalénien final de la grotte du Rond du Barry, commune de Polignac (Haute-Loire)' in *Congrès Préhistorique de France, XIX° session, Auvergne, 1969, 1972*. Ed. Société Préhistorique Française, 37–69.

Bayle Des Hermens (De), R. (1986) 'Découverte d'un nouveau bâton perforé dans la grotte du Rond du Barry', *Bulletin de la Société Préhistorique Française* 83, 36–46.

Bouchud, J. (1966) *Essai sur le renne et la climatologie du Paléolithique moyen et supérieur*, Imprimerie Magne, Périgueux, 1966.

Chauvière, F. -X. and Rigaud, A. (2005) 'Les "sagaies" à "base raccourcie" ou les avatars de la typologie : du technique au "non-fonctionnel" dans le Magdalénien à navettes de la Garenne (Saint-Marcel, Indre)' in *Industrie osseuse et parure du Solutréen au Magdalénien en Europe*, ed. V. Dujardin, Actes de la table ronde sur le Paléolithique supérieur récent, Mémoires de la Société Préhistorique Française XXXIX, 233–242.

Costamagno, S. (1999) *Stratégies de chasse et fonction des sites au Magdalénien dans le sud de la France*, Thèse de Doctorat de l'Université de Bordeaux I, 2 vol.

Fontana, L. and Chauvière, F.-X. (2009). 'The total exploitation of Reindeer at the site of Les Petits Guinards: what's new about the annual cycle of nomadism of

Magdalenian groups in the French Massif Central?' in *In Search of Total Animal Eexploitation. Case studies from the Upper Palaeolithic and Mesolithic*, eds L. Fontana, A. Bridault and F.-X. Chauvière, Proceedings of the XVth UISPP Congress, Session C61, vol. 42, Lisbon, 4–9 September 2006, BAR International Series 2040, Archaeopress, Oxford, 101–111.

Fontana, L., Digan, M., Aubry, T., Llach, J.-M. and Chauvière, F.-X. (2009) 'Exploitation des ressources et territoire dans le Massif central français au Paléolithique supérieur: approche méthodologique et hypothèses', in *Le concept de territoires dans le Paléolithique supérieur européen*, eds F. Djindjian, J. Kozlowski and N. Bicho, Proceedings of the XVth UISPP Congress, Lisbon, 4–9 September 2006, BAR International Series 1938, Archaeopress, Oxford, 201–215.

Goutas, N. (2004) '*Caractérisation et évolution du Gravettien en France par l'approche techno-économique des industries en matières dures animales (étude de six gisements du Sud-Ouest)*', Thèse de l'Université Paris I Panthéon Sorbonne, 2 vol.

Julien, M. (1999) 'Une tendance créatrice au Magdalénien: à propos des stries d'adhérence sur quelques harpons', *Préhistoire d'os, recueil d'études sur l'industrie osseuse préhistorique offert à Henriette Camps-Fabrer*. Publications de l'Université de Provence, Provence, France, 133–142.

Lafrage, A. (2007) '*Modéles stochastiques pour la resconstruction tridimensionnelle d'environments urbains*', Unpublished PhD dissertation. Ecole des Mines de Paris.

Lafarge, A., Delvigne, V., Remy, D., Fernandes, P. and Raynal, J.-P. (2012) 'Ancient Magdalenian of the French Massif Central revisited: a reappraisal of unit F2 of the Rond du Barry cave (Polignac, Haute-Loire, France)', in *Unravelling the Palaeolithic. Ten years of Research at the Centre for the Archaeology of Human Origin (CAHO, University of Southampton)*, eds K. Ruebens, I. Romanowska and R. Bynoe, BAR International Series 2400, Archaeopress, Oxford, 109–129.

Mons, L. (1988) 'Fiche sagaie à base raccourcie', in *Fiches typologiques de l'industrie osseuse préhistorique. Cahier I: Sagaies*, ed. H. Camps-Fabrer, Publications de l'Université de Provence, Provence, France, 1–9.

Pétillon, J. -M. (2006) '*Des Magdaléniens en armes. Technologie des armatures de projectile en bois de cervidé du Magdalénien supérieur de la grotte d'Isturitz (Pyrénées atlantiques)*', *Artefacts 10*, Editions du Centre d'études et de documentation archéologiques, Treignes, Belgium.

Pétillon, J.-M. and Ducasse, S. (2012) 'From flakes to grooves: A technical shift in antlerworking during the last glacial maximum in southwest France', *Journal of Human Evolution* 62, 435–465.

Rigaud, A. (2001) 'Les bâtons percés. Décors énigmatiques et fonction possible', *Gallia Préhistoire* 43, 101–151.

Robert, F. (1937) 'Mémoire géologique sur le bassin du Puy', *Annales de la Société d'agriculture, sciences, arts et commerce du Puy* 7–8, 53–78.

Spiess, A. E. (1979) *Reindeer and Caribou Hunters: An Archaeological Study*, New York Academic Press.

12

Deer (*Rangifer tarandus* and *Cervus elaphus*) Remains from the Final Gravettian of the Abri Pataud and their Importance to Humans

Carole Vercoutère, Laurent Crépin, Dorothée G. Drucker,
Laurent Chiotti, Dominique Henry-Gambier
and Roland Nespoulet

Introduction

Traditionally, researchers of prehistory have approached the study of faunal remains from butchery processes and objects made of animal hard tissues separately. This partitioning of analyses has prevented a holistic understanding of the exploitation and role of species within the prehistoric human groups studied. Thus, over the past decade, priority has been given to integrated, multidisciplinary studies. Moreover, new disciplines, such as biogeochemistry, provide additional data about the contribution of meat to the human diet and in the ethology of hunted animals.

Since 2005, a new multidisciplinary research project has been conducted at the Abri Pataud (Dordogne, France; Figure 12.1) on the Final Gravettian level (archaeological level two) under the direction of R. Nespoulet and L. Chiotti. The main goal of this project is to re-examine the archaeological context and the nature of the human remains from level two through new excavations and studies of previously made collections and archives (Nespoulet *et al.* 2008; Chiotti *et al.* 2009; Nespoulet *et al.* 2013). This project has allowed us, among others, to carry out an integrated study of faunal remains. This paper presents preliminary results of the archaeozoological, osseous artefacts and biogeochemical studies to discuss the role of deer in human culture and subsistence during the Final Gravettian at the Abri Pataud.

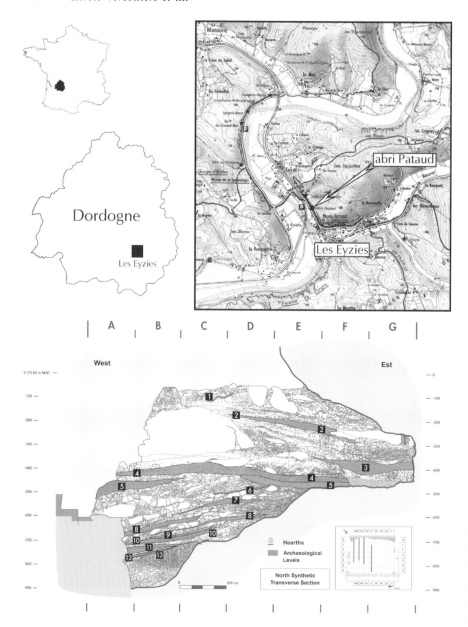

FIGURE 12.1.
Geographical location and stratigraphical sequence (Synthetic North Section) of the abri Pataud (Les Eyzies-de-Tayac, Dordogne, France).

Archaeological context

The Abri Pataud, located at Les Eyzies-de-Tayac (Dordogne, France), is a rock-shelter at the base of a cliff. Excavation of this site was mainly conducted by H. L. Movius from 1958 to 1964 (Movius 1977; Bricker 1995). Movius divided the filling of the rock-shelter into fourteen main archaeological levels, in stratigraphic continuity, from the Early Aurignacian to the beginning of the Solutrean, after an intermediate phase in the Gravettian (Figure 12.1).

The archaeological material from level two was attributed to the Final

Gravettian (historically named 'Protomagdalenian'), dated around 22,000 years before present (BP). This cultural facies, well defined at the Abri Pataud has been confidently recognised in only four archaeological sites in France: Abri Pataud and Laugerie-Haute Est in Dordogne, Les Peyrugues in Lot and Le Blot in Haute-Loire. Therefore, Pataud is a key site for improving knowledge of human behaviour at the end of the Gravettian period. At Pataud, level two corresponds to recurrent seasonal camp type occupations. It has yielded thousands of archaeological objects (faunal remains, lithic and bone industries, painted and engraved wall fragments from an ancient parietal decoration, charcoals), among which were more than four hundred human remains. Within the rock-shelter, the human remains were all located in the rear zone, behind a line of fallen blocks. But, after Movius' excavations in 1958 and 1963, the question about the nature of these human remains still has to be dealt with. More precisely, the question of whether it is possible to interpret their presence in the rock-shelter as the result of funerary activities needs to be addressed. This question is the focus of the new research program initiated in 2005 and is addressed via three approaches: new excavations adjoining those of Movius, study of Movius' collections, and analysis of Movius' archives (Nespoulet *et al.* 2008; 2013). Therefore, as far as archaeological data are concerned, two corpuses, before and after 2005, are now available in our database.

Human (*Homo sapiens*) remains from level two at Pataud represent one of the most important series of human remains for the French Gravettian and for the European Final Gravettian to date. The human remains were located in two concentrations at level two, one in the southern part and one in the northern part of the site, which could be identified as a mortuary deposition (Nespoulet *et al.* 2008; Henry-Gambier *et al.* 2013). Some artefacts from the same level were qualified as 'extra-ordinary' because of their scarcity in comparison with other types of remains from level two, their spatial arrangement (near the north human remains concentration), the raw material they were made of and their high technical investment and apparently long use. Some of these 'extraordinary' objects were associated with the human remains, and could thus be considered as mortuary deposits (Chiotti *et al.* 2009).

Holistic investigation of deer exploitation

Among the large game of taxa represented in level 2, two species of deer have been recognised: the reindeer (*Rangifer tarandus*) and the red deer (*Cervus elaphus*). While previous archaeozoological studies considered only the faunal remains produced from butchery process (Cho 1998), an integrated study of the complete assemblage of faunal remains, including products of butchery wastes and osseous artefacts was carried out. Moreover, remains were analysed using three approaches: archaeozoology, osseous typo-technology and biogeochemistry. Our aim is to have a holistic understanding of the exploitation of deer by people who lived at the site during the Final Gravettian period.

Archaeozoological data

The archaeozoological analysis carried out on finds from the new excavations (2005–2008: Crépin 2013) established a faunal spectrum dominated by herbivores (fifteen herbivores and only two carnivores), in particular reindeer, with 291 remains (75.8 percent of the NISP: Number of Identified Specimens) corresponding to six individuals (one juvenile between five and eight month, two young between one and two years old and three pregnant females). Young and adult reindeer were killed from autumn to the middle of winter, based on dental eruption, the presence of foetal bones and female antler. During this cold season, they were likely 'easy prey' to target because of the ease with which they could be located: females and calves are together in small groups, whereas single males are scattered in landscape. Females are also the most nutritious preys at this period of the year. They have more fat than males after rut period (Spiess 1979; Gerhart *et al.* 1996; Couturier *et al.* 2009). All skeletal elements are represented in this bone assemblage (particularly cranial and limb ones) with an important deficit of axial bones (ribs and vertebrae probably used as fuel). Therefore, the complete carcasses of reindeer were likely brought to the Abri Pataud, where the butchery process was undertaken. Reindeer remains show a high frequency of marks linked with this dietary treatment (dismembering, filleting and intensive breakage activities). These marks indicate relatively intensive use of carcasses with a significant recovery of meat, fat and marrow. Red deer on the other hand was less important as a source of meat in human's diet: only fourteen remains (cranial and limb bones) were discovered that show few cut marks. It seems that only some pieces of one old red deer were brought to the shelter, thus it is not possible to determine the season of hunting.

Typo-technological studies

In both corpuses (before and after 2005), reindeer and red deer provided raw materials (antlers, bones, teeth) for bone industry and body ornamentation (Table 12.1). As far as industry is concerned, a few antler objects, whose manufacture demands high technical and time investment, were found (Vercoutère 2007; Vercoutère 2013). Both collections include informative antler waste products,

Taxa	NR		Total	%
	Movius' *collection*	*2005–2010* *excavations*		
Rangifer tarandus	118	56	174	50.29
Cervus elaphus	6	1	7	2.02
Cervids	5	-	5	1.45
Bos or *Bison*	-	3	3	0.87
Bos or *Bison* or *Equus*	3	4	7	2.02
Mammuthus primigenius	12	11	23	6.65
Undetermined	106	21	127	36.71
Total	250	96	**346**	

TABLE 12.1. Taxa that supplied raw materials for osseous artefacts from Pataud level 2 (NR: Number of Remains).

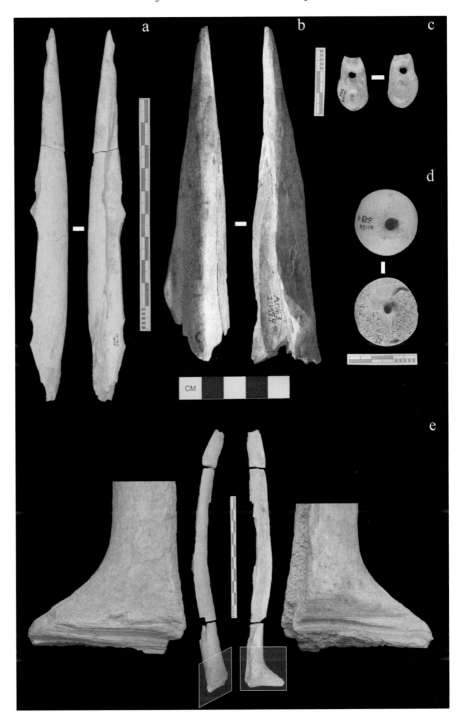

FIGURE 12.2. Osseous artefacts from Pataud Level 2. a: AP/08-W76C-147, awl made on reindeer metapodal (© L. Chiotti); b: AP/63-2-1974, awl made on red deer metapodal (© C. Vercoutère); c: AP/58-2-885, red deer canine pierced (© L. Chiotti); d: AP/58-2-529, reindeer femoral head pierced (© L. Chiotti); e: AP/08-V76A-469, waste product made on reindeer tine (© L. Chiotti).

which suggested that at least a part of the antler tool manufacturing took place at the site (Figure 12.2). Bone tools are more varied and are associated with treatment of skins, e.g. awls and smoothers. It was noticed that there was a high exploitation of reindeer metapodials to make awls. Moreover, the only

	Dietary exploitation	Non-dietary exploitation			
		Industry	*Ornamentation*	*Art*	*Funerary context*
Rangifer tarandus	■	■	■	■	■
Cervus elaphus	■	■	■	■	?
Bos/Bison	■	?		■	■
Equus sp.	■	?			■
Mammuthus primigenius			■		■

TABLE 12.2. Animal status for the Final Gravettian level of the Abri Pataud.

two tools made on red deer remains were also manufactured using metapodials (Figure 12.2).

A 'bâton percé' made of reindeer antler is among the artefacts that could be considered to be a mortuary deposit. It consists of a non-shed reindeer antler, without any decoration or other modifications, except for a perforation. Its large size – *c.*60 cm long – is noticeable.

A 'deer skull complex' is probably also associated with human remains but its nature is less clear. It consists of three red deer skulls: two from females (fourteen and two years old) and one from a juvenile (six months old) found in association. Movius (1977) noticed that these skulls were quite complete when they were discovered, only the lower jaws were lacking. That is quite exceptional both because there are a few cranial remains in level two and the faunal remains are generally highly broken. Moreover, new excavations provided a right maxilla, with all cheek teeth, from a red deer of more than ten years old. This bone bares some ochre microtraces and was also found near human remains.

The results of the analyses carried out on faunal remains from Pataud level two are consistent with the main role of reindeer in food resources and osseous industry. No taboo has been noticed in reindeer use (Table 12.2). This cervid species is even represented within body ornamentation and mortuary context. But, while previous research concluded that the economy of the Final Gravettian people was based on reindeer, new studies demonstrate the important contribution of other species such as large bovids (*Bos* or *Bison*) to the meat diet and mammoth (*Mammuthus primigenius*) to body ornamentation (Vercoutère 2007; Vercoutère 2013; Vercoutère *et al.* 2011; Crépin 2013). Even if these species are less represented in the faunal spectrum than reindeer, red deer were exploited in the same way as reindeer, with the exception that no manufactured object was made of red deer antler. The association of those two cervids, with a high proportion of reindeer, as a main source of meat and raw material, and a low representation of red deer, was also seen at Laugerie-Haute Est (Cho 1998) and Le Blot (Chauvière and Fontana 2005), taking into account that those two sites seem to be summer camps in contrast to the few evidences available for Pataud.

Biogeochemical analyses

In order to go further in understanding cervid exploitation by human groups that lived at the Abri Pataud during the Final Gravettian, biogeochemical analyses on human, wolf (*Canis lupus*), bovid, horse (*Equus* sp.) and reindeer bones were also conducted. Sampling of red deer has been postponed due to ongoing investigations (checking for possible matches among specimens in the old and new collections). Stable isotope analyses of fossil remains have been routinely used for the last decades to reconstruct the environment and diet of Palaeolithic animal and human specimens (review in Koch 2007).

Materials and Method

The stable isotope carbon-13 abundances ($\delta^{13}C$) of herbivore tissues reflect those of plants, which depend on the source of atmospheric carbon and the photosynthetic process (review in Tieszen 1991; Heaton 1999). In Western Europe, ancient and modern indigenous plants exhibit a C_3 photosynthesis pattern and their $\delta^{13}C$ values vary according to environmental conditions. The nitrogen isotopic composition of bone collagen ($\delta^{15}N$) from herbivores depends on dietary, physiological and climatic parameters. Bone collagen reflects the long-term isotopic trend as bone is continuously remodelled with a turnover of several years over life (review in Koch 2007). Thus, bone $\delta^{15}N_{coll}$ values of mature individuals of a given species can reflect environmental change through long time periods. In Western Europe during the last 40,000 years, the chronological and geographical variation in $\delta^{15}N_{coll}$ values of large herbivore seemed to follow the intensity of soil activity driven by temperature and glacier occurrence (Drucker *et al.* 2003a; Drucker *et al.* 2003b; Stevens *et al.* 2004; Stevens *et al.* 2008). In a given trophic chain, the isotopic signature of nitrogen is significantly enriched in bone collagen compared to its average diet, typically by 3 to 5‰ (Bocherens and Drucker 2003). Therefore, the nitrogen isotopic signature of a given individual depends on the isotopic signature at the base of the food web to which it belongs (in the plants), and on the position of the specimen within the food web (herbivore or predator). For human and animal predator diet, the relative contribution of different preys can be estimated when they exhibit contrasting stable isotope signature in their bone collagen (e.g. Drucker and Henry-Gambier 2005; Bocherens *et al.* 2005; Bocherens *et al.* 2011).

Isotopic analysis of carbon and nitrogen were performed on bone remains of human, wolf, reindeer, horse and large bovids (*Bos* or *Bison*) from the level two of Abri Pataud (Table 12.3). In order to augment the sample size, data previously obtained on horse and reindeer at the sites of Laugerie-Haute Est and Les Peyrugues were also added. These sites are located in south-western France and delivered archaeological levels contemporaneous with the level two of Abri Pataud (Drucker *et al.* 2003a).

Collagen was extracted following a protocol based on Longin (1971) and modified by Bocherens *et al.* (1997). The extraction process includes a step of soaking in 0.125M NaOH between the demineralisation and solubilisation steps to achieve the elimination of lipids. Elemental analysis (C_{coll}, N_{coll}) and isotopic analysis ($\delta^{13}C_{coll}$, $\delta^{15}N_{coll}$) were conducted at the Department of Geosciences of Tübingen University using a NC2500 CHN-elemental analyser coupled to a Thermo Quest Delta+XL mass spectrometer. The standard, internationally defined, is a marine carbonate (PDB) for $\delta^{13}C$ and atmospheric nitrogen (AIR) for $\delta^{15}N$. Analytical error, based on within-run replicate measurement of laboratory standards (albumen, modern collagen, USGS 24, IAEA 305A), was ±0.1‰ for $\delta^{13}C$ values and ±0.2‰ for $\delta^{15}N$ values. Reliability of the $\delta^{13}C_{coll}$ and $\delta^{15}N_{coll}$ values can be established by measuring its chemical composition, with C/N_{coll} atomic ratio ranging from 2.9 to 3.6 (DeNiro 1985), and percentage of C_{coll} and N_{coll} above 8 and 3 percent, respectively (Ambrose 1990).

Results and discussion

The $\delta^{13}C_{coll}$ and $\delta^{15}N_{coll}$ values of the large herbivores varied from -21.0 to -18.7‰ and from 2.6 to 6.0‰, respectively (Table 12.3). The horse showed $\delta^{13}C_{coll}$ values (-21.0 to -20.4‰) significantly lower than those of reindeer (-19.5 to -18.8‰), while large bovid exhibited intermediate $\delta^{13}C_{coll}$ values compared to horse and reindeer (-20.0 to -19.6‰). This pattern of $\delta^{13}C_{coll}$ distribution between the three herbivores groups was evidenced during the Upper Palaeolithic in south-western France (Drucker *et al.* 2003a) as well as in eastern and northern France and Belgium (Bocherens *et al.* 2011). Diet specialisation such as high lichen consumption for reindeer was considered to be the cause of this constant pattern in $\delta^{13}C_{coll}$ values (e.g. Drucker *et al.* 2003a). In contrast, important variations in the $\delta^{15}N_{coll}$ values were observed over time for each considered species in Western Europe (Drucker *et al.* 2003a; Stevens *et al.* 2004; Stevens *et al.* 2008). It was thus important to choose additional samples that were provided by other Final Gravettian context of the same region to complete the isotopic data of horse and reindeer of Pataud level two. Interestingly, this chronological period was characterised by comparable $\delta^{15}N_{coll}$ mean values among reindeer, large bovid and horse in south-western France. The new results from Abri Pataud were consistent with this observation since average $\delta^{15}N_{coll}$ values of horse (4.2‰), reindeer (4.5‰) and large bovid (4.4‰) were similar (Figure 12.3).

The wolf and the human of Abri Pataud had comparable $\delta^{13}C_{coll}$ values (-19.2‰ and -19.6‰, respectively) but contrasted $\delta^{15}N_{coll}$ values of (9.0 and 10.1‰, respectively). The $\delta^{15}N_{coll}$ value of the wolf was 4.5 to 4.8‰ higher than those of the large herbivores of Abri Pataud, as expected between a predator and its potential prey. The human had a higher $\delta^{15}N_{coll}$ values than the 9.2 to 9.5‰ calculated as the maximum values for a diet based on the associated large herbivores of the region, when we consider their averaged $\delta^{15}N_{coll}$ values. Interestingly, the $\delta^{13}C_{coll}$ value of the human was too low to include a significant part of reindeer in the diet, while it could be explained by a major contribution of horse. As a matter of fact, horse was limited to four samples in this study due to the rather low occurrence of this species in the site of Abri Pataud. It could be speculated that the mean $\delta^{15}N_{coll}$ value of horse was higher than the one calculated so far and would fit the expected range for the human's

No. lab	Species	Anatomical part	No. excavation	Level	C_{coll} (%)	N_{coll} (%)	C/N_{coll} (%)	$\delta^{13}C_{coll}$ (‰)	$\delta^{15}N_{coll}$ (‰)
PAT-1	Human	vertebra	AP89 no. 288	2 trench F/G VIII	27.7	10.1	3.2	-19.6	10.1
PAT-30	Wolf	metatarsal	Pataudo8 U77C 123G	2	35.8	12.9	3.2	-19.2	9.0
PAT-4	Reindeer	left metacarpal	AP58 B64-2	trench 2 or 3	45.7	16.2	3.3	-19.3	4.7
PAT-5	Reindeer	tibia	AP58 B64-1	trench 2 or 3	45.9	16.3	3.3	-19.2	3.8
PAT-6	Reindeer	radius/metacarpal	AP58 B64-3	trench 2 or 3	43.3	14.9	3.4	-19.5	4.4
PAT-7	Reindeer	left metacarpal	AP58 B59-5	trench V or VI	44.7	15.5	3.4	-19.1	4.9
PAT-8	Reindeer	tibia/femur	AP58 B59-6	trench V or VI	43.7	14.7	3.5	-18.7	4.3
PRG4200	Reindeer	antler	46PRG 9E 18 T763	18	40.0	14.6	3.2	-19.5	3.8
PRG4300	Reindeer	skull	46PRG 8C 18 377	18	41.9	15.1	3.3	-19.5	5.6
PRG4400	Reindeer	right femur	46PRG 6C 18 77	18	42.6	15.3	3.2	-19.2	5.4
PRG4500	Reindeer	femur	46PRG 9B 18 323	18	42.2	14.9	3.3	-19.1	4.8
PRG4600	Reindeer	tibia	46PRG 7C 18 205	18	42.5	15.4	3.2	-18.7	3.8
PRG4700	Reindeer	metacarpal	46PRG 12C 18 464	18	40.8	14.8	3.2	-19.2	3.3
LGH2200	Reindeer	metapodal	LHE F9	36	43.4	15.9	3.2	-19.5	4.2
							Mean	**-19.2**	**4.4**
							SD	**0.3**	**0.6**
PAT-15	Large bovid	left femur	Pataudo7 U75B 97G	2	33.7	10.9	3.6	-19.9	4.4
PAT-16	Large bovid	left femur	Pataudo8 U76B 358G	2	30.9	10.9	3.3	-20.0	4.6
PAT-18	Large bovid	left femur	Pataudo6 V76A 126D	2	40.4	13.3	3.5	-19.6	4.4
							Mean	**-19.8**	**4.5**
							SD	**0.2**	**0.1**
LGH1900	Horse	3rd phalanx	LHE F8	36	43.6	16.0	3.2	-20.8	
PAT-26	Horse	tibia	Pataudo6 T75C 176F	2	40.0	14.2	3.3	-20.4	2.6
PAT-28	Horse	tooth root	Pataudo6 S75C 193E	2	40.0	14.3	3.3	-21.0	4.3
PAT-29	Horse	tooth root	Pataudo6 S75D 156E	2	40.3	14.3	3.3	-20.7	3.8
							Mean	**-20.7**	**4.2**
							SD	**0.2**	**1.2**

TABLE 12.3. Results of the isotopic analysis of collagen ($\delta^{13}C_{coll}$, $\delta^{15}N_{coll}$) of human (*Homo sapiens*) and wolf (*Canis lupus*) of the level 2 of abri Pataud compared the averaged $\delta^{13}C_{coll}$ and $\delta^{15}N_{coll}$ values of reindeer (*Rangifer tarandus*), horse (*Equus* sp.) and large bovid (*Bos* or *Bison*) from Pataud level 2, Laugerie-Haute Est level 36 and Les Peyrugues level 18 (Drucker *et al.*, 2003a; this work). The carbon and nitrogen composition of the collagen is given through elemental composition (Ccoll, Ncoll) and atomic ratio (C/Ncoll).

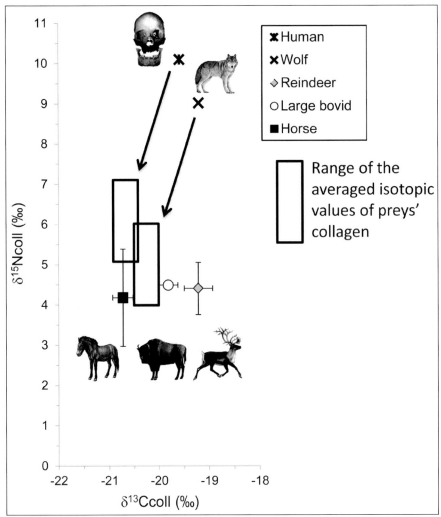

FIGURE 12.3. The $\delta^{13}C_{coll}$ and $\delta^{15}N_{coll}$ values of human (*Homo sapiens*) and wolf (*Canis lupus*) of Pataud level 2 compared the averaged $\delta^{13}C_{coll}$ and $\delta^{15}N_{coll}$ values of reindeer (*Rangifer tarandus*), horse (*Equus* sp.) and large bovid (*Bos* or *Bison*) from Pataud level 2, Laugerie-Haute Est level 36 and Les Peyrugues level 18 (Drucker *et al.*, 2003a; this work). The rectangles stand for the range of averaged $\delta^{13}C_{coll}$ and $\delta^{15}N_{coll}$ values of preys bone collagen.

diet. Another herbivore species that showed similar $\delta^{13}C_{coll}$ and higher $\delta^{15}N_{coll}$ amounts compared to horse of the same ecosystem was the woolly mammoth (e.g. Bocherens *et al.* 2011). Remains of mammoth, mainly ivory, were found in Abri Pataud but there is no evidence of mammoth exploitation for subsistence purpose. However, the mammoth seems to have been a significant part of the human subsistence during the early Palaeolithic in south-western France (Bocherens *et al.* 2005). Red deer could be also investigated for stable isotopes, but to date their isotopic signature was found to be very close to those of large bovid (e.g. Bocherens *et al.* 2011). Thus, red deer did not probably contribute to a large part of the diet of the human of Abri Pataud. Finally, input of freshwater fish in the human diet needs to be considered, but very few remains of fish have been evidenced in the site of Abri Pataud. Freshwater fish are known to exhibit a large range of $\delta^{13}C_{coll}$ and $\delta^{15}N_{coll}$ values that overlap those found for terrestrial herbivores (Dufour *et al.* 1999). Thus, the isotopic signature of freshwater fish should be defined but, so far, the available material is scarce for the study region.

Further investigations are in process to better decipher the human diet composition at the Abri Pataud. The low input of reindeer is rather surprising in comparison the archaeozoological data (Crépin 2013). Reindeer could have contributed more to the lipid part of the human diet, which is not reflected by the isotopic composition of the collagen because fat does not contain any nitrogen element. A second hypothesis is that people may have periodically visited the Abri Pataud site to kill reindeer, but their primary diet was derived from other species. Horse and mammoth appear as the most likely terrestrial contributors, while red deer is highly unlikely based on the known isotopic signature of this species.

Conclusion

Thanks to the new holistic study of faunal remains from Pataud level two, including archaeozoological, typo-technological and biogeochemical analyses, we now have a more precise understanding of the exploitation of deer during the Final Gravettian. Reindeer and red deer appear to have been exploited for multiple uses. Even if the former is more commonly represented in faunal remains, both are consumed and used as raw materials for bone industry and body ornamentation and are also represented in the funerary context.

Without challenging previous faunal data, it is clearly demonstrated that the role of deer, in particular reindeer, must be balanced. Reindeer remains are clearly the most numerous and this cervid unquestionably provided a high proportion of dietary and non-dietary resources. But, other herbivores took an important part within the meat diet (i.e. bovids) and in a more social/symbolic field (i.e. mammoth, very well represented in body ornamentation).

Moreover, biogeochemical data raise new questions about the role of deer in the human daily life as far as diet is concerned. Therefore, we need now to pursue synchronic and diachronic research to better interpret the role and importance of deer for the Final Gravettian people. Use of multidisciplinary approaches as were initiated in the study of level two at Pataud are a promising way to explore this question.

Acknowledgements

This study was conducted as a part of the archaeological project: 'The occupation of the Abri Pataud 22,000 years ago', supported by the French Ministry of Culture (DRAC Aquitaine), the Conseil Général of the Dordogne, and the National Museum of Natural History.

The authors would like to acknowledge the reviewers, who provided comments which improved the manuscript. We thank Walter Joyce (Geoscience, University of Tübingen) for his kind reviewing of the English.

References

Ambrose, S. H. (1990) 'Preparation and characterization of bone and tooth collagen for isotopic analysis', *Journal of Archaeological Science* 17, 431–451.

Bocherens, H., Billiou, D., Patou-Mathis, P., Bonjean, D., Otte, M. and Mariotti, A. (1997) 'Paleobiological implications of the isotopic signature (^{13}C, ^{15}N) of fossil mammal collagen in Scladina cave (Sclayn, Belgium)', *Quaternary Research* 48, 370–380.

Bocherens, H. and Drucker, D. (2003) 'Trophic level isotopic enrichments for carbon and nitrogen in collagen: case studies from recent and ancient terrestrial ecosystems', *International Journal of Osteoarchaeology* 13, 46–53.

Bocherens, H., Drucker, D., Billiou, D., Patou-Mathis, M. and Vandermeersch B. (2005) 'Isotopic evidence for diet and subsistence pattern of Saint-Césaire I Neanderthal', *Journal of Human Evolution* 49, 71–87.

Bocherens, H., Drucker, D. G., Bonjean, D., Bridault, A., Conard, N. J., Cupillard, C., Germonpré, M., Höneisen, M., Münzel, S. C., Napierala, H., Patou-Mathis, M., Stephan, E., Uerpmann, H. -P. and Ziegler R. (2011) 'Isotopic evidence for dietary ecology of cave lion (*Panthera spelaea*) in North-Western Europe: prey choice, competition and implications for extinction', *Quaternary International* 245, 249–261.

Bricker, H. M. (1995) *Le Paléolithique supérieur de l'abri Pataud (Dordogne): les fouilles de H. L. Movius Jr.*, Éditions de la maison des sciences de l'homme, Documents d'archéologie française, Paris, 50.

Chauvière, F.-X. and Fontana, L. (2005) 'Modalités d'exploitation des rennes dans le Protomagdalénien du Blot (Haute-Loire, France) : entre subsistance, technique et symbolique', in *Industries Osseuses et Parures du Solutréen au Magdalénien en Europe*, ed. V. Dujardin, actes de la table ronde sur le Paléolithique supérieur récent, Angoulême (28–30 mars 2003). Société préhistorique française, Paris, Mémoires de la SPF, XXXIX, 137–147.

Chiotti, L., Nespoulet, R., Henry-Gambier, D., Morala, A., Vercoutère, C. and coll. (2009) 'Statut des objets "extra-ordinaires" du Gravettien final de l'abri Pataud (Les Eyzies, Dordogne): objets abandonnés dans l'habitat ou dépôt intentionnel?' in *Du Matériel au Spirituel. Réalités Archéologiques et Historiques des "dépôts" de la Préhistoire à nos Jours*, S. Bonnardin, C. Hamon, M. Lauwers and B. Quilliec (dir.), XXIXèmes Rencontres internationales d'archéologie et d'histoire d'Antibes, Juan-les-Pins, APDCA, Antibes, 29–46.

Cho, T. S. (1998) *Etude Archéozoologique de la Faune du Périgordien Supérieur (couches 2, 3 et 4) de l'abri Pataud (Les Eyzies, Dordogne): Paléoécologie, Taphonomie, Paléoéconomie*, Thèse de Doctorat, Muséum National d'Histoire Naturelle, Paris.

Couturier, S., Côté, S. D., Huot, J. and Otto, R. D. (2009) 'Body-condition dynamics in a northern ungulate gaining fat in winter', *Canadian Journal of Zoology* 87, 367–378.

Crépin, L. (2013) 'Données archéozoologiques des grands mammifères, fouilles 2005–2008', in *Le Gravettien final de l'abri Pataud (Dordogne, France). Résultats des Fouilles et des Etudes 2005–2009*, eds R. Nespoulet, L. Chiotti and D. Henry-Gambier, BAR International Series 2458, Archaeopress, Oxford, 63–88.

DeNiro, M. J. (1985) 'Postmortem preservation and alteration of in vivo bone collagen isotope ratios in relation to palaeodietary reconstruction', *Nature* 317, 806–809.

Drucker, D. G., Bocherens and H., Billiou, D. (2003a) 'Evidence for shifting environmental conditions in south-western France from 33,000 to 15,000 years ago derived from carbon-13 and nitrogen-15 natural abundances in collagen of large herbivores', *Earth and Planetary Science Letters* 216, 163–173.

Drucker, D., Bocherens, H., Bridault, A. and Billiou, D. (2003b) 'Carbon and nitrogen isotopic composition of red deer (*Cervus elaphus*) collagen as a tool for tracking palaeoenvironmental change during the Late-Glacial and Early Holocene in the northern Jura (France)', *Palaeogeography, Palaeoclimatology, Palaeoecology* 195, 375–388.

Drucker, D. G. and Henry-Gambier, D. (2005) 'Determination of the dietary habits of a Magdalenian woman from Saint-Germain-la-Rivière in south-western France using stable isotopes', *Journal of Human Evolution* 49, 19–35.

Dufour, E., Bocherens, H. and Mariotti, A. (1999) 'Palaeodietary implications of isotopic variability in Eurasian lacustrine fish', *Journal of Archaeological Science* 26, 627–637.

Gerhart, K. L., White, R. G., Cameron, R. D. and Russell, D. E. (1996) 'Body composition and nutrient reserves of artic caribou', *Canadian Journal of Zoology* 74, 136–146.

Heaton, T. H. E. (1999) 'Spatial, species, and temporal variations in the $^{13}C/^{12}C$ ratios of C_3 plants: Implications for palaeodiet studies', *Journal of Archaeological Science* 26, 637–649.

Henry-Gambier, D., Villotte, S., Beauval, C., Bruzek, J. and Grimaud-Hervé, D. (2013) 'Les vestiges humains: un assemblage original', in *Le Gravettien final de l'abri Pataud (Dordogne, France) Résultats des fouilles et des études 2005–2009*, eds R. Nespoulet, L. Chiotti, and D. Henry-Gambier, BAR International Series 2458, Archaeopress, Oxford, 135–177.

Koch, P. L. (2007) 'Isotopic study of the biology of modern and fossil vertebrates', in *Stable Isotopes in Ecology and Environmental Science*, eds R. Michener and K. Lajtha, Wiley-Blackwell, Oxford, 99–154.

Longin, R. (1971) 'New method of collagen extraction for radiocarbon dating', *Nature* 230, 241–242.

Movius, H. L. (1977) *Excavation of the abri Pataud, Les Eyzies (Dordogne): Stratigraphy*, Cambridge, Massachussetts, Peabody Museum, Harvard University, American School of Prehistoric Research Bulletin, 31.

Nespoulet, R., Chiotti, L., Agsous, S., Guillermin, P., Grimaud-Hervé, D., Henry-Gambier, D., Lenoble, A., Marquer, L., Morala, A., Patou-Mathis, M., Pottier, C., Vannoorenberghe, A., Vercoutère, C. and Vérez, M. (2008) 'L'occupation humaine de l'abri Pataud il y a 22 000 ans: problématique et résultats préliminaires des fouilles du niveau 2', in *Les Sociétés Paléolithique dans un Grand Sud-Ouest de la France: Nouveaux Gisements, Nouveaux Résultats, Nouvelles Méthodes*, eds J. Jaubert, J.-G. Bordes and I. Ortega, Mémoire de la Société Préhistorique Française XLVII, Paris, 325–334.

Nespoulet, R., Chiotti, L. and Henry-Gambier, D. eds (2013) *Le Gravettien Final de l'abri Pataud (Dordogne, France). Résultats des Fouilles et des Etudes 2005–2009.* BAR International Series 2458, Archaeopress, Oxford.

Spiess, A. (1979) *Reindeer and Caribou Hunters: An Archaeological Study*, Academic Press, New York.

Stevens, R. E. and Hedges, R. E. M. (2004) 'Carbon and nitrogen stable isotope analysis of north-west European horse bone and tooth collagen, 40,000 BP–present: Palaeoclimatic interpretations', *Quaternary Science Review* 23, 977–991.

Stevens, R. E., Jacobi, R., Street, M., Germonpré, M., Conard, N. J., Münzel, S. C. and Hedges, R. E. M. (2008) 'Nitrogen isotope analyses of reindeer (*Rangifer tarandus*),

45,000 BP to 9,000 BP: Palaeoenvironmental reconstructions', *Palaeogeography, Palaeoclimatology, Palaeoecology* 262, 32–45.

Tieszen, L. L. (1991) 'Natural variations in the carbon isotope values of plants: Implications for archaeology, ecology, and paleoecology', *Journal of Archaeological Science* 18, 227–248.

Vercoutère, C. (2007) 'De la viande à la pendeloque – Exemple de l'exploitation du Renne dans l'occupation Gravettienne du niveau 2 de l'abri Pataud (Dordogne, France)', in *Les Civilisations du Renne d'hier et d'aujourd'hui. Approches Ethnohistoriques, Archéologiques et Anthropologiques*, eds S. Beyries and V. Vaté, Actes des XXVIIe Rencontres internationales d'archéologie et d'histoire d'Antibes, Juan-les-Pins, 19–21 octobre 2006. Editions APDCA, Antibes, 325–343.

Vercoutère, C. (2013) 'Les objets en matières dures d'origine animale', in *Le Gravettien Final de l'abri Pataud (Dordogne, France). Résultats des Fouilles et des Études 2005–2009*, eds R. Nespoulet, L. Chiotti and D. Henry-Gambier, BAR International Series 2458, Archaeopress, Oxford, 89–109.

Vercoutère, C., Müller, K., Chiotti, L., Nespoulet, R., Staude, A., Riesemeier, H. and Reiche, I. (2011) 'Rectangular beads from the Final Gravettian level of the abri Pataud: raw material identification and its archaeological implications', in *Dossier Thématique: The International ArBoCo Workshop Towards a Better Understanding and Preservation of Ancient Aone Materials*, ed. I. Reiche, *ArchéoSciences* 35, 259–271.

Deer Stones and Rock Art in Mongolia during the Second to First Millennia BC

Kenneth Lymer, William Fitzhugh and Richard Kortum

Introduction

The evocative image of the deer has been drawn upon time and time again by numerous prehistoric societies during the Bronze and Iron Ages in Central Asia and South Siberia, *c.* second to first millennia BC. Depictions of cervids are found carved in exposed bedrock at many rock art sites within the region but also conspicuously adorn Late Bronze Age monoliths and menhirs, commonly referred to as deer stones. These two differing contexts for the imagery provide tantalising clues about the beliefs and worldviews of past societies within the greater geographical region. Moreover, recent archaeological field research has not only contributed to the growing corpus of data about deer stones but has also supplemented new discoveries of rock art that bring fresher perspectives to persisting problems concerning their function and meanings.

This paper focuses mainly upon data collected from northern, central and western Mongolia and draws upon relevant correspondences with the neighbouring South Siberian and Central Asian republics of the Altai, Buryatia, Tuva, Kazakhstan and Kyrgyzstan. In particular, images of deer were carved in natural rock surfaces that spanned across this territory during the second millennium BC. Furthermore, during the later Bronze Age in Mongolia, there emerged significant practices characterised by erecting plinths and standing stones carved with stag motifs that gave rise to the category of deer stone in archaeological scholarship. These were interconnected with the construction of funerary monuments known as khirigsuurs and together form a unique mortuary and ceremonial tradition, the Deer Stone-Khirigsuur Complex (DSKC). Therefore, the following brief examination of the contexts of rock art and deer stones importantly highlights the multi-faceted nature of cervid imagery in the art, beliefs and worldviews of prehistoric societies in Central Asia and South Siberia from second to first millennium BC.

Rock art

Established archaeological chronologies of the region traditionally characterise the second millennium BC as belonging to the Bronze Age. During this period numerous scenes featuring deer were cut into natural stone (petroglyphs) at sites ranging from Kazakhstan and Kyrgyzstan to the Altai, Tuva and Mongolia (e.g. Pomaskina 1976, figs 18–29; Novgorodova 1978, figs 5–6; Maksimova *et al.* 1985, pls 61, 63; Kubarev and Jacobson 1996, figs 141, 143, 156; Devlet 1998 pls 4, 11, 14; Kortum and Tserendagva 2007, figs 4–5). Similar petroglyph scenes have been recorded since the late nineteenth century. A good example is the detailed pen and ink drawing (1888) of deer in flight engraved on a rock surface along the Kemshik River, Tuva by the Finnish archaeologist J. R. Aspelin (Appelgren-Kivalo 1931, 63). The chronology for rock art was developed in the twentieth century by Soviet archaeologists through the methodology of relative dating. Scientific dating techniques for petroglyphs have yet to produce viable results; thus age assignments have mainly been derived through stylistic comparisons based upon established archaeological sequences. Though the term 'Bronze Age' used in the context of rock art is an imperfect one (Lymer 2010), it provides a general framework that is broadly valid. Even so, we must be mindful of its inherent problems.

The deer, like all other petroglyph images, were pecked, carved and incised and take part in a wide variety of scenes, whether depicted in motion or standing still (Figure 13.1A–D). They are also associated in varying combinations with other elements such as humans, animals or various geometric shapes and motifs (Figure 13.1A–B). Moreover, the images of deer feature a wide variety of artistic styles with varying degrees of recognisable anatomical details.

Some images have distinguishable antler morphologies of adult males that indicate they may be representations of the Central Asian red deer (*Cervus elaphus*), which is frequently referred to as a maral in the literature. However, the present day distribution of the maral is considered by some zoologists to lie westward in the Caucasus and Asia Minor, as it is a subspecies of the European red deer and known as the Caucasian deer, *Cervus elaphus maral* (Geist 1998, 183; Wrobel 2007, 90). Overall, scenes commonly feature depictions of mature stags, as older male cervid antlers start growing in the spring immediately after shedding the previous ones and develop into impressive multi-tined beams by late summer (see Geist 1998). Antlers begin to emerge as two beams on a young adult by its second year and as the stag grows older, generally, its antlers grow bigger and produce more tines with each successive year. On occasion rock art images are found of antlerless deer that could be viewed as hinds (females), hummels (adult males that are unable to produce antlers) or juveniles of either sex; even so, for the most part the image of the stag with branching antlers predominates.

The scenes in Figure 13.1 are more than simple encounters with beasts of prey. Hunting activities in traditional cultures typically involve interactions entangled

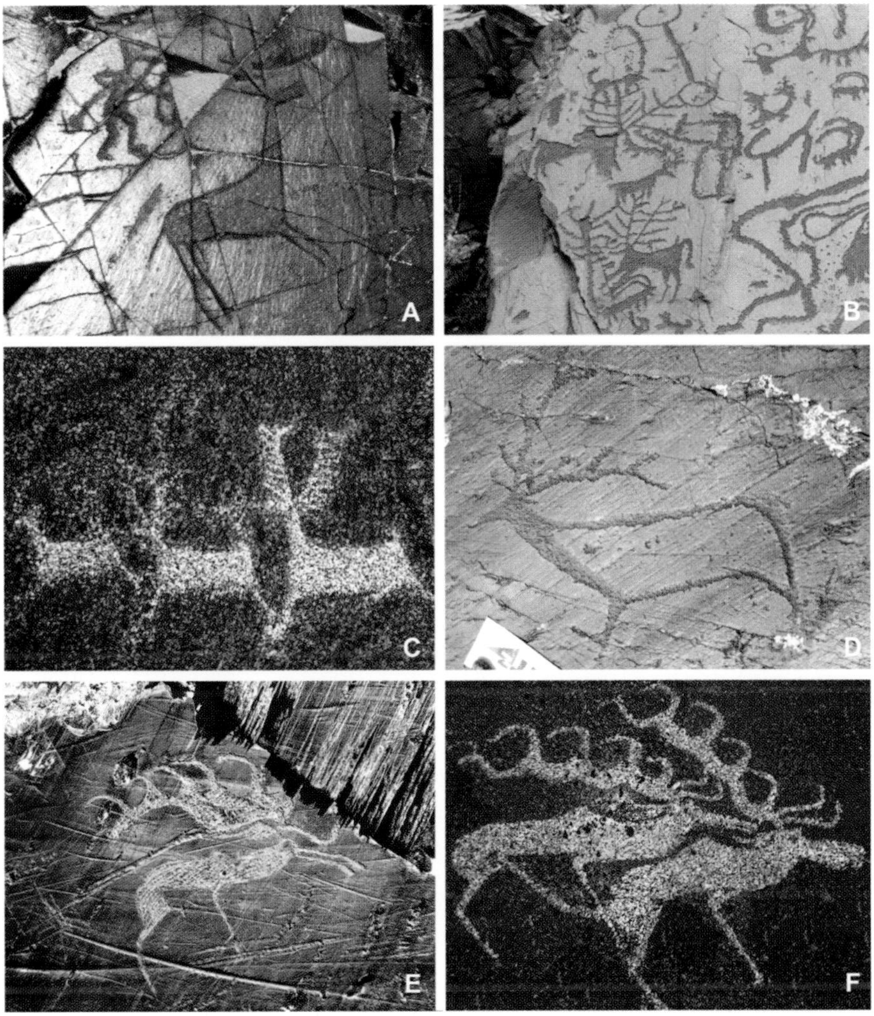

FIGURE 13.1. Rock art images of deer: A. Bronze Age scene from Tamgaly, Kazakhstan (photograph K. Lymer); B. Bronze Age scene from Saimaly Tash, Kyrgyzstan (photograph K. Lymer); C. Bronze Age scene from Boregtiin Gol, Bayan Ölgii, Mongolia (photograph by R. Kortum); D. Bronze Age deer from Biluut, Bayan Ölgii, Mongolia (photograph R. Kortum); E. Mongolian deer (LBA) from Biluut, Bayan Ölgii, Mongolia (photograph K. Lymer); F. Mongolian deer (LBA) from Boregtiin Gol, Bayan Ölgii, Mongolia (photograph R. Kortum).

within religious practices. Among Siberian peoples, the shamanic practitioner is assisted by spirit-helpers in the form of animals, which sometimes appear as deer (Basilov and Zhukovskaya 1989, 162). Furthermore, spirit masters found in the landscape preside over wild animals; these must be negotiated by the shaman on behalf of the community to continue the ongoing access to deer and other game. Thus, hunting needs to be considered as part and parcel of a host of other activities engaged in by a community in its interactions with the landscape, seeing that hunting is itself embedded within complex social and religious interactions (Lymer 2009, 233). Moreover, these ancient rock art scenes of stags may have been carved as part of an ancient society's daily activities – practices that assisted in maintaining the well-being of the wild deer, while at the same time they negotiated with the spirit masters of the environment. Hence, the petroglyph image of the deer constitutes a powerful medium through which people and spirits interact within the landscape (Lymer 2002, 91)

Deer stones

In the later Bronze Age in Mongolia there emerged the practice of decorating standing stones with a distinctive stylised image of the stag. The majority of these deer stones are concentrated in central and northern Mongolia and extend north into the Lake Baikal region as well as west into western Mongolia and the Altai Republic (Bayarsaikhan 2011, 21). A small number have also been reported in the Altai Mountains of Xinjiang province, China (e.g. Mu Shunying *et al.* 1994, 131). Decorated standing stones have been erected by prehistoric societies in South Siberia since at least the third millennium BC (Okunevo archaeological culture); however, the direct antecedents of Mongolian stelae in the later Bronze Age are still unknown (Fitzhugh 2009a, 196). Some suggest they may have been originally executed in wood (Ol'khovskii 1989, 60; Jettmar 1994, 4).

Deer stones have been recognised by researchers in this part of the world since the nineteenth century. One of the earliest discoveries was the Ivolga Stone collected in 1856 by D. P. Davydov not far from present day Ulan Ude, Buryatia. Its significance, however, was not realised until A. P. Okladnikov wrote his famous article in *Sovetskaya Arkheologiya* (1954), which rediscovered this deer stone residing in the State Historical Museum, Irkutsk. In addition, J. R. Aspelin recorded two 'stone pillars' engraved with deer standing on tiptoes during his travels around Tuva in 1888 (Appelgren-Kivalo 1931, figs 329, 331), while V. V. Radlov (1892) published drawings of standing stones with 'classic' Mongolian deer figures found in the Orkhon River valley, northern Mongolia.

During the Soviet period, archaeologist V. V. Volkov (1981) developed the broadly accepted typology, which recognises three distinct regional variations of deer stones:

1. The *Mongolian-Transbaikal* or *Mongolian type* (Figures 13.2A and 13.3B) spreads from Mongolia to Lake Baikal, with the most famous examples found clustered in northern and central Mongolia. All four sides of the stone are carved with images, especially the distinct stylised form of a deer with swirling antlers and folded legs. Other elements accompany these figures, including weapons, shields and wild animals. A handful of stelae have human faces sculpted out of the top of the stone (Figure 13.3C).

2. The *Sayan-Altai type* (Figure 13.2B) is distributed across the Sayan and Altai Mountain ranges; it clusters mainly within Tuva and the Altai Republic. These characteristically have fewer engravings, especially when compared to the Mongolian stones; images of stags and wild animals are usually depicted with legs extended, as if standing on tiptoes. Images of weapons, shields, circular shapes and geometric designs are also typical of these stones.

3. The *Eurasian type* of standing stone (Figure 13.2C) is classified as a deer stone, but in fact has no images of cervids or other animals. It features a minimal amount of engravings in the form of simple belt-lines with hanging weapons, necklaces, parallel slashes and circular rings. They are also found in the Sayan

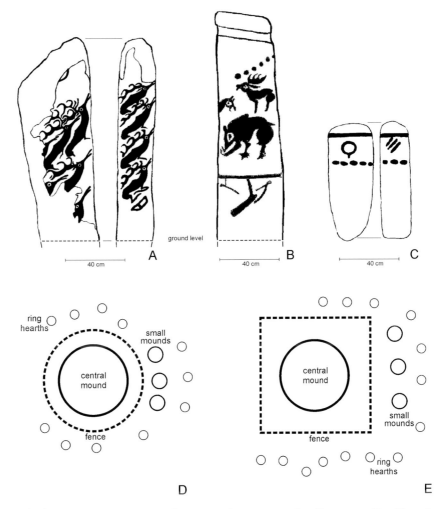

FIGURE 13.2. Deer stone and khirigsuur illustrations:
A. Mongolian type deer stone from Khushuutiin Devseg, Khövsgöl;
B. Sayan-Altai type deer stone from Urd Khuurai, Arkhangai; C. Eurasian type stone from Sairiin Tal, Bayan Ölgii;
D. Diagram of a circular 'fenced' khirigsuur;
E. Diagram of a square 'fenced' khirigsuur.

and Altai Mountain regions, but extend west into the Caucasus. The Eurasian stones from the Caucasus, however, bear little resemblance to the 'classic' Mongolian deer stones other than being menhirs with some combination of belts, necklaces and diagonal slashes. Cultural connections between the two are debatable (Fitzhugh 2009b, 386).

It is important to note that there are areas of considerable overlap, as all three deer stone types can be found in the same region – sometimes at the very same site. In particular, numerous examples of all three styles have recently been found in different localities in the Altai Mountains of Mongolia's westernmost Bayan Ölgii province, which borders China (Fitzhugh and Bayarsaikhan 2009; Kortum 2009). Researchers have also recognised for years that this wide geographical spread of deer stones from Mongolia to the Altai Republic and Tuva demonstrates they were not exclusively associated with just one archaeological culture (e.g. Volkov 1967, 69).

FIGURE 13.3. Deer stone and khirgisuur photographs: A. Uushikiin Övör, Khövsgöl (photograph W. Fitzhugh); B. Uushikiin Övör deer stone menhir (photograph W. Fitzhugh); C. Uushikiin Övör deer stone stele with human face (photograph W. Fitzhugh); D. Circular 'fenced' khirigsuur near the Altai Mountains at Biluut, Bayan Ölgii (photograph K. Lymer).

The 'classic' Mongolian type can be either a rectangular plinth shaped from granite (Figure 13.3C) or a menhir derived from a plate of slate-like greywacke (Figure 13.3B). Their sizes generally range around 1.8 m tall with some 1 m or less, although larger stones can reach 3–4 m, such as deer stone no. 2 at Ulaan Tolgoi and deer stone no. 14 at Uushikiin Övör (Figure 13.3C), both from Khövsgöl province.

Mongolian deer stones are renowned for their evocative carvings of the Mongolian stag – a highly stylised depiction of a cervid in profile that features an unusually long snout, a bird-like head with widened circular eye and massive antlers with curled tines flowing over its back, while its vestigial legs are tucked under the torso. This iconic image is most likely based on the red deer (Erdeii 1978, 139); however, some scholars believe it represents the reindeer, *Rangifer tarandus* (e.g. Jacobson 1993, 157, 169; Vitebsky 2005, 6). Gryaznov (1984, 76), on the other hand, asserts that this latter proposition is not plausible.

Nevertheless, this highly stylised representation is neither a photographic snapshot nor an anatomically correct drawing. The swirling antlers, saucer eye and long thin snout in essence illustrate the embodiment of a cosmological worldview about deer for societies of the Late Bronze Age.

While the Mongolian deer image is highly standardised, its application to the stone surface varies considerably in terms of size, position, direction/ orientation and number of cervids depicted together. Characteristically, these 'classic' Mongolian deer are placed tightly side by side; their upwards tilt across the surface of the stone imparts a definite sense of movement. This has been interpreted as representing the spirit of the deer sprinting towards the heavens, thereby making an association with a shaman's flight into other spirit realms (Magail 2005, 178).

In addition to the Mongolian deer, these standing stones feature circles, zigzags, stripes, tools, knives, bows, axes, shields, wild animals (such as caprids, suids and felid predators) and occasionally equids. The Mongolian deer stone's surface can be densely covered with these images, especially when compared to the Sayan-Altai type, which is sparser in its choice of deer and other icons. Significantly, Sayan-Altai stones exhibit a different style of stag: the cervid, with a blunt snout, is shown in profile with elongated legs, as though it were standing on tiptoes. Overall, each individual menhir or monolith possesses a unique composition regardless of type. Designs vary not only from stone to stone but from site to site across the greater geographical region.

The assemblage of specific weapon and tool icons on deer stones strongly suggests that deer stones represent specific individuals. It has been argued by many scholars that images of weapons indicate these decorated plinths functioned as cenotaphs for chiefs or warriors (e.g. Okladnikov 1954; Dikov 1958; Chlenova 1962; Erdeii 1978). Animal figures carved onto the sides of deer stones are believed by many to replicate ornaments embroidered or sewn onto a warrior's costume and/or tattoos emblazoned upon the human body (Erdeii 1978, 140; Kubarev 1979, 82). Additionally, deer stones with sculptured human faces give rise to the suggestion that these monuments were representations of special ancestors (Maidar 1981, 39) or, because of their rounded open mouths, chanting shamans (Fitzhugh 2009b, 388).

The aforementioned repertoire of images has also provided the basis for dating them by comparison with artefacts derived from the later Bronze Age (Karasuk) or Iron Age (Scytho-Siberian) (e.g. Borovka 1927, Okladnikov 1954; Dikov 1958; Maidar 1981; Volkov 1981). It was not until the beginning of the twenty-first century, however, that deer stones were reliably dated using radiocarbon techniques. Hundreds of deer stones have been found across northern, central and western Mongolia, but very few excavations have examined their archaeological contexts in close detail. The Smithsonian Institution coordinated several seasons of intensive archaeological fieldwork in a joint Mongolian-American research project that produced new data and suitable contexts for radiocarbon dating. Since 2002 the project has scientifically dated over twenty horse-head burials directly associated with Mongolian deer stones, which range between 1200 and 700 cal BC (see Fitzhugh 2009c, table 1). Moreover, this has brought new clarity to the relationship between deer stones and other Late Bronze Age monuments known as khirigsuurs. These dates demonstrate that these two distinct forms are, in reality, interconnected facets of a singular mortuary and ceremonial tradition.

Khirigsuurs

Researchers have known since the nineteenth century that deer stones were closely associated with ancient stone mounds called khirigsuurs. V. V. Radlov's drawings in his *Atlas of Ancient Mongolia* (1892) clearly demonstrate that Mongolian deer stones are associated with various kinds of constructions, including graves and circular mounds. These mounds are referred to in Soviet studies as *khereksurs*. Unfortunately, there is no definitive explanation for the origin of this term, although the contemporary Mongolian form of the word, 'khirigsuur', can be translated as 'Kyrgyz burial' (Maidar 1981, 46; Honeychurch *et al.* 2009, 80).

Khirigsuurs are stone structures featuring central burial mounds surrounded by stone 'fences' (Figure 13.2D–E). Mounds can reach up to 40 m in diameter, while their bordering 'fences' have circular or square perimeters. The central mound is made of a pile of stones and usually covers a human burial placed in a shallow pit or stone slab box. Most burials are found within the upper 0.5 m of the ground, or were laid directly upon the surface. The poor preservation of organics – mostly the result of unfavourable soil conditions – could conceivably account for the lack of human remains within some mounds. Grave goods are rarely found and only small items, such as bronze buttons or belt buckles, have been recovered. This type of burial practice contrasts sharply with later kurgan structures of the Iron Age (seventh to third centuries BC in Central Asia), where large funerary chambers made of logs contain lavishly interred bodies beneath stone mounds, such as the Early Saka kurgans at Shilikty, eastern Kazakhstan (Chernikov 1965) and the Pazyryk kurgans at the famous site of Pazyryk in the Altai Republic (Rudenko 1970).

Typically, in northern and central Mongolia, khirigsuurs are directly associated with small stone rings apparently used as hearths and smaller peripheral mounds that contain sacrificial offerings of horses (Fitzhugh 2009c, 78). The hearth stone ring is usually 1–1.5 m in diameter and contains charcoal and calcined goat or sheep bones. The small horse mounds usually contain an east-facing equine skull with articulated maxilla and mandibles. Cervical vertebrae and hooves usually accompany the skull; more rarely, ceramic fragments and bones of other domesticated animals are present. Larger khirigsuur structures can have several rows of horse-head burial mounds clustered in a dense concentration on the east side, while the largest ones, measuring 200–400 m across, may have hundreds. Urt Bulagyn, for example, in Mongolia's Arkhangai province, possesses over 1,700 horse-head mounds (Allard and Erdenebaatar 2005, 549).

Khirigsuurs differ in size, construction and terrestrial design; they range from small mounds and 'fences' to large groups of mounds amidst multi-component stone constructions. The largest khirigsuurs and most complex structures are situated on valley floors, while smaller and simpler constructions are found along the flanks of hills and upper slopes or else constructed in valley margins closer to hillsides (Frohlich and Bazarsad 2005; Fitzhugh 2009a, 191). Khirigsuur complexes possess a high degree of architectural intricacy, involving an arrangement of satellite mounds, hearths, stone pavements ('plazas') and other stone structures that are distinct Late Bronze Age phenomena unique to Mongolia.

The Smithsonian sponsored Mongolian-American project produced radiocarbon dates of features directly associated with Mongolian deer stones that clearly demonstrate their contemporaneity with khirigsuurs (Fitzhugh 2009b; Fitzhugh 2009c). The Deer Stone-Khirigsuur Complex, therefore, is a Late Bronze Age mortuary and ceremonial tradition which dates *c.*1350–750 cal BC (see Fitzhugh and Bayarsaikhan 2009, table 13.1). It is important to note, however, that not all deer stones are directly associated with khirigsuurs. On the other hand, deer stones connected with khirigsuurs are erected as singular stones or in groups (with other stones without imagery), which are sometimes arranged in rows. The most famous example is found at Uushikiin Övör, Khövsgöl province (Figure 13.3A–C), where deer stones are arranged on a north–south axis in two rows directly to the west of a large khirigsuur (Volkov and Novgorodova 1975, fig. 1). Additionally, recent field surveys in western Mongolia have found that on occasion deer stones, menhirs and small upright stones are embedded in khirigsuur mounds (Kortum 2009, figs 8, 10, 15, 23–26).

Deer stones and khirigsuurs possess a structural similarity: for instance, both have associated periphery features of horse-head burials and feasting hearths (Fitzhugh 2009a, 191). Smithsonian excavations reconfirmed the earlier observation that, characteristically, deer stones are not directly associated with human remains; they also demonstrate conclusively the relative absence of associated artefacts (Fitzhugh 2009b, 403). Oval hearths accompanying deer stones may be the remains of feasts associated with dedication ceremonies, while khirigsuur hearths seem to have played a role in elaborate funeral rituals. The number of horse deposits roughly correlates with the size of a khirigsuur. There is, thus, an indication that higher status individuals occupy larger khirigsuurs and are accompanied by a greater number of offerings. The frequent occurrence of deer stones directly associated with khirigsuur complexes strongly suggests that they belong to a cultural practice of honouring significant individuals in society. Hence, they can be seen as two types of ceremonial activity operating within a greater socio-religious system. One commemorated the dead interred under stone mounds, while the other honoured ancestors through an association with deer stelae.

Other contexts

In addition to the DSKC, evidence from rock art sites provides another important clue to the contextual complexity of deer motifs. The distinct forms of the Mongolian stag (Figure 13.1E–F) and deer on tiptoes were not limited to deer stones; they are also found in the natural landscape on exposed bedrock surfaces at petroglyph sites from Kazakhstan and Kyrgyzstan to the Altai Republic, Tuva and Mongolia (e.g. Gaponenko 1963, figs 5–6; Kubarev 1979, pl. XI; Devlet 1998, fig. 15; Kortum and Tserendagva 2007, fig. 8; Kortum 2009, figs 17, 20, 29, 30; Kortum 2012, fig. 9). Furthermore, the 'classic' Mongolian deer stone has not yet been found in Kazakhstan, but there are a few known rock art depictions of Mongolian deer (Samashev 2001, figs 33–36).

These acts of carving petroglyphs in hillsides indicate not only that deer imagery was connected to intimate personal encounters with special places in the landscape, but that, in addition, this was also part and parcel of public clan or lineage rituals embodied by deer stones and khirigsuurs. These pecked and incised scenes of stags may have been created as individual responses to a wider sacred landscape involving activities that included negotiating with spirits of the environment (Lymer 2002, 89–93). There is evidence to suggest that deer may have also performed the role of an intermediary. For example, among some Mongol groups the shaman's drum is called the 'black stag' (Heissig 1944, 47; see also Bayarsaikhan 2005). This instrument serves as an important go-between between the practitioner and the world of the spirits. Moreover, the peculiar stylisation of the Mongolian deer, with its bird-like head, wide circular eyes, open mouth (perhaps calling or singing), flowing antlers and legs folded, as if in flight, strongly suggests the form of a spirit transformation figure – a clearly recognisable aspect of the world of Siberian and circumpolar shamans (Fitzhugh 2009a, 192).

Archaeological excavations have unearthed very few artefacts associated with the Late Bronze Age DSKC in Mongolia. Yet the evidence from the rock art shows that these motifs were not exclusive to standing stones (Figure 13.1E–F). It is certainly conceivable that cervid imagery decorated many other types of items, as well as objects made of perishable materials, such as wood, leather and textiles. It is important to note that, conversely, there is substantial evidence of deer images being used to adorn objects of material culture during the early part of the first millennium BC in Central Asia and South Siberia. A striking example is a bronze mirror from the Frolov Collection, *c.* seventh century BC, found in the Altai Republic, which features six outlined figures of tiptoed deer radiating from around the centre (Kiryushin and Tishkin 1997, fig. 66). Another fine example was found with a richly outfitted couple who were buried together in the great kurgan of Arzhan-2, which belongs to the Aldybel archaeological culture of Tuva, *c.* seventh to sixth centuries BC (Chugunov *et al.* 2004). The male wore a headdress adorned with an exquisite golden plaque of a stag on tiptoes, while the female had a golden headdress pin with a tiny version of a similar figure mounted on its pinnacle (Čugunov *et al.* 2006, pls 5, 49–52).

Objects recovered from Arzhan-2 also demonstrate the prominent role that the stag plays in biographical objects intimately associated with the lives of their owners. Thus, these stylised images of deer were not only images of power, but must have functioned also as symbols of rank and prestige.

Closing remarks

The image of the deer has been drawn upon repeatedly, if not continuously, by prehistoric societies across Central Asia, South Siberia and Mongolia over a very long period of time, from second to first millennium BC, at least. Moreover, this deer imagery was deeply embedded in many socio-cultural contexts, from personal decorations and objects of material culture to rock art, deer stones and khirigsuur ceremonial and funerary structures. Petroglyphs, standing stones and khirigsuurs all played an active and meaningful part in the lives of individuals. They were undoubtedly created in specific socio-religious contexts; but at the same time, they were essential components of activities performed within and in response to the landscape on the part of different communities.

The deer stones of Mongolia, Tuva and the Altai Mountains display a variety of distinguishable styles that are situated within varying architectural contexts and archaeological cultures. Deer stones and khirigsuurs appeared in northern Mongolia around 1300 BC; they flourished for several hundred years, with variations extending north into Tuva and west into the high Altai. By the beginning of the early first millennium BC changes and shifts in cultures, values and practices led ultimately to the disappearance of the iconic Mongolian deer motif. However, as the practices involving the 'classic' deer stone come to an end, an abundance of evidence points to an entanglement of a modified stag motif with people's identities through its use to decorate and/or empower personal objects and the human body. In their twilight the artistic legacy of deer stones was nevertheless profound. Their considerable reverberations spread widely as stylised images of cervids travelled across Siberia and Central Asia with subsequent Iron Age (Scytho-Siberian) cultures that spanned the Eurasian steppes. This array of interconnections not only demonstrates the power of the deer imagery, but also shows how it was inextricably interwoven into cultural, social and religious beliefs and practices throughout the Bronze and Iron Ages of Central Asia and South Siberia.

References

Allard, F. and Erdenebaatar, D. (2005) 'Khirigsuurs, ritual and mobility in the Bronze Age of Mongolia', *Antiquity* 79, 547–563.

Appelgren-Kivalo, H. (1931) *Alt-Altaische Kunstdenkmäler: Briefe und Bildermaterial von J. R. Aspelins Reisen in Sibirien und der Mongolei 1887–1889*, K. F. Puromies Buchdruckerei, Helsingfors.

Basilov, V. N. and Zhukovskaya, N. L. (1989) 'Religious beliefs', in *Nomads of Eurasia*, ed. V. N. Basilov, University of Washington Press, Seattle and London, 160–181.

Bayarsaikhan, J. (2005) 'Shamanistic elements in Mongolian deer stone art', in *The Deer Stone Project: Anthropological Studies in Mongolia 2002–2004*, eds W. W. Fitzhugh, J. Bayarsaikhan and P. K. Marsh, Arctic Studies Center and the National Museum of Mongolian History, Washington DC and Ulaanbaatar, 41–45.

Bayarsaikhan, J. (2011) 'Brief introduction of the Mongolian deer stone culture', in *Deer stones of the Jargalantyn am*, eds Ts. Turbat, J. Bayarsaikhan, D. Batsukh and N. Bayarhuu, Mongolian Tangible Heritage Association, Ulaanbaator, 18–41.

Borovka, G. (1927) 'Arkheologicheskoe obsledovanie srednego techeniya r. Toly', in *Severnaya Mongoliya, Tome 2. Predvaritel'nye otchety lingvisticheskoi i arkheologicheskoi ekspeditsii o rabetakh, proizvedennykh v 1925 godu*, Leningrad, 43–44.

Chernikov, S. S. (1965) *Zagadka zolotogo kurgana*, Nauka, Moscow.

Chlenova, N. L. (1962) 'Ob olennykh kamnyakh Mongolii i Sibiri', in *Mongolskii Arkheologicheskii Sbornik*, ed. S. V. Kiselev, Akademii Nauk SSSR, Moscow, 27–35.

Chugunov, K. V., Parzinger, H. and Nagler, A. (2004) 'Chronology and cultural affinity of the kurgan Arzhan-2 complex according to archaeological data', in *Impact of the Environment on Human Migration in Eurasia*, eds E. M. Scott, A. Alekseev and G. Zaitseva, Nato Science Series: IV: Earth and Environmental Sciences, Volume 42. Kluwer Academic Publishers, Dordrecht, 1–7.

Čugunov, K. V., Parzinger, H. and Nagler, A. (2006) *Der Goldschatz von Aržan. Ein Fürstengrab der Skythenzeit in der südsibirischen Steppe*, Schirmer/Mosel Verlag GmbH, München.

Devlet, M. A. (1998) *Petroglify na dne Sayanskogo Morya (gora Aldy-Mozaga)*, Pamyatniki Istoricheskoi Mysli, Moscow.

Dikov, N. N. (1958) *Bronzovyi vek Zabaikal'ya*, Nauka, Ulan-Ude.

Erdeii, I. (1978) 'Nekotorye itogi rabot Mongol'sko-Vengerskoi ekspeditsii', in *Arkheologiya i Etnografiya Mongolii*, ed. A. P. Okladnikov, "Nauka" Sibirskoe Otdelenie, Novosibirsk, 136–151.

Fitzhugh, W. W. (2009a) 'The Mongolian Deer Stone-Khirigsuur Complex: Dating and Organization of a Late Bronze Age Menagerie', in *Current Archaeological Research in Mongolia*, eds J. Bemmann, H. Parzinger, E. Pohl and D. Tseveendorzh. Bonn Contributions to Asian Archaeology Volume 4. Vor- und Frühgeschichtliche Archäologie, Rheinische Friedrich-Wilhelms-Universität, Bonn, 183–199.

Fitzhugh, W. W. (2009b) 'Deer stones and khirigsuurs: pre-Scythian Bronze Age ceremonialism and art in northern Mongolia', in *Social Complexity in Prehistoric Eurasia: Monument, Metals, and Mobility*, eds B. Hanks and K. Linduff, CUP, Cambridge, 378–411.

Fitzhugh, W. W. (2009c) 'Stone shamans and flying deer of Northern Mongolia: Deer goddess of Siberia or chimera of the Steppe?', *Arctic Anthropology* 46, 72–88.

Fitzhugh, W. W. and Bayarsaikhan, J. (2009) *American-Mongolian Deer Stone Project: Field Report 2008*, Arctic Studies Center, Smithsonian Institution and National Museum of Mongolia, Washington DC and Ulaanbaatar.

Frohlich, B. and Bazarsad, N. (2005) 'Burial mounds in Hovsgol aimag, northern Mongolia: Preliminary results from 2003 and 2004', in *The Deer Stone Project. Anthropological Studies in Mongolia 2002–2004*, eds W. Fitzhugh, J. Bayarsaikhan and P. K. Marsh, Arctic Studies Center, Smithsonian Institution and National Museum of Mongolia, Washington DC and Ulaanbaatar, 57–88.

Gaponenko, V. M. (1963) 'Naskal'nye izobrazheniya Talasskoi doliny', in *Arkheologicheskie pamyatniki Talasskoi doliny*, ed. P. N. Kozhemyako, Akademii Nauk Kirgizskoi SSR, Frunze, 101–110.

Geist, V. (1998) *Deer of the World: Their evolution, behaviour, and ecology*. Stackpole Books, Mechanicsburg, Pennsylvania.

Gryaznov, M. P. (1984) 'O monumental'nom iskusstve na zare skifo-sibirskikh kul'tury v stepnoi azii', *Arkheologicheskii Sbornik* 25, 76–82.

Heissig, W. (1944). 'Schamanen und Geisterbeschwörer im Küriye-Banner', *Folklore Studies* 3, 39–72.

Honeychurch, W., Fitzhugh, W. W. and Amartuvshin, C. (2009) 'Precursor to empire: Early cultures and Prehistoric peoples', in *Genghis Khan and the Mongol Empire*, eds W. W. Fitzhugh, M. Rossabi and W. Honeychurch, Dino Inc., the Mongolian Preservation Foundation and Arctic Studies Center, Smithsonian Institution, Washington DC, 74–83.

Jacobson, E. (1993) *The Deer Goddess of Ancient Siberia: A Study in the Ecology of Belief*, E. J. Brill, Leiden.

Jettmar, K. (1994) 'Body-painting and the roots of the Scytho-Siberian animal style', in *The Archaeology of the Steppes: Methods and Strategies*, ed. B. Genito, Istituto Universitario Orientale, Dipartimento di Studi Asiatici, Naples, 3–15.

Kiryushin, Yu. F. and Tishkin, A. A. (1997) *Skifskaya epokha Gornogo Altaya, Chast 1. Kul'tura naseleniya v ranneskifskoe vremya*, Altaiskii Gosudarstvennyi Universitet, Barnaul.

Kortum, R. (2009) 'An initial survey of southern Bayan Ulgii Aimag, June 2–8, 2008', in *American-Mongolian Deer Stone Project: Field Report 2008*, eds W. Fitzhugh and J. Bayarsaikhan, Arctic Studies Center, Smithsonian Institution and National Museum of Mongolia, Washington DC and Ulaanbatar, 166–193.

Kortum, R. (2012) 'Latest rock art research at Khoton Lake, summer 2011', in *Rock Art & Archaeology: Investigating Ritual Landscape in the Mongolian Altai, Field Report 2011*, eds W. Fitzhugh, R. Kortum and J. Bayarsaikhan, Arctic Studies Center, Smithsonian Institution and National Museum of Mongolia, Washington DC and Ulaanbatar, 108–114.

Kortum, R. and Tserendagva, Y. (2007) 'Boregtiin Gol: A new petroglyph site in Bayan Olgii Aimag, western Mongolia', *International Newsletter on Rock Art* 47, 8–15.

Kubarev, V. D. (1979) *Drevnie izvayaniya Altaya (Olennye kamni)*, "Nauka" Sibirskoe Otdelenie, Novosibirsk.

Kubarev, V. D. and Jacobson, E. (1996) *Répertoire des Pétroglyphes d'Asie Centrale Fascicule No. 3: Sibérie du Sud 3: Kalbak-Tash I (République de l'Altai)*, Mémoires de la Mission Archéologique Francaise en Asie Centrale Tome v.3, Diffusion de Boccard, Paris.

Lymer, K. (2002) 'The deer petroglyphs of Arpauzen, South Kazakhstan', in *Spirits and Stones: Rock art and shamanism in Central Asia and Siberia*, eds A. Rozwadowski and M. M. Kośko, Instytut Wschodni, Uniwersytet im. Adama Mickiewicza, Poznań, 80–95.

Lymer, K. (2009) 'Maps and Visions: shamanic voyages and rock art images in the Republic of Kazakhstan', in *Proceedings of the Ninth Conference of the European Society for Central Asian Studies*, eds T. Gacek and J. Pstrusińska, Cambridge Scholars Publishing, Newcastle-Upon-Tyne, 226–246.

Lymer, K. (2010) 'Rock Art and Religion: The percolation of landscapes and permeability of boundaries at petroglyph sites in Kazakhstan', *Diskus* 11, http://www.basr.ac.uk/diskus/diskus11/lymer.htm

Maidar, D. (1981) *Pamyatniki istorii i kul'tury Mongolii*, Tome VIII. Mysl', Moscow.

Magail, J. (2005) 'L'art des "pierres à cerfs" de Mongolie', *Arts Asiatiques* 60, 172–180.

Maksimova, A., Ermolaeva, A. and Mar'yashev, A. (1985) *Naskal'nye izobrazheniya urochishcha Tamgaly*, Öner, Alma-Ata

Mu Shunying, Qi Xiaoshan and Zhang Ping (1994) *Zhongguo Xinjiang gudaiyishu*, Xinjiang Meishu sheying chubanshe, Urumqi.

Novgorodova, E. A. (1978) 'Drevneishie izobrazheniya kolesnits v gorakh Mongolii' *Sovetskaya Arkheologiya* 4, 192–206.

Okladnikov, A. P. (1954) 'Olennyi kamen reki Ivolgi', *Sovetskaya Arkheologiya* 19, 207–220.

Ol'khovskii, V. S. (1989) 'Olennye kamni (k semantike obraza)', *Sovetskaya Arkheologiya* 1, 48–62.

Pomaskina, G. A. (1976) *Kogda bogi na zemle… (Naskal'naya galereya Saimaly-Tasha)*, "Kyrgyzstan", Frunze.

Radlov, V. V. (1892) *Altas drevnostei Mongolii*, Trudy Orkhonskoi Ekspeditsii, Imperatorskoi Akademii Nauk, Sankt-Peterburg.

Rudenko, S. I. (1970) *Frozen Tombs of Siberia: The Pazyryk burials of Iron Age horsemen*, J. M. Dent and Sons, London.

Samashev, Z. (2001) 'Petroglyphs of Kazakhstan', in *Petroglyphs of Central Asia,* ed. K. Tashbayeva, Bishkek, 151–219.

Vitebsky, P. (2005) *The Reindeer People: Living with Animals and Spirits in Siberia*, HarperCollins, London.

Volkov, V. V. (1967) *Bronzovyi i rannyi zheleznyi vek severnoi Mongolii*, Shinjlekh Ukhaany Akademiin Khevlel, Ulaanbaatar.

Volkov, V. V. (1981) *Olennye kamni Mongolii*, An MNR, Ulan-Bator.

Volkov, V. V. and Novgorodova, E. A. (1975) 'Olennye kamin Ushkiin-Uvera (Mongoliya)', in *Pervodytnaya arkheologiya Sibiri*, ed. A. M. Mandel'shtam, "Nauka" Leningradskoe Otdelenie, Leningrad, 78–84.

Wrobel, M. (2007) *Elsevier's Dictionary of Mammals: in Latin, English, German, French and Italian*, 1st edition, Elsevier, Amsterdam and London.

Zooarchaeological Analyses
from the Roman and Medieval UK

Chasing Sylvia's Stag: Placing Deer in the Countryside of Roman Britain

Martyn G. Allen

Introduction

Roman attitudes to deer were complex. Stags played a prominent role in contemporary literature (Virgil, *Aeneid* 7.475–571; Ovid, *Metamorphosis* 3.177–252), they were representative of the Goddess Diana, symbolic of baptism in later Christian contexts (Gilhus 2006, 104) and the keeping of live deer was an important signifier of elite status (Anderson 1985, 83–121; Starr 1992). Deer, in the classical worldview, were animals which could transcend the boundary separating wildness and domesticity, and they were used in a number of literary contexts as a device to emphasise conflicting anthropological dichotomies: love/hate, sex/violence (see Greene 1996). The importance of deer to Roman literature led Wofford (1992, 441) to proclaim that 'the stag [w]as an emblem of a dangerous duality (of nature and culture) in the Italian world'.

As the Roman Empire expanded into new territories such as Britain, the exchange of contrasting worldviews impacted on existing cultures, stimulating new ideas surrounding the social roles played by different animals (Sykes 2009). For example, the introduction of fallow deer (*Dama dama*) to the high-status settlement at Fishbourne Palace in southern England during the first century AD has been interpreted as evidence for a deer park (a *vivarium*), a landscape feature which must have been very alien to local people, perpetuating imperial ideals and demonstrating elite power (Sykes *et al.* 2006). It also articulates more broadly with Roman attitudes towards nature, where political conquests and the social power of the elite were intimately tied to control and domination over the natural world, from the siting and construction of the villa estate (Purcell 1987, 187–203) to the death and destruction of the wild beast in the arena (Lindstrøm 2010).

It is perhaps surprising then, considering the prominence of deer in Roman culture, that cervid remains are rarely viewed with reference to wider social contexts. When identified, deer bones generally inspire little discussion in faunal reports regarding their importance to people, at best eliciting comments noting the presence of nearby woodland (for useful discussions see Grant 1989, 144; Maltby 2010, 269–272). The paucity of cervid bones from Romano-British sites has led Grant (1981, 205–207) to argue that deer were exploited as a risk-buffering strategy: venison was consumed when economic and ecological pressures called for it, i.e. where agricultural systems had 'broken down' leading to 'a need to supplement the meat supply'. Such a hypothesis is firmly grounded within traditional economic zooarchaeology, where animals are cleaved from the social realm and their remains are limited to answering a select range of time-honoured research questions (see Dobney 2001, 36).

I do not subscribe to the view that deer-hunting was solely a risk-buffering strategy of Romano-British society because such perspectives ignore the possibility that venison was not automatically classified as food (Hamilakis 2003, 239). It also neglects wider cultural perceptions and categories that may have been attached to deer as living animals, affecting the ways people interacted with them, be they from legal, religious or environmental perspectives. In this paper, close attention will be paid to the context of venison consumption in Roman Britain, whilst still recognising that eating meat was only one part of the human-deer relationship. Indeed, there is evidence to suggest that hunting deer provided a means of signifying social inequality and cultural identity through specific symbolic displays. To contextualise this argument, a zooarchaeological dataset of 366 assemblages from across England and Wales (see Allen 2011 for full dataset) with contemporary iconography and historical texts will be examined and integrated. This paper is not intended to be a universal answer for all examples of deer exploitation in Roman Britain but rather to demonstrate that evidence from faunal remains can be useful when placed alongside other streams of evidence to interpret wider social contexts.

Patterns of deer exploitation

Very few assemblages of Late Iron Age date contain high frequencies of deer bones. Where present, specimens tend to be of antler, demonstrating that venison was rarely consumed prior to the Roman Conquest. However, this does not mean that deer were unimportant to Iron Age society. The presence of whole red deer (*Cervus elaphus*) bodies interred, without dismemberment or other sign of butchery, in sealed pits at Danebury (Grant 1991, 478, 480) and Winklebury (Jones 1977, 64) hillforts imply that these animals were deliberately buried for reasons not associated with meat production. Such deposits resonate with late prehistoric burials of red deer excavated from various Scottish Islands and are suggested to have signified the red deer as a totemic animal (Morris 2005, 14). These deer were not eaten prior to deposition, an aspect which led Morris (*ibid.*) to argue that wealth and power were being symbolised through conspicuous non-consumption of the meat (an important economic resource). It remains possible that the red deer burials at the hillforts in southern Britain entailed similar meanings.

Settlement Type	1st C BC –1st C AD	1st–2nd C AD	2nd–3rd C AD	3rd–4th C AD
religious	5	6	7	9
nucleated *(hillfort/oppidum/small town)*	11	19	16	19
urban	0	23	11	27
military	0	13	5	9
small rural	64	47	16	26
villa	0	7	10	16
% presence *Cervus*	56.8	57.4	64.6	75.5
% presence *Capreolus*	24.7	44.4	44.6	55.2

TABLE 13.1. Number of assemblages considered in analysis by phase, and the percentage of sites by phase containing red deer and roe deer remains. N.B. includes only assemblages with NISP >100 from cattle, sheep/goat and pig remains. Source: Allen (2011).

After the Roman Conquest, there appears to have been an increase in deer exploitation. Red deer remains occur on 57 percent of late Iron Age sites, a figure which rises to 76 percent on late Roman sites; the frequency of sites producing roe deer (*Capreolus capreolus*) remains more than doubles across the same period (Table 13.1). Viewing the data by site-type reveals apparent differences between social groups, as greater mean frequencies of cervid remains are recovered from villa and military sites, possibly reflecting differential access to deer (Figure 13.1). A socio-economic dichotomy between 'villa' and 'non-villa' rural sites does not follow a strict separation, but even at this level a difference between overtly high-status rural settlements and other settlement types is observable. The presence of deer does not necessarily reflect hunting practices, as shed antler may be collected from the countryside. The ratio of antler to bone specimens recovered from differing site-types varies, possibly indicating patterns in the exchange of raw materials (Figure 13.2). Higher proportions of non-antler elements are recovered from villa sites compared to non-villa rural sites, though the sample size from the latter is frustratingly small, due to a general lack of detailed description of deer remains in many faunal reports.

At the elite settlement at Fishbourne, West Sussex, a range of red deer elements from different phases of occupation (*c.*AD 43–280) were commonly found to come from pre-adult individuals: the majority of femorae identified exhibited unfused proximal and distal epiphyses, indicating that very few animals were older than five years (Table 13.2). The bone sizes and bone surface morphology were in agreement with characteristics of juveniles/young adults; no neonatal individuals were found (Allen 2011, 162). It would seem that red deer at Fishbourne were consistently killed prior to maturity, raising the possibility that particular age groups were a focus. As Cartmill (1993, 29) points out, hunting is not a random matter of going out and killing any animal; it must be specified, chosen to be chased and killed in a particular fashion. In conjunction, comparative analysis of red deer distal tibia widths showed that the Fishbourne specimens were consistent with biometric data from contemporary sites at the upper end of the size range (Figure 13.3), possibly indicating that stags were singled out. Such evidence suggests that the hunting of deer was deliberately carried out by local elites as an expression of status; perhaps the consumption of venison was also a display of social exclusivity.

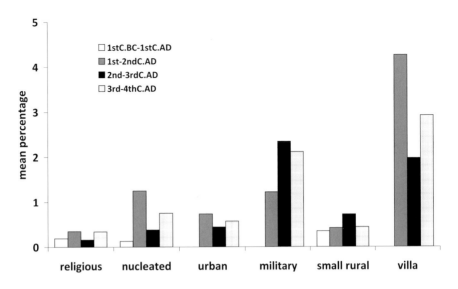

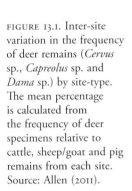

FIGURE 13.1. Inter-site variation in the frequency of deer remains (*Cervus* sp., *Capreolus* sp. and *Dama* sp.) by site-type. The mean percentage is calculated from the frequency of deer specimens relative to cattle, sheep/goat and pig remains from each site. Source: Allen (2011).

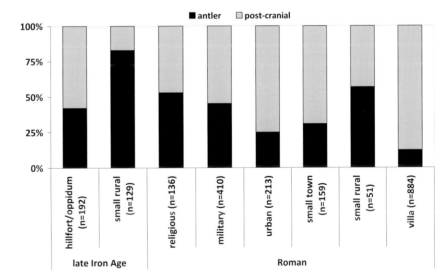

FIGURE 13.2. Ratio of antler and bone from different site types in late Iron Age and Roman Britain, n = number of specimens. N.B. 'post-cranial' includes mandible specimens. Source: Allen (2011).

The social role of venison consumption

Cool (2006, 114) argues that venison was eaten in 'unusual circumstances' in Roman Britain which would suggest that deer meat was highly-prized, though the meaning of its consumption may well have been context-specific. Whilst large and well-studied faunal assemblages from Romano-British villas are comparatively scarce, a good number have been found to contain quantities of butchered post-cranial deer bones, such as at Fishbourne (Allen 2011), Keston (Locker 1991) and Shakenoak (Cram 2005), displaying clear evidence for venison consumption at these sites. Cervid bones were also more often associated with high-status areas in early Roman phases at Colchester (Luff 1993, 9, 25)

Age at fusion	Element	F	UF	Total	%F
*c.*8 months	d. metapodials	14	5		
	p. phalanges	14	0		
	Total	28	5	33	84.8
*c.*30 months	pelvis	11	2		
	p. radius	2	0		
	p. femur	1	6		
	Total	14	8	22	63.0
*c.*36 months	atlas	1	0		
	calcaneus	2	0		
	p. ulna	0	1		
	Total	3	1	4	75.0
*c.*48 months	d. radius	13	3		
	d. ulna	0	1		
	p. tibia	5	1		
	d. tibia	8	3		
	Total	26	8	34	76.5
*c.*60–72 months	d. femur	1	5		
	Total	1	5	6	16.7

TABLE 13.2. Epiphyseal fusion data from red deer specimens at Fishbourne Palace, West Sussex, *c.*AD 43–280. Ages of epiphyseal fusion based on Habermehl (1985). Source: Allen (2011).

and Southwark (Reilly 2005, 160), suggesting that intra-site variation in deer assemblages reflected socio-economic difference within urban contexts.

The Roman military also exhibited a preference for venison, as suggested by relatively high deer frequencies. Since unofficial hunting was punishable by military law, this was probably a penchant of the higher-ranking officers (Davies 1971, 128). Certainly, Caesar was recorded to have favoured the meat from roe deer, stating that it provided the venison of choice for the discerning individual (Dalby 2000, 248). Organised hunting and meat-supply may have been highly regulated, as insinuated by one of the Vindolanda tablets which states that roe venison was ordered for the Praetorium at the fort (Bowman and Thomas 1994, 157). Military garrisons around the Empire carried specialist teams of trackers (*vestigiatores*) and hunters (*venatores immunes*) (Epplett 2001, 217), and epigraphic evidence demonstrates that such groups were operating in Britain (*ibid.* 213). Supporting archaeological evidence derives from excavations on the Tribune's house at Caerleon which produced remains of butchered red and roe deer haunches alongside the bones of hare and wild boar (Hamilton-Dyer 1993, 133; Maltby and Hambleton, this volume); contrastingly, deer bones were absent from areas associated with legionaries and other low-ranking personnel (O'Connor 1986). Higher frequencies of deer remains at elite and military sites after the Conquest imply that socio-political changes impacted on wild resource exploitation, a change which cannot be explained by venison consumption alone.

Fishbourne

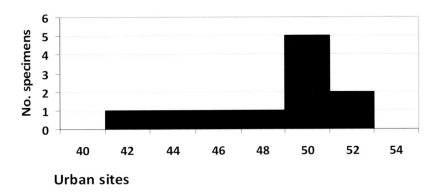

Urban sites

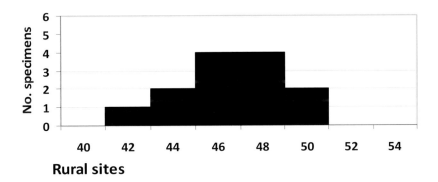

Rural sites

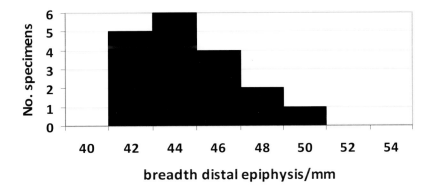

FIGURE 13.3. Red deer tibia measurements from Fishbourne Palace compared to contemporary urban and rural sites. Source: Allen (2011).

Deer in Roman law and the provincial landscape

Hunting in pastoral societies has long been argued to be intimately tied to landscape organisation and power relationships, particularly where shifting political circumstances unfold (Helms 1993, 84; Hamilakis 2003). Therefore, deer exploitation may have had specific relevance for Roman provinces since they were politically-dominated landscapes. From a legal perspective, it is known that deer were imbued with a status which differentiated them from domesticated animals (Starr 1992, 438). In the wild, deer were not owned; they were nobody's property (*res nullius*) and thus could be taken at will. Accordingly, deer were free to enter or leave a person's property as they wished. However, when residing on private property, deer could be seized by the landowner as *occupatio* (see McLeod 1989 for further discussion of these and other related legal concepts). Consequently, the exploitation of deer in Roman Britain must also be viewed in terms of landownership.

The existence of major estates, both private and public/imperial, ranged across southern and midland Britain, whilst the introduction of Imperial law meant that land around urban settlement and areas of productive farmland became increasingly defined and delineated (Mattingly 2006, 354–358, 454–456). For example, we know that woodland could be purchased and owned during this period, potentially for the enclosure of deer, as attested by the financial transaction of a five-acre wood in Kent detailed on a second century AD writing tablet from London (Tomlin 1996). Indeed, Columella (*De Re Rustica* IX, 1.1–9) highlights the importance of including woodland and the animals associated therein on private villa estates, suggesting that deer were contained within purpose-built wooded enclosures rather than roaming free on unbounded land (see also Starr 1992, 436). As already noted, the identification of fallow deer at Fishbourne has provided the suggestion that a *vivarium* existed at the site during the first century AD, just after the Roman Conquest. Consideration of other *Dama* specimens in Roman Britain is now indicating that the phenomenon may not have been an isolated incident (Sykes 2010). Only in special circumstances were hunting restrictions in place outside legal boundaries (Davies 1971, 128; Anderson 1985, 105), though by erecting physical enclosures around property, people were able to take possession over deer and, in effect, stop others from hunting what was once a 'free' resource. These changes suggest an imposed restriction or regulation in the access to wider landscape resources in some areas.

It thus seems probable that the act of possessing deer was an expression of landownership in Roman Britain. Marvin (2001, 204) argues that hunted animals can be raised and cared for by people, but they must act like wild animals, a point which resonates with the status of deer in Roman law – they can be edible, but must be killed according to certain ritual practices. Halstead (1999, 84–86) suggests that the consumption of non-domestic animals in a public arena would generally be avoided in pastoral societies, since a wild beast is public property and brings with it a stronger obligation to share meat. In private households however, hunted animals could be restricted and redistributed amongst a select range of people. This concept, Hamilakis (2003, 242) suggests, works well in places with unequal socio-political organisations of space and it is tempting to see this with regards to deer-hunting in Roman Britain. If wild resources were a focus of elite attentions, their enclosure may have said more about dominant cultural views of landscape than simply being demonstrations of financial power. The specific ways deer were experienced and exploited must then be key to further understanding their importance.

Deer symbolism

Whilst deer remains are generally rare in Iron Age assemblages, they featured in British iconography prior to the Roman Conquest, mainly as live, free-living animals (e.g. Green 1992, 168). Similarly, antler was commonly depicted in Iron Age imagery, particularly on human figures usually attributed to the 'stag-god' Cernunnos (Ross 1996, 176–182). Such artwork seems to imbue deer with a meaning which belies their rarity in the zooarchaeological record, perhaps symbolic of liminality or fertility. Whatever the case, deer seem to have been revered with a status which meant they were not perceived as sources of meat but were fundamental elements of a, possibly sacred, worldview.

After the Roman Conquest, deer deposits of a ritual character continue, but in places seem to have become more complex. A prime example is the burial of a horse with the skull and foot bones from a red deer stag in a second century AD pit at Fishbourne (Allen 2011, 238–239), a deposit which may represent an elaborate burial rite, perhaps involving the skin of the deer. In human burials too, roe deer bones have been recovered from first and second century AD cremations respectively at Winnall, Hampshire, and Youngsbury, Hertfordshire (Philpott 1991, 198, 252), whilst a burnt red deer radius was deposited in association with a first century AD cremation at Baldock, Hertfordshire (Stead and Rigby 1986, 66). Antlers were also used in religious contexts as evidenced by second/third century AD deposits at Witham Romano-Celtic temple, Essex. Many tines had been sawn from the beams and included knife and shaving marks for the removal of small flakes along with much evidence of burning (Luff 1999, 218). Within such contexts we must give credence to the sensory aspect of such practices and how these might have been experienced by the temple's congregants. The Elder Pliny, for example, noted how antlers contained a healing drug and the smell of their burning was thought to combat epilepsy (*Hist. Nat.* 8.41).

One pit at Witham contained sawn-off antler tines buried with a human skull, horse and dog bones, and was argued by Green (1999, 255–256) to represent a hunting cult. Similarly, the remains of stags interred with dogs in shafts is known from a number of sites in southern Britain (Bird 2008, 80), 'special deposits' which Rudling (2008, 130–13) argues are linked to long-established religious beliefs, possibly in worship of local, or even classical, deities. By the later Roman period, the deposition of whole red deer in wells at high-status settlements is known from a number of sites. Instances include twin carcasses of calves from fourth century deposits at Baldock (Chaplin and McCormick 1986, 396) and Rudston villa (Chaplin and Barnetson 1980, 158–160), as well as an adult stag in a late third-mid fourth century well at Bays Meadow villa (Noddle 2006, 218). Antler from red deer stags were also excavated at Rudston and Bays Meadow villas which were shown to have been deliberately removed

Site	Site type	Mosaic type/deer description	Date	Design group	Source
Leicester, High Cross Street	town house (urban)	Cyparissus and the Stag	*c.* fourth century AD		Cosh and Neal 2002, 86–89; no. 25.3
Winterton, Lincolnshire	villa	Orpheus and the Animals	*c.* AD 350	Northern group	Cosh and Neal 2002, 199; no. 68.1
Malton, Yorkshire	town house (small town)	Four Seasons	mid-fourth century AD	Northern group	Cosh and Neal 2002, 344; no. 137.1
Rudston, Yorkshire	villa	Nude Venus and *bestiarii* (stag in woodland)	early fourth century AD		Cosh and Neal 2002, 353–356; no. 143.2
York, Micklegate Bar	town house (urban)	2 fawns in octagon (mosaic now lost – deer image possibly embellished by engraver)	probably fourth century AD		Cosh and Neal 2002, 369–370; no. 149.1
Frampton, Dorset	villa	Scenes from Virgil's *Aeneid* and Ovid's *Metamorphoses* – stag and hind hunted in woodland by	mid-fourth century AD	Durnovarian group	Cosh and Neal 2005, 130–134; no. 168.1
Frampton, Dorset	villa	Chi-Rho mosaic/Bellerophon and Pegasus slaying the Chimera (stag in woodland)	mid-fourth century AD	Durnovarian group	Cosh and Neal 2005, 134–137; no. 168.2
East Coker, Somerset	villa	Two huntsmen with killed hind on pole	2nd half of fourth century AD		Cosh and Neal 2005, 209–210; no. 198.4
Newton St Loe, Somerset	villa	Orpheus and the Animals	2nd half of fourth century AD		Cosh and Neal 2005, 274–278; no. 209.2
Littlecote, Wiltshire	villa	Orpheus and the Animals	post-AD 360	Dubonnic group	Cosh and Neal 2005, 351–355; no. 248.1
St Albans, Hertfordshire	town house (urban)	Lion with stag's head in mouth	late second century AD		Cosh and Neal 2009, 341–343; no. 348.40
Withington, Gloucestershire	villa	Orpheus mosaic with hunting and aquatic scenes	fourth century AD		Cosh and Neal 2010, 204–209; no. 455.4
Woodchester, Gloucestershire	villa	'The Great Pavement' – Orpheus and the Animals	fourth century AD		Cosh and Neal 2010, 214–223; no. 456.1

TABLE 13.3. Examples of deer iconography on mosaics from Romano-British villa houses.

from the skulls of hunted animals, whilst the fourth century well at Baldock also included a group of arrow and spearheads (Manning and Scott 1986, 145–149), further suggesting a votive event related to deer-hunting.

The repeated nature of these types of deposit and their dating to the late Roman period places them in a broader allegorical context, which is also demonstrated by contemporary deer imagery seen within Romano-British houses (see also Cool 2006, 116–118). Cervid iconography survives from at least thirteen polychrome mosaics, all from the late third/fourth centuries and deriving mostly from rural villas, but also town houses (Table 13.3). Some mosaics are associated with Christian imagery, whilst others are concerned with classical stories and myth, but all seem to depict red deer. The mosaic from Frampton villa, which includes portrayals of Virgil's *Aeneid* and Ovid's *Metamorphoses* in its central motifs, has scenes skirting the main design of deer-hunting with dogs through woodland. Such use of natural imagery at villa estates also formally links them with Imperial worldviews where visual depictions of wild landscapes were common on the continent (Greene 1996, 229–230). Indeed, Starr (1992, 437) argues that Virgil's *Aeneid*, where Sylvia and her stag reside, 'exaggerates and extends' the relationship between people and deer in the countryside into an idealised cultural landscape (see also Anderson 1985, 95).

Deer were a common element in Orphic imagery – the most common theme on Romano-British mosaics where the poet Orpheus controls wild animals with music – which is said to have been a theatrical element, providing a context for social encounters in villas for feasts and story-telling (Scott 2004, 47–48). The perception of the red deer as a truly wild animal thus becomes an emblem of wilderness, but one which could be demonstratively controlled by elite groups through a combination of hunting, venison consumption, and harmonious imagery. In truly elite contexts, such as Fishbourne, the keeping of fallow deer allowed such imagery to take place in the landscape itself, representing a wild deer in its fully tamed form. Villas then, were not simply economic production centres, but also social arenas where environmental resources could be viewed as 'civilised' landscapes, ready to be exploited and enjoyed by people in safety, whilst perpetuating social power and ideology.

Conclusion

As stated in the introduction, the central discourse presented here is not intended to be universally representative of human-deer relationships in Roman Britain. However, I would suggest that the evidence for deer-hunting and venison consumption entailed complex meanings beyond those traditionally cited. Moreover, I have sought to argue that the social changes seen in Britain from the Late Iron Age and across the Roman period impacted upon attitudes and behaviour of people towards deer. Considering their symbolism in art and in structured deposits, the act of killing a deer seems to have entailed a multifaceted set of communicative phenomena, interconnecting people's ideas of landownership with religious/mythological perspectives of nature. If nothing else, I hope to have raised awareness for the need to be more explicit and integrative in zooarchaeological work, not only for Roman Britain but in general, as deer continue to sit in the shadow of domestic animals, a position which truly belies their importance to pastoral societies.

References

Allen, M. G. (2011) *Animalscapes and Empire: New Perspectives on the Iron Age/Romano-British Transition*. Unpublished PhD dissertation. University of Nottingham.

Anderson, K. (1985) *Hunting in the Ancient World*, University of California, Berkeley.

Bird, D. (2008) 'Roman period temples and religion in Surrey' in *Ritual Landscapes of Roman South-East Britain*, ed. D. Rudling, Heritage Marketing, Great Dunham, 63–86.

Bowman, A. K. and Thomas, D. (1994) *The Vindolanda Writing Tablets*, British Museum, London.

Cartmill, M. (1993) *A View to a Death in the Morning: Hunting and Nature through History*, Harvard University Press, Cambridge.

Chaplin, R. E. and Barnetson, L. P. (1980) 'Animal bones' in *Rudston Roman Villa*, ed. I. M. Stead, Yorkshire Archaeological Society, York, 149–161.

Chaplin, R. E. and McCormick, F. (1986) 'The animal bones' in *Baldock: The Excavation of a Roman and Pre-Roman Settlement 1968–72*, ed. I. M. Stead and V. Rigby, Society for the Promotion of Roman Studies, London, 396–415.

Columella, (1955), *De Re Rustica,* trans E. S. Forster and E. H. Heffner, Heinemann, London.

Cool, H. E. M. (2006) *Eating and Drinking in Roman Britain*, CUP, Cambridge.

Cosh, S. R. and Neal, D. S. (2002; 2005; 2009; 2010) *Roman Mosaics of Britain, Volumes I–IV*, Society of Antiquaries of London, Barham/London.

Cram, C. L. (2005) 'Animal bones' in *The Roman Villa at Shakenoak Farm, Oxfordshire: Excavations 1960–76*, eds A. C. C. Brodribb, A. R. Hands and D. R. Walker, BAR British Series 395, Archaeopress, Oxford, 384–401, 498–528.

Dalby, A. (2000) *Empire of Pleasures*, Routledge, London.

Davies, R. W. (1971) 'The Roman military diet', *Britannia* 2, 122–142.

Dobney, K. (2001) 'A place at the table: the role of vertebrate zooarchaeology within a Roman research agenda for Britain' in *Britons and Romans: Advancing the Research Agenda*, eds S. James and M. Millet, CBA Research Report 125, Council for British Archaeology, London, 36–45.

Epplett, C. (2001) 'The capture of animals by the Roman military', *Greece and Rome* 48, 210–222.

Gilhus, I. S. (2006) *Animals, Gods and Humans: Changing Attitudes to Animals in Greek, Roman and Early Christian Ideas*, Routledge, London.

Grant, A. (1981) 'The significance of deer remains at occupation sites of the Iron Age to the Anglo-Saxon period' in *The Environment of Man: The Iron Age to the Anglo-Saxon Period*, eds M. Jones and G. W. Dimbleby, BAR British Series 87, Archaeopress, Oxford, 205–13.

Grant, A. (1989) 'Animals in Roman Britain' in *Research on Roman Britain 1960–1989*, ed M. Todd, Britannia Monograph 11, Society for the Promotion of Roman Studies, London, 135–146.

Grant, A. (1991) 'The animal bones' in *Danebury: An Iron Age Hillfort in Hampshire, Vol. 5, The Excavations, 1979–88: The Finds*, eds B. Cunliffe and C. Poole, CBA Research Report 73, Council for British Archaeology, London, 447–487.

Green, M. (1992) *Animals in Celtic Life and Myth*, Routledge, London.

Green, M. (1999) 'Religion and deities' in *Excavations of an Iron Age Settlement and Roman Religious Complex at Ivy Chimneys, Witham, Essex 1978–83, Chelmsford*, ed. R. Turner, EAA Report 88, East Anglian Archaeology, Chelmsford, 255–257.

Greene, C. M. C. (1996) 'Did the Romans hunt?' *Classical Antiquity* 15, 222–260.

Habermehl, K. H. (1985) *Altersbestimmung bei Wild- und Pelztiere: Möglichkeiten und Methoden*. Ein praktischer Leitfaden für Jäger, Biologen und Tierärzte, Hamburg, Parey Verlag.

Halstead, P. (1999) 'Neighbours from hell? The household in Neolithic Greece' in *Neolithic Society in Greece*, ed. P. Halstead, Sheffield Academic, Sheffield, 77–95.

Hamilakis, Y. (2003) 'The sacred geography of hunting: wild animals, social power and gender in early farming societies' in *Zooarchaeology in Greece: Recent Advances*, eds E. Kotjabopoulou, Y. Hamilakis, P. Halstead, C. Gamble and V. Elefanti, British School at Athens, London, 239–247.

Hamilton-Dyer, S. (1993) 'The animal bone' 132–136 in eds J. D. Zienkiewicz, J. Hillam, E. Besly, B. M. Dickinson, P. V. Webster, S. A. Fox, S. Hamilton-Dyer, A. E. Caseldine and P. A. Busby, 'Excavations in the *Scamnum Tribunorum* at Caerleon: The Legionary Museum Site 1983–5', *Britannia* 24, 27–140.

Helms, M. W. (1993) *Craft and the Kingly Ideal: Art, Trade and Power*, University of Texas, Austin.

Jones, R. (1977) 'Animal bones' 58–66, in ed. K. Smith, 'The excavation of Winklebury Camp, Basingstoke, Hampshire', *Proceedings of the Prehistoric Society* 43, 31–130.

Lindstrøm, T. C. (2010) 'The animals of the arena: how and why could their destruction and death be endured and enjoyed?' *World Archaeology* 42, 310–323.

Luff, R. M. (1993) *Animal Bones from Excavations in Colchester, 1971–85*. CA Report 12, Colchester Archaeological Trust, Colchester.

Luff, R. M. (1999) 'Animal and human bones' in *Excavations of an Iron Age Settlement and Roman Religious Complex at Ivy Chimneys, Witham, Essex 1978–83, Chelmsford*, ed. R. Turner, EAA Report 88, East Anglian Archaeology, Chelmsford, 204–223.

Maltby, M. (2010) *Feeding a Roman Town: Environmental Evidence from Excavations in Winchester 1972–1985*, Winchester Museums Services, Winchester.

Manning, W. H. and Scott, I. R. (1986) 'Iron Objects' in *Baldock: The Excavation of a Roman and Pre-Roman Settlement, 1968–72*, eds I. M. Stead and V. Rigby, Britannia Monograph 7, Society for the Promotion of Roman Studies, London, 145–162.

Marvin, G. (2001) 'The problem of foxes: legitimate and illegitimate killing in the English countryside' in *Natural Enemies: People-Wildlife Conflict in Anthropological Perspective*, ed. J. Knight, Routledge, London.

Mattingley, D. (2006) *An Imperial Possession: Britain in the Roman Empire, 54 BC–AD 43*, Penguin, London.

McLeod, G. (1989) 'Wild and tame animals and birds in Roman law' in *New Perspectives in the Roman Law of Property: Essays for Barry Nicholas*, ed P. Birks, Clarendon, Oxford.

Morris, J. T. (2005) 'Red deer's role in social expression on the isles of Scotland' in *Just Skin and Bones: New Perspectives on Human-Animal Relations in the Historic Past*, ed. A. G. Pluskowski, BAR International Series 1410, Archaeopress, Oxford, 9–18.

Noddle, B. (2006) 'Animal bone' in 'Bays Meadow Villa, Droitwich: excavations 1967–77', ed. L. Barfield, 216–220 in *Roman Droitwich: Dodderhill Fort, Bays Meadow Villa, and Roadside Settlement*, ed D. Hurst, CBA Research Report 146, Council for British Archaeology, York, 78–242.

O'Connor, T. P. (1986) 'Animal bones' in *The Legionary Fortress Baths at Caerleon, vol. II: The Finds*, ed. J. D. Zienkiewicz, Alan Sutton, Gloucester, 224–248.

Ovid, (1986) *Metamorphoses*, trans A. D. Melville, OUP, Oxford.

Locker, A. (1991) 'The animal bone' in *The Roman Villa Site at Keston. Kent*, ed. B. Philp, Kent Archaeological Rescue Unit, Dover Castle, 285–292.

Philpott, R. A. (1991) *Burial Practices in Roman Britain: A Survey of Grave Treatment and Furnishing AD 43–410*, BAR British Series 219, Archaeopress, Oxford.

Pliny, (1940) *Natural History*, Trans H. Rackham, Heinemann, London.

Purcell, N. (1987) 'Town in country and country in town' in *Ancient Roman Villa Gardens*, ed. E. B. MacDougall, Dumbarton Oaks, Washington DC, 185–204.

Reilly, K. (2005) 'Animal remains' in *A Prestigious Roman Building Complex on the Southwark Waterfront: Excavations at Winchester Palace, London 1983–90*, ed. B. Yule, MOLAS Monograph 23, MOLAS, London, 158–167.

Ross, A. (1996) *Pagan Celtic Britain*, Academy Chicago Publishers, Chicago.

Rudling, D. (2008) 'Roman-period temples, shrines and religion in Sussex' in *Ritual Landscapes of Roman South-East Britain*, ed. D. Rudling, Heritage Marketing, Great Dunham, 95–138.

Scott, S. (2004) 'Elites, exhibitionism and the society of the late Roman villa' in *Landscapes of Change: Rural Evolutions in Late Antiquity and the Early Middle Ages*, ed. N. Christie, Ashgate, Aldershot, 39–65.

Starr, R. J. (1992) 'Sylvia's Deer (Vergil, Aeneid 7.479–502): game parks and Roman law', *The American Journal of Philology* 113, 435–439.

Stead, I. M. and Rigby, V. (1986) *Baldock: The Excavation of a Roman and Pre-Roman Settlement, 1968–72*, Britannia Monograph 7, Society for the Promotion of Roman Studies, London.

Sykes, N. J. (2009) 'Worldviews in transition: the impact of exotic plants and animals on Iron Age/Romano-British landscapes', *Landscapes* 10, 19–36.

Sykes, N. J. (2010) 'European fallow deer' in *Extinctions and Invasions: a Social History of British Fauna*, ed. T. P. O'Connor and N. J. Sykes, Windgather, Oxford.

Sykes, N. J. White, J. Hayes, T. and Palmer, M. (2006) 'Tracking animals using strontium isotopes in teeth: the role of fallow deer (*Dama dama*) in Roman Britain', *Antiquity* 80, 948–959.

Tomlin, R. S. O. (1996) 'A five-acre wood in Roman Kent' in *Interpreting Roman London: Papers in Memory of Hugh Chapman*, eds J. Bird, M. Hassall and H. Sheldon, Oxbow, Oxford, 209–215.

Virgil, (1969) *Aeneid* VII–XII, trans H. R. Fairclough, Heinemann, London.

Wofford, S. L. (1992) *The Choice of Achilles: the Ideology of Figure in the Epic*, Stanford University, Stanford.

Deer and Humans in South Wales during the Roman and Medieval Periods

Mark Maltby and Ellen Hambleton

Introduction

Discussions of documentary and archaeological evidence pertaining to the history of the exploitation of deer in Britain have tended to focus on the evidence from Medieval England, from where most of the recent research has been carried out (e.g. Birrell 2006; Sykes 2006; Rotherham 2007). There have, however, been several recent papers that have incorporated documentary and topographical evidence from Wales and the Welsh Marches, mainly concerned with Medieval parks, forests and chases (e.g. Silvester 2010; Langton 2011; Smith, this volume). To complement these discussions, this paper will consider zooarchaeological evidence for the exploitation of deer in south Wales from a number of important Roman and post-Roman excavations carried out during the last 30 years. The evidence relies heavily on recent, as yet unpublished, analyses of animal bones from Medieval and Early post-Medieval deposits at Laugharne Castle and the Roman town at Caerwent (Hambleton and Maltby 2004a; Hambleton and Maltby 2004b; Hambleton and Maltby 2009). The locations of these settlements and the other major sites discussed in the paper are shown in Figure 15.1. Raw data of species counts from all sites are provided in Table 15.1. Counts are restricted to those of the major domestic mammal food–producing species (cattle (*Bos* sp.); sheep/goat (*Ovis/Capra*); pig (*Sus* sp.)) and the three species of deer that have been recorded in Wales during the periods involved. In most cases the counts are of the number of individual specimens (NISP). In a few cases the counts have been derived from selected bone counts. Although both methods tend to bias counts towards large mammals, the results from the two methods are usually compatible (Maltby 2010).

Deer remains from Roman period sites

Archaeological evidence for the exploitation of deer is limited to a handful of sites in southern Wales, although large assemblages have been analysed at the

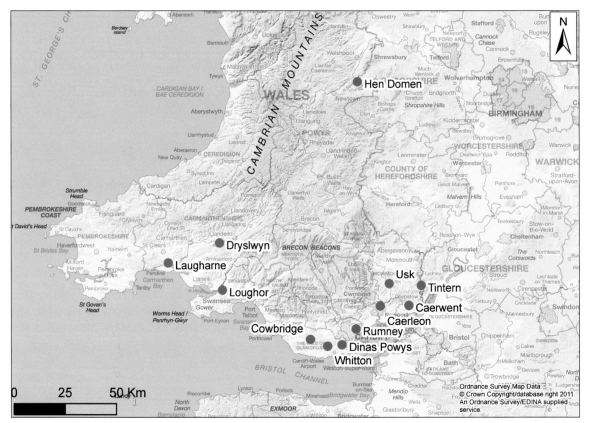

FIGURE 15.1. Map of south Wales showing principal sites discussed in this survey.

town of Caerwent, the fortress at Caerleon and the auxiliary fort at Loughor. The largest assemblage in this survey was obtained from the area around the Basilica at the *civitas* capital at Caerwent, Gwent (Hambleton and Maltby 2004b). Red deer (*Cervus elaphus*) provided only 0.3 percent of the selected mammal assemblage (Table 15.1). This is typical of Romano-British urban assemblages, in which red deer elements provide less than 1% of the total cattle, sheep/goat, pig and deer counts (Maltby 2010). There was no marked chronological variation in their relative abundance, although they were marginally better represented in late Roman deposits (Hambleton and Maltby 2004b).

There was no significant bias towards elements from particular parts of the body, although vertebrae were poorly represented. Notably, there was no bias towards hindlimb elements. Indeed forelimb bones (apart from the metacarpal) were better represented. At least five scapulae were represented (Table 15.2).

Butchery marks on red deer bones included superficial chop marks on the medial aspect of a proximal radius and heavy axial blade marks along the edges of the proximal articulation of a tibia. Both are characteristic of marks commonly inflicted on cattle upper limb bones on Roman urban settlements (Maltby 2007; Maltby 2010), indicating that some deer were acquired and processed by specialist butchers. Three scapulae had been chopped transversely where the shoulder had been segmented from the upper forelimb. It is feasible

Site	Settlement Type	Date	Cattle	Sheep	Pig	Red	Roe	Fallow	NISP*	% Red	% Roe	% Fallow	Source
Caerwent Basilica	Town	Roman	7460	5128	5920	50	49		18607	0.3	0.3	0.0	Hambleton and Maltby 2009
Caerwent NW Tower	Town	Roman	387	103	62	16	2		570	2.8	0.4	0.0	Noddle 1983
Caerwent House	Town	Roman	3640	1843	550	9			6042	0.1	0.0	0.0	Burnett n.d.
Caerleon Baths	Legionary Fort	Roman	4470	966	654	24	1		6115	0.4	0.0	0.0	O'Connor 1986
Caerleon BT Site	Legionary Fort	Roman	500	98	127	1	3		729	0.1	0.4	0.0	Hamilton-Dyer n.d.
Caerleon Garden	Legionary Fort	Roman	362	112	213	11	19		717	1.5	2.6	0.0	Hamilton-Dyer 1993
Caerleon S Defences	Legionary Fort	Roman	338	134	395	3			870	0.3	0.0	0.0	Jones 2010
Loughor (Leucarum)	Auxiliary Fort	Roman	1834	546	728	60	63		3231	1.9	1.9	0.0	Sadler 1997
Cowbridge, Bear Field	Roadside	Roman	311	51	15	1	1		379	0.3	0.3	0.0	Sadler 1996
Cowbridge, High Street	Roadside	Roman	778	507	155	7			1447	0.5	0.0	0.0	Jones 1996
Thornwell Farm	Rural settlement	Roman	166	141	47				354	0.0	0.0	0.0	Pinter-Bellows 1996
RAF St Athan	Rural settlement	Roman	50	47	16				113	0.0	0.0	0.0	Higbee 2006
Whitton	Rural settlement	IA/Roman	2185	2465	1008	106	75		5839	1.8	1.3	0.0	Kinnes 1989
Rumney Wharf	Coastal settlement	Roman	50	78	8	1			137	0.7	0.0	0.0	Hamilton-Dyer 1994

TABLE 15.1. Counts of animal bones from Roman and post-Roman sites in south Wales. Counts are derived from Number of individual specimen (NISP) counts in call cases except Thornwell Park and RAF St Athan where selected element counts were employed. Continues pp. 190–191.

Site	Settlement Type	Date	Cattle	Sheep	Pig	Red	Roe	Fallow	NISP*	% Red	% Roe	% Fallow	Source
Laugharne (Inner Ward)	English Castle	12th	169	227	282	298	8		984	30.3	0.8	0.0	Hambleton and Maltby 2004a
Laugharne (Inner Ward)	English Castle	L12–E13th	2195	2590	2575	470	175		8005	5.9	2.2	0.0	Hambleton and Maltby 2004a
Laugharne (Inner Ward)	English Castle	14th	145	153	58	6	3	1	366	1.6	0.8	0.3	Hambleton and Maltby 2004a
Laugharne (Outer Ward)	English Castle	L12–14th	118	155	68	4	1		346	1.2	0.3	0.0	Hambleton and Maltby 2004a
Laugharne (Inner Ward)	Castle/Mansion	Tudor	52	466	222	1	6		747	0.1	0.8	0.0	Hambleton and Maltby 2004b
Laugharne (Inner Ward)	Castle/Mansion	17th	280	301	64	2	1		648	0.3	0.2	0.0	Hambleton and Maltby 2004b
Dryslwyn Area F	Welsh Castle	1220–1287	691	369	510	33	121		1724	1.9	7.0	0.0	Gidney 2007
Dryslwyn Other Areas	Welsh Castle	1220–1287	3247	1985	2737	88	209	1	8267	1.1	2.5	0.0	Gidney 2007
Dryslwyn Other Areas	English Castle	1287–1400	1866	1792	1451	133	115	3	5360	2.5	2.1	0.1	Gidney 2007
Dryslwyn Other Areas	Decomissioning	1400–1430	541	450	336	13	33		1373	0.9	2.4	0.0	Gidney 2007
Loughor	Castle	E12th	227	97	170	89	4		587	15.2	0.7	0.0	Noddle 1993
Loughor	Castle	L12–E13th	421	58	114	124	5		722	17.2	0.7	0.0	Noddle 1993
Loughor	Castle	13th	221	168	134	80	7		610	13.1	1.1	0.0	Noddle 1993
Loughor	Castle	14–16th	258	87	69	29	1		444	6.5	0.2	0.0	Noddle 1993
Rumney	Castle	12th	100	20	51	6	2		179	3.4	1.1	0.0	Jones 1992 and *pers. comm.*

* Counts are derived from Number of individual specimen (NISP) counts in call cases except Thornwell Park and RAF St Athan where selected element counts were employed.

TABLE 15.1. continued.

Site	Settlement Type	Date	Cattle	Sheep	Pig	Red	Roe	Fallow	NISP*	% Red	% Roe	% Fallow	Source
Rumney	Castle	13th	211	74	163	7	1	1	457	1.5	0.2	0.2	Jones 1992 and *pers. comm.*
Hen Domen	Timber Castle	11–13th	178	14	619	34	31		876	3.9	3.5	0.0	Browne 2000
Caerleon	Small town	12–14th	1112	325	271	23		6	1737	1.3	0.0	0.3	O'Connor 1986
Tintern	Abbey	13th	160	170	58	2	2		392	0.5	0.5	0.0	Jones 1989
Tintern	Abbey	15–16th	136	48	40	3	1		228	1.3	0.4	0.0	Jones 1989
Rhossili	Settlement	12–13th	21	33	4	1			59	1.7	0.0	0.0	Jones 1987
Usk	Town	Mainly 17th	1139	313	190			5	1647	0.0	0.0	0.3	Jones 1981

TABLE 15.1. continued.

* Counts are derived from Number of individual specimen (NISP) counts in call cases except Thornwell Park and RAF St Athan where selected element counts were employed.

that these were shoulder joints preserved by smoking and/or salting, again a practice that has parallels in Roman cattle butchery (Dobney 2001; Maltby 2007).

Roe deer (*Capreolus capreolus*) also contributed 0.3 percent of the identified mammal assemblage at the Caerwent Basilica site (Table 15.1). Their rarity is typical of most Romano-British urban assemblages (Maltby 2010). There was a bias towards larger, denser bones with no records of tarsals, phalanges, ribs or vertebrae. Forelimbs were slightly better represented than hindlimbs. At least six mandibles and humeri were represented (Table 15.2).

A courtyard house in the north-west corner of Caerwent has also produced a substantial assemblage. Red deer were very poorly represented and no roe deer remains were identified (Burnett nd). A much smaller assemblage was obtained from excavations near the town wall of Caerwent (Noddle 1983). This produced an unusually high percentage of red deer (2.8 percent).

Four sites from the legionary fortress at Caerleon, 13 km to the west of Caerwent, are included in this survey (Table 15.1). Three assemblages contained very small percentages of red and roe deer (both species under 0.5 percent). The fourth, from the Museum Garden site, produced very high percentages of roe (2.6 percent) and red deer (1.5 percent) (Hamilton-Dyer 1993). These bones were largely derived from a well, which also included significant numbers of wild boar (*Sus scrofa*) and crane (*Grus* sp.) bones, and other luxury foods such as grapes and figs. The well had partly been filled with kitchen waste from the residence of a high-ranking officer (Zienkiewicz 1993, 77). Excluding this deposit, the percentages of red deer (0.6 percent) and roe deer (0.5 percent) are at levels more typical of other assemblages from the fortress (Hamilton-Dyer 1993).

	Caerwent				Leucarum		Laugharne Castle			
	Basilica Site (Hambleton and Maltby 2009)				(Sadler 2007)		Medieval deposits from Inner Ward (Hambleton and Maltby 2004a)			
	Red		Roe		Red	Roe	Red		Roe	
Element	*NISP*	*MNE*	*NISP*	*MNE*	*NISP*	*NISP*	*NISP*	*MNE*	*NISP*	*MNE*
Antler	4	3	1	1	17	5	20	nc	5	nc
Maxilla	1	1	1	1	1		11	5	4	4
Skull frag			1	1			11	3		
Mandible	3	3	6	6	2	4	7	4	7	6
Teeth	5	nc			2	1	18	nc	9	nc
Scapula	6	5	1	1	3	1	5	4	7	7
Humerus	1	1	7	6	5	7	8	6	9	7
Radius	6	4	4	4	2	9	10	6	9	8
Ulna	4	3	4	3	1	6	6	4	2	2
Pelvis	2	1	5	5	3	3	42	19	9	9
Femur	3	1	1	1	5	3	112	39	4	3
Patella							1	1		
Tibia	3	2	6	4	1	8	125	61	21	13
Carpals					2		1	1		
Astragalus					1	2	40	39		
Calcaneus					2		45	43	6	6
Cuboid							7	7		
Tarsals							9	9		
Metacarpal			8	5	2	3	2	2	5	4
Metatarsal	4	2	4	2	3	8	3	3	19	8
Metapodial									1	nc
Phalanx 1	3	1			2	3	2	1	1	1
Phalanx 2	4	1			1		1	1		
Phalanx 3							1	1		
Atlas	1	1							1	1
Axis							1	1	2	2
Lumbar V					5					
Sacral V							1	1		
Total	50		49		60	63	489		121	

NISP = number of individual specimens; MNE = minimum number of element; nc = MNE not calculated

TABLE 2. Deer elements in assemblages recorded in detail. NISP = number of individual specimens; MNE = minimum number of element; nc = MNE not calculated. Caerwent = Basilica Site (Hambleton and Maltby 2009); Leucarum (Sadler 2007); LC = Laugharne Castle Medieval deposits from Inner Ward (Hambleton and Maltby 2004a).

Both red and roe deer were also (by Romano-British standards) well represented at another military site, the auxiliary fort of *Leucarum* situated at Loughor, on the outskirts of Swansea. They each provided 1.9 percent of the assemblage (Sadler 1997), although the red deer numbers were slightly inflated by a high antler count and possibly an associated group of five lumbar vertebrae (Table 15.2).

Assemblages from the roadside settlement of Cowbridge, South Glamorgan, had low numbers of red and roe deer (Jones 1996; Sadler 1996) (Table 15.1). Only one fragment of red deer was found at Rumney Wharf (Hamilton-Dyer 1994) and two small assemblages from rural settlements at Thornwell Farm

(Gwent) and RAF St Athan (Vale of Glamorgan) included no deer remains at all (Pinter-Bellows 1996; Higbee 2006). The large assemblage from Whitton (Vale of Glamorgan) produced higher percentages of both red (1.8 percent) and roe deer (1.3 percent) (Kinnes 1989). Unfortunately the assemblage from this high status Late Iron Age and Romano-British settlement was not subdivided by period and no further details of the deer assemblages were published.

Information about mortality profiles of the deer found on Roman sites is limited, because of small sample sizes and lack of discussion in some reports. However, the great majority of the red and roe deer from Caerwent, Caerleon and *Leucarum* were adult animals.

Deer remains from post-Roman sites

Evidence for the hunting of deer is very poorly documented in the archaeological record of the period between the fifth and eleventh century. The hillfort of Dinas Powys, Vale of Glamorgan produced a large bone assemblage. Unfortunately, only part of it was retained. The assemblage analysed from this high status site was dominated by pig, followed by cattle and sheep/goat. Only fifteen elements of deer were identified but not to species level (Haglund-Calley and Cornwall 1963). Gilchrist's (1988) reanalysis of the bones did not consider wild species in the published discussion. No other Dark Age sites from south Wales have produced a faunal sample worthy of detailed analysis.

Nearly all of the few large animal bone assemblages from the period after 1066 have been obtained from castles. The two main assemblages that have been examined in detail are located at Laugharne and Dryslwyn, Dyfed (Figure 15.1).

Laugharne Castle was occupied for most of the Medieval period by Anglo-Norman families. Red deer bones were particularly abundant (30.3 percent) in the earliest twelfth century deposits from the Inner Ward outnumbering all the domestic species in the assemblage. They were also found commonly (5.9 percent) in the substantial assemblages obtained from late twelfth and early thirteenth century deposits. They became less abundant in Late Medieval features (1.6 percent) and virtually disappeared in the post-Medieval deposits (0.1–0.3 percent). There were intra-site variations, however. In the Outer Ward, red deer fragments were relatively uncommon (1.2 percent) throughout the Medieval deposits (Table 15.1) (Hambleton and Maltby 2004a; Hambleton and Maltby 2004b).

The high percentages of red deer in the earlier deposits, were remarkable considering only a small range of their bones were commonly deposited in the Inner Ward. The assemblages were dominated by bones of the haunches and adjacent areas (femur, tibia, pelvis, astragalus and calcaneus). Metatarsals, phalanges and all forelimb bones were very poorly represented. Cranial elements consisted mainly of worked antler (Table 15.2). Disarticulation of the feet was evidenced by knife cuts and, less commonly, by chop marks on the astragalus, calcaneus and distal tibia. Chop marks on the pelvis and the proximal femur indicate segmentation of the carcass around the hip joint.

Roe deer bones were less well represented than red deer in the Medieval deposits, but were nevertheless found in larger proportions than in most of the Roman assemblages. There was a much more even representation of forelimb and hindlimb elements, but cranial elements were under-represented. Only a single pelvis from a fourteenth century context was identified as fallow deer (*Dama dama*).

The largest faunal assemblage from a Medieval site in south Wales has been obtained from Dryslwyn Castle. This is situated about 23 km to the north-east of Laugharne (Figure 15.1). In contrast to Laugharne, Dryslwyn was a Welsh castle from its foundation in the 1220s until its capture by the English in 1287. Thereafter, a constable held the castle on behalf of the English king or lords (Gidney and Caple 2007). Red deer percentages from thirteenth and fourteenth century samples were slightly lower (1.1–2.5 percent) than contemporary assemblages from Laugharne (Gidney 2007). Indeed, roe deer outnumbered red deer bones at Dryslwyn in most periods, particularly in the earliest phase (7.0 percent). Fallow deer identifications were restricted to four bones, three from the initial phase of English occupation in the fourteenth century (Table 15.1). No detailed information about butchery and body part representation has been published, although Gidney (2007, 307) did note that red deer hindlimbs were better represented than the forelimbs.

Another Anglo-Norman foundation at Loughor Castle, 30 km east of Laugharne, was located within the boundaries of the Roman auxiliary fort of *Leucarum* discussed above. The assemblage of over 2,000 bones produced very high percentages of red deer throughout its twelfth to thirteenth century phases (13.1–17.2 percent). Although percentages decreased in the fourteenth to sixteenth century deposits (6.5 percent), red deer bones were substantially more abundant than at Laugharne and Dryslwyn. Unfortunately, no information about body part representation was published. Roe deer were present in smaller but consistent numbers (0.2–1.1 precent) but no fallow deer bones were identified (Noddle 1993).

Excavations at the Anglo-Norman Rumney Castle, South Glamorgan, produced a small twelfth century assemblage dominated by cattle, but with a relatively high percentage of red deer (3.4 percent). Roe deer was also present (1.1 percent). A larger sample from thirteenth century levels produced lower percentages of red deer (1.5 percent) and single occurrences of both roe and fallow deer (Jones 1992 and *pers. comm.*).

Further north-east, the timber castle at Hen Domen, Montgomery, produced an assemblage mainly derived from one pit from the bailey. This was dominated by pig. Both red (3.9 percent) and roe deer (3.5 percent) were well represented. The limited evidence for body part representation showed a bias towards the hindlimb (Browne 2000, 131). No bones of fallow deer were identified.

Unfortunately, assemblages from non-castle sites in southern Wales are extremely scarce. Medieval deposits from Caerleon have produced the largest sample of over 1,700 specimens (O'Connor 1986). Red deer were found in smaller amounts (1.3 percent) than in nearly all the castle samples and no roe deer bones were identified at all. Perhaps surprisingly, several fallow deer bones were present in demolition layers dated to the thirteenth century, a period when they were not recorded in any of the castles discussed above.

A fairly small sample was obtained from Tintern Abbey. Modest quantities of red deer (0.5–1.3 percent) and roe deer bones (0.4–0.5 percent) were recovered from both Medieval and Early post-Medieval deposits but no fallow deer bones were identified (Jones 1989). The rural settlement at Rhossili, West Glamorgan, produced a very small faunal sample that included a single bone of red deer (Jones 1987).

Apart from Laugharne, the only post-Medieval assemblage of significant size has come from the town of Usk, Gwent. Most of the assemblage was of seventeenth century date and was dominated by cattle, whereas deer bones were rare. The only species identified was fallow (0.3 percent) (Jones 1981).

Discussions related to post-Roman mortality patterns of deer have been limited. It seems, however, that the majority of venison consumed belonged to adult animals of all three species. At Laugharne, 85 percent of the latest-fusing epiphyses of red deer limb bones had fused, indicating the focus on the acquisition of fully-grown animals. Adults also dominated the smaller roe deer assemblage (Hambleton and Maltby 2004a). Sub-adult roe deer were quite common in the Dryslwyn assemblage (Gidney 2007, 307).

Discussion

The survey has confirmed that only two species of deer were definitely present in southern Wales during the Roman period. The absence of fallow deer is unsurprising, as the presence of this imported species has hitherto only been authenticated on a small number of high status sites in southern England (Sykes *et al.* 2011). Although the percentages of red and roe deer are generally low on the Welsh sites, there are some indications that they were more likely to be found on sites associated with people of high status. The clearest example is the presence of both species amongst other 'luxury' foods in kitchen waste associated with high-ranking officers at the legionary fortress of Caerleon. The deposition of these together within a disused well suggests that they were consumed at a special banquet hosted by the officer involved (Hamilton-Dyer 1993, 136). If it was derived from one meal, it included the meat from at least four cranes, a wild boar, two roe deer and possibly two red deer. The remaining assemblages from this site and others in Caerleon were much less diverse with deer bones present in only small quantities.

Variations in dietary breadth within Roman military sites have also been observed at the fort of South Shields on Hadrian's Wall. Here, assemblages associated with the commandant's residence produced a much broader range of species, including red and roe deer, than in deposits associated with the barrack blocks (Stokes 2000). The high percentages of deer remains at *Leucarum* may also reflect the fact that most of the bones were recovered from the vicinity of the *praetorium*, and perhaps also associated with the garrison commanders (Sadler 1997).

Significantly higher percentages of red deer remains were found in the assemblage from near the north-west defences in Caerwent than from elsewhere in the town, which may indicate the presence of a high status residence nearby. There are indications from the butchery evidence that some of the red deer in Caerwent were acquired and processed by the urban specialist butchers.

The presence of large quantities of deer on high status Medieval sites in Wales is to be expected given their prevalence at castles and manors in England (Sykes 2006). The main difference in comparison with southern England is that fallow deer were present, if at all, only in small numbers. The absence of fallow from twelfth century deposits is understandable, as their frequency tended to increase from the mid-twelfth century onwards as deer parks stocked with fallow became established. However, they are also virtually absent from later Medieval deposits from Laugharne, Dryslwyn, Loughor and Rumney. They were found as infrequently as in urban deposits at Caerleon. There is therefore only tentative evidence that fallow deer were introduced to Wales during this period. One could account for the presence of the few fallow deer as imports of salted venison (Birrell 2006).

There may well have been sufficient numbers of red and roe deer in the forests and chases of south Wales to meet the needs of the Welsh or Anglo-Norman gentry. The percentages of deer on the castle sites tended to decrease during the Medieval period (Table 15.1), which might reflect pressures on diminishing resources. However, percentages also vary according to areas within the castles excavated. At Laugharne in particular, much higher percentages of deer remains were associated with the kitchen waste from the Inner Ward than from peripheral parts of the castle. High percentages of deer in the earliest phases of these castles may reflect the desire of the elite to demonstrate their wealth, legitimise their status and strengthen their position in the area by commonly hosting feasts.

The relative abundance of red and roe deer varies between Medieval sites. At the Anglo-Norman Castles of Laugharne, Loughor and Rumney, red deer outnumbered roe. The opposite was the case in the period of Welsh occupation at Dryslwyn. Whether this reflects variations in local availability is unclear, but it is interesting to note that red deer outnumbered roe at Dryslwyn during the initial phase of English occupation. At this time, documentary evidence indicates that provisioning of the castle garrison relied more on purchases from market towns (Gidney and Caple 2007). Isotopic analysis of pigs has raised the possibility that they may have been procured from another source at this time (Millard *et al.* 2011). It is feasible that at least some of the red deer may have been acquired from outside the local area.

Only at Laugharne (Table 15.2) has it been possible to demonstrate the expected dominance of haunches of red deer that are such a prominent feature of assemblages of red and fallow deer from high status sites in England (Sykes 2006; Thomas 2007). It is likely that similar strict rules regarding the hunting of large deer and the distribution of their body parts were in operation in these Anglo-Norman castles. The same may have been true during the Welsh occupation at Dryslwyn but details are lacking.

This brief survey has demonstrated both similarities and differences regarding the acquisition and consumption of venison between south Wales and southern England, particularly in the Medieval period. It provides the basis for more detailed comparisons with future assemblages. Re-examination of some of the analysed assemblages would provide further information about element representation, butchery practices and isotopic signatures. There is also potential for a comprehensive survey of the stature of red deer and possible changes related to variations and changes in habitat in the region.

Acknowledgements

We would like to acknowledge the kind assistance of Gill Jones and Sheila Hamilton-Dyer for additional information that was unavailable from their published works, and for pointing us towards several reports that we were unaware of. Astrid Caseldine also kindly supplied references to other published works. Thanks too to Richard Brewer for allowing us to use material from excavations in Caerwent in advance of publication. Mark Dover created Figure 15.1. The paper is dedicated to the memory of Richard Avent, who provided us with the opportunity to study the material from his excavations at Laugharne Castle.

References

Birrell, J. (2006) 'Procuring, preparing and serving venison in late Medieval England', in *Food in Medieval England*, eds C. Woolgar, D. Serjeantson and T. Waldron, OUP, Oxford, 176–188.

Browne, S. (2000) 'The animal bones', in *Hen Domen, Montgomery: a Timber Castle on the English-Welsh Border: a Final Report*, eds R. Higham and P. Barker, University of Exeter Press, Exeter, 126–134.

Burnett, D. (n.d.) 'The animal bones from excavations of a courtyard house in the NW corner of Caerwent', Consultancy Report for National Museum of Wales.

Dobney, K. (2001) 'A place at the table: the role of vertebrate zooarchaeology within a Roman research agenda', in *Britons and Romans: advancing an Archaeological agenda*, eds S. James and M. Millett, CBA Research Report, London, 36–45.

Gidney, L. (2007) 'Animal and bird bones', in *Excavations at Dryslwyn Castle 1980–95*, ed. C. Caple, Society for Medieval Archaeology Monograph, Leeds, 295–314.

Gidney, L. and Caple, C. (2007) 'Food production and consumption in Wales: 13th to 15th centuries', in *Excavations at Dryslwyn Castle 1980–95*, ed. C. Caple, Society for Medieval Archaeology Monograph, Leeds, 283–295.

Gilchrist, R. (1988) 'A reappraisal of Dinas Powys: local exchange and specialized livestock production in 5th- to 7th-century Wales', *Medieval Archaeology* 32, 50–62.

Haglund-Calley, L. and Cornwall, I. (1963) 'Report on the Dinas Powys animal bones', in *Dinas Powys: An Iron Age, Dark Age, and Medieval Settlement in Glamorgan*, ed. L. Alcock, University of Wales Press, Cardiff, 192–194.

Hambleton, E. and Maltby, M. (2004a) 'Animal Bones from Medieval Contexts at Laugharne Castle, Dyfed, Wales', Consultancy Report for CADW, Bournemouth University.

Hambleton, E. and Maltby, M. (2004b) 'Animal Bones from Post-medieval Contexts at Laugharne Castle, Dyfed, Wales', Consultancy Report for CADW, Bournemouth University.

Hambleton, E. and Maltby, M. (2009) 'Animal Bones from Caerwent Forum-Basilica, Wales', Consultancy Report for National Museum of Wales, Bournemouth University.

Hamilton-Dyer, S. (1993) 'The animal bones', in ed. V. Zienkiewicz, 'Excavations in the *Scamnum Tribunorum* at Caerleon', *Britannia* 24, 132–136.

Hamilton-Dyer, S. (1994) 'The animal bone', in ed. M. Fulford, J. Allen and S. Rippon, 'The settlement and drainage of the Wentlooge Level, Gwent: excavation and survey at Rumney Great Wharf, 1992', *Britannia* 25, 197–200.

Hamilton-Dyer, S. (n.d.) 'The animals bones from excavations at the British Telecom site, Caerleon'. Consultancy Report.

Higbee, L. (2006) 'Animal bone', in ed. A. Barber, S. Cox and A. Hancocks, 'Vale of Glamorgan: evaluation and excavation 2002–3', *Archaeologia Cambrensis* 155, 91–94.

Jones, G, (1981) 'The animal bone', in ed. V. Metcalfe Dickenson, 'Excavations at Old Market Street, Usk', *The Monmouthshire Antiquary* 4, 33–35.

Jones, G. (1987) 'Animal bones', in ed. H. Owen John, 'Excavations at the sand covered Medieval settlement at Rhossili, West Glamorgan', *The Bulletin of the Board of Celtic Studies* 34, 268–269.

Jones, G. (1989) 'The animal bones' in ed. P. Courtney, 'Excavations in the outer precinct of Tintern Abbey', *Medieval Archaeology*, 33, 138–141.

Jones, G. (1992) 'The animal bone' in. ed K. Lightfoot. 'Rumney Castle, a ringwork and manorial centre in South Glamorgan', *Medieval Archaeology* 36, 151–155.

Jones, G. (1996) 'The animal bones from 75 High Street', in *Excavations at Cowbridge, South Glamorgan, 1977–88*, eds J. Parkhouse and E. Evans, BAR British Series 245, Archaeopress, Oxford, 226–232.

Jones, G. (2010) 'Animal bones', in *The Excavation of the Southern Defences of the Caerleon Legionary Fortress 1982*, eds H. Mason, P. Macdonald and H. Cool, Archaeology Data Service, York (doi:10.5284/1000161).

Kinnes. I. (1989) 'The animal bones', in *Whitton: an Iron Age and Roman Farmstead in South Glamorgan*, eds M. Jarrett and S. Wrathmell, University of Wales Press, Cardiff, 232.

Langton, J. (2011) 'Forest and chases in Wales and the Welsh Marches: an exploration of their origins and characteristics', *Journal of Historical Geography*, 37, 263–272.

Maltby, M. (2007) 'Chop and change: specialist cattle carcass processing in Roman Britain', in *TRAC 2006: Proceedings of the 16th Annual Theoretical Roman Archaeology Conference, Cambridge 2006*, eds B. Croxford, N. Ray, R. Roth and N. White, Oxbow, Oxford, 59–76.

Maltby. M. (2010) *Feeding a Roman Town: Environmental Evidence from Excavations in Winchester, 1972–1985*, Winchester Museums Service, Winchester.

Millard, A., Jimenez-Cano, N., Lebrasseur, O. and Saki, Y. (2011) 'Isotopic investigation of animal husbandry in the Welsh and English periods at Dryslwyn Castle,

Camarthenshire, Wales', *International Journal of Osteoarchaeology* DOI:10.1002/oa.1292

Noddle, B. (1983) 'The animal bones', in ed. P. Casey, 'Caerwent (Venta Silarum): the excavations of the north-west corner tower and analysis of the structural sequence of the defences', *Archaeologia Cambrensis* 132, 63–69.

Noddle, B. (1993) 'The mammal bones', in J. M. Lewis, 'Excavations at Loughor Castle, West Glamorgan 1969–73', *Archaeologia Cambrensis* 142, 159–169.

O'Connor, T. (1986) 'The animal bones', in *The Legionary Fortress Baths at Caerleon: Volume II. The Finds*, eds J. D. Zienkiewicz, National Museum of Wales/CADW, Cardiff, 225–248.

Pinter-Bellows, S. (1996) 'The animal bone', in *The Excavation of a Late Prehistoric and Romano-British Settlement at Thornwell Farm, Chepstow, Gwent, 1992*, ed. G. Hughes, BAR British Series 244, Archaeopress, Oxford, 81–84.

Rotherham, I. (2007) 'The ecology and economics of medieval deer parks', *Landscape Archaeology and Ecology* 6, 86–102.

Sadler, P. (1996) 'The animal bones from Bear Field', in *Excavations at Cowbridge, South Glamorgan, 1977–88*, eds J. Parkhouse and E. Evans, BAR British Series 245, Archaeopress, Oxford, 224–227.

Sadler, P. (1997) 'Faunal remains', in *Leucarum: excavations at the Roman auxiliary fort at Loughor, West Glamorgan 1982–84 and 1987–88*, eds A. Marvell and H. Owen-John, Britannia Monograph, London, 396–409.

Silvester, R. (2010) 'Historical concept to physical reality: forests in the landscape of the Welsh borderlands', in *Forests and Chases of Medieval England and Wales, c.1000 to 1500*, eds J. Langton and G. Jones, St John's College, Oxford, 163–178.

Stokes, P. (2000) 'A cut above the rest?' Officers and men at South Shields Roman fort', in *Animal Bones, Human Societies*, ed. P. Rowley-Conwy, Oxbow, Oxford, 146–151.

Sykes, N. (2006) 'The impact of the Normans on hunting practices in Britain', in *Food in Medieval England*, eds C. Woolgar, D. Serjeantson and T. Waldron, Oxford University Press, Oxford, 162–175.

Sykes, N. (2010) 'Deer, land, knives and halls: social change in Early Medieval England', *Antiquaries Journal* 90, 175–191.

Sykes, N., Baker, K., Carden, R. F., Higham, T., Hoelzel, R. and Stevens, R. (2011) 'New evidence for the establishment and management of European fallow deer (*Dama dama dama*) in Roman Britain', *Journal of Archaeological Science* 38, 156–165.

Thomas, R. (2007) 'Maintaining social boundaries through the consumption of food in medieval England', in *The Archaeology of Food and Identity*, ed. K. Twiss, Southern Illinois University Centre for Archaeological Investigations Occasional Paper, Carbondale, 130–151.

Zienkiewicz, V. (1993) 'Excavations in the *Scamnum Tribunorum* at Caerleon', *Britannia* 24, 27–140.

Making a Fast Buck in the Middle Ages: Evidence for poaching from Medieval Wakefield

Matilda Holmes

Introduction

Animal bones are often one of the most abundant finds recovered from archaeological sites. They can represent domestic waste or that resulting from industrial or craft-based processes. However, occasionally they are the result of events that are harder to classify. In the deposit discussed in this paper, a large number of fallow deer bones were recovered from a single cess pit, context (20141), in Wakefield, northern England dated to the fifteenth or sixteenth centuries. This paper will consider the likely events that led to the deposition of this group of bones, and what this can tell us of those living in the associated property.

Methods

Site excavation

During November 2008, Birmingham Archaeology excavated a site to the west of the centre of Wakefield, northern England (NGR SE 328 207) in advance of a multi-use development (McNicol and Hewitson 2009). Excavation revealed three burgage plots dating from the Early Medieval period that fronted onto Westgate, a street which is still in use today. The backyards were used for industrial processes, most likely cloth manufacture, and the tenements at the front of these plots likely housed the workers and their families who would have been of relatively modest status. The group of bones central to this discussion came from a domed, stone-lined cess pit, situated at the south-western edge of the middle tenement, dated by other artefacts from the same fill to the Late Medieval period (fifteenth to sixteenth century). The cess deposits were in a clear sequence: the bottom was lined with clay, and then a twenty centimetre layer of organic fill was sealed by another layer of organic waste which contained a large amount of animal bone that forms the basis of this discussion. The plant macrofossil evidence (Grinter 2009) suggests that there was an area of stabling nearby (probably in the backyard of a neighbouring tenement), and open common or agricultural land directly to the north-west.

Results

Although fallow deer (*Dama dama*) bones predominate in the fill of the cess pit, fragments from a cattle (*Bos* sp.) mandible and maxilla and a pig (*Sus* sp.) ulna were also recovered, in similar condition to the deer remains. The inclusion of these bones from other species is most likely coincidental, and it is the group of fallow deer bones that will be explored further in this paper.

The bones were in good condition with no refits and only one fresh break suggesting good preservation coupled with limited, if any, post depositional movement. There was no evidence of canid or rodent gnawing, indicating they were buried quickly. No bones had been burnt or exhibited craft working modifications, although butchery marks were consistent with the removal of antlers from the skull and subsequent chopping and sawing of the antlers, as well as the disarticulation of the femur and pelvis with a knife. Jointing of the carcass was evidenced by transverse chops through a femur and a cervical vertebra that is presumed to be fallow deer, but could only be identified to the level of cervid species.

As the majority of bones appear to have been buried rapidly after deposition, and were from one context, it is suggested that they result from one activity that occurred over a short time.

The fallow deer assemblage was dominated by fragments from the metapodials (lower leg) as well as a few other limb bones as Table 16.1 shows. There were no girdle elements (pelves or scapulae) or countable vertebrae, with the exception of the aforementioned vertebra. One complete skull, without antlers, and one partial skull with antlers, two mandibles and two antlers with the frontal bone attached were also recovered, as well as various other antler fragments. Foot bones (phalanges) were recorded, but not in large enough quantities to suggest they were deposited articulating with the lower leg bones. Although this could be an example of recovery bias (i.e. being missed during hand excavation), fallow deer phalanges are robust and fairly large, and with no sieved samples to

Element	MNE
Antler	2
Skull	2
Mandible*	2
Radius P	1
Radius D	1
Ulna	1
Femur P	1
Femur D	1
Tibia D	1
Calcaneus	1
Metacarpal P	24
Metacarpal D	22
Metapodial D	1
Metatarsal P	18
Metatarsal D	18
1st phalange**	2
2nd phalange**	1
3rd phalange**	1
Total	100

** adjusted for frequency bias
* mandibles with molars
P = proximal end; D = distal end

TABLE 16.1. Minimum number of fallow deer elements recorded from Wakefield.

compare with, the degree to which this explains their apparent absence cannot be confirmed.

Based on the greatest number of left proximal metacarpals at least thirteen animals are represented in the assemblage.

Work carried out on fallow deer bones from Dudley Castle has shown that measurements from metapodia can be used to investigate sexual dimorphism with some success, as there is a notable difference between the relatively short and slender female (doe) and the longer and more robust male (buck) metapodia (Thomas 2005, 61). These measurements (greatest length against smallest diameter) have been plotted in Figure 16.1 using

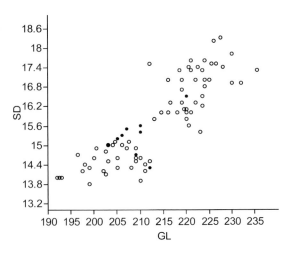

FIGURE 16.1. Sexual dimorphism of fallow deer metatarsals. GL = greatest length; SD = smallest diameter of shaft. Open circle = data from Dudley Castle (Thomas 2005, fig. 134); filled circle = data from Wakefield.

the most abundant complete bones present (the metatarsals). These elements show a clear separation between a large grouping similar to metatarsi of does from Dudley Castle and a single element corresponding with those of bucks. There are also two ambiguous measurements, which may be bucks or does. The presence of bucks is also apparent from at least two antlers with parts of the skull connected, meaning they were not from chance finds of cast antler.

The most abundant data available for investigating the age at death of the fallow deer was from bone fusion (Carden and Hayden 2006). Figure 16.2 shows that, although most bones were fully fused, the animals were not selected for a cull at a particular age. Instead, there were a mixture of new born animals and those which died before reaching 23 months, 29 months and 42 months of age.

The presence of bucks with sets of antlers narrows the season of death for these animals to between early summer, when the antlers are nearly full grown until after the rut when they are shed during March and April (Chapman and Chapman 1997, 106). However, the best indication for seasonality comes from the presence of fawns which are traditionally born from May to late June (Chapman and Chapman 1997, 139). Thus, the presence of neonatal animals implies that the likely season in which these animals were killed was between May and July.

Discussion

Origins of the assemblage

The assemblage is characterised by a predominance of metapodials, which are bones of little meat value. When combined with the scarcity of limb bones and vertebrae this suggests that the deposit did not represent food waste as these elements are commonly removed with joints of meat, or filleted out as part of the secondary butchery process. Antler and metapodials are also useful sources of raw material for craftsmen; however, although the antler has been heavily chopped and sawn there are no signs of offcuts or working on these elements, so their disposal as a result of craft-work may also be discounted.

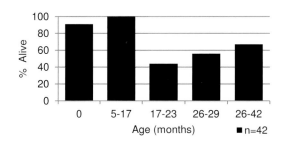

FIGURE 16.2. Fusion ages of fallow deer (after Carden and Hayden 2006). Showing the proportion of bones representing live deer beyond each age range.

Once food and craft-based origins have been disregarded, the implication is that the assemblage derives from some other form of processing waste. Two closely linked modes of processing frequently provide skeletal patterns observed within this deposit – those of primary butchery and skin processing. The particular elements discarded from each of these processes vary with the need of the individual practitioner and the anatomy of the animal. For example, when domestic animals were skinned and the pelts processed by the tanner or whittawyer, they often included the feet, tail and sometimes horn cores and metapodials of the animal, but generally not the skull, which would have made the skin heavy and cumbersome to move (Serjeantson 1989, 136–139; Armitage 1990, 84; Albarella 2003, 77). However, a group of roe deer (*Capreolus capreolus*) bones were recovered from a tannery in London, which included the antlers, suggested by the authors to have been left on to aid identification (Armitage and Butler 2005). Indeed this assemblage is similar to contemporary skin processing deposits (e.g. Harman 1996; Holmes 2009), except for the inclusion of skull and antler fragments, and the paucity of phalanges. The other possibility, that the bones originated from primary butchery activities, should also be considered. The butcher may be expected to remove the skin and pass it on to the tanner or tawyer, as well as dismembering the carcass, thereby removing the limb bones and vertebrae with the joints of meat. The skull may also have been passed on as a source of cheek meats, tongue and brawn, although it is not unreasonable to suggest that antlers may have been removed to make the head more portable.

Given the relative absence of phalanges and caudal (tail) vertebrae, elements that are commonly removed with the skin and taken elsewhere, it is therefore most likely that this group of bones were deposited following processing of the carcass – after skinning and jointing for meat – the metapodials and antlers were disposed of as waste products of the primary butchery process.

Procurement of venison

Legally, the right to hunt deer during the fifteenth and sixteenth centuries was reserved for the aristocracy, and the house within which the fallow deer were recovered would not have accommodated a high status household. It is also unlikely that the inhabitants of this part of Wakefield could have legitimately bought this quantity of venison. In an attempt to keep the status of venison as a highly desirable meat, a proscription was placed upon its sale or purchase until the early nineteenth century (Munsche 1981, 177; Manning 1993, 11), thereby restricting the availability, reinforcing social obligations through gift giving and making obvious to law enforcers that lower classes in possession of the venison were not doing so legally (Birrell 1996, 85).

Carcass Parts	Dudley Castle	Sandal Castle	Okehampton Castle	Launceston Castle	Wakefield
Mandible	-	14	28	-	8
Vertebrae	2	-	-	-	0
Upper Foreleg	7.3	14.5	11.3	5.7	2
Lower Foreleg	19	34	45	17	100
Upper Hindleg	55.2	28.5	54	51.3	2.4
Lower Hindleg	100	100	100	100	75
Phalanges	16	-	45	12	5.3
Total Number	570.5	622	396	163	96

TABLE 16.2. Relative proportion of carcass parts recorded at Wakefield and selected high-status sites referred to in the text. Due to the way data were recorded in the comparanda, this is given as the mean proportion of the most common anatomical element. Upper foreleg = scapula, humerus, radius and ulna; upper hindleg = pelvis, femur, tibia, astragalus and calcaneus; lower legs = metapodials.

As well as rules regarding who could hunt deer or have access to venison, there were also rules of a more practical nature. Not only was hunting deer prohibited during the fawning (birthing) season, rights of way through parks were often restricted at this time to minimise disturbance to the does (Almond 2003, 86). In this instance this was not the case – there is clear evidence that the deer were hunted during this time from the presence of new born fawns. Additionally, legitimate hunts would be carried out in seasons specific to the prey – for example, bucks would be hunted between June and September, and does between September and February (Cummins 1988, 33; Almond 2003, 87). The presence of both bucks and does in the Wakefield assemblage again points to nefarious origins, as does the likely season of death (May to July), which is outside the official hunting seasons for both sexes.

Furthermore, the representation of carcass parts is unusual. Traditionally, fallow deer were hunted in the same way as red deer (*Cervus elaphus*), either on horseback with a pack of hounds, or by being rounded up and driven towards hunters who would have killed many at once (Almond 2003, 22). Following the hunting and killing of a deer on a legitimate hunt, there was a strict 'unmaking' ceremony, described in Medieval hunting manuals, whereby specific portions of the carcass were redistributed in particular ways. As a result, the hind legs were taken to high-status sites, while hunt servants were given the foreleg (Sykes 2007, 150–153; Thomas 2007, 137). This process is reflected archaeologically, where a predominance of hind limb bones has been observed at many high status sites such as the nearby Sandal Castle (Griffith *et al.* 1983) as well as Dudley Castle (Thomas 2005), Okehampton Castle (Maltby 1982) and Launceston Castle (Albarella and Davis 1996). This bias can be directly related to the breaking up of the deer carcass in a highly ritualised manner. If the body parts from Wakefield are compared with those recorded at contemporary high-status sites (Table 16.2), it is clear that the bones from the cess pit are atypical of the patterns recorded at more likely legitimate locations as there are considerably fewer meat-bearing bones, and greater numbers of lower fore leg bones (Figure 16.3).

The evidence implies that the assemblage excavated at Wakefield was not one of legitimate gain: (i) the remains of so many deer outside of a high-status

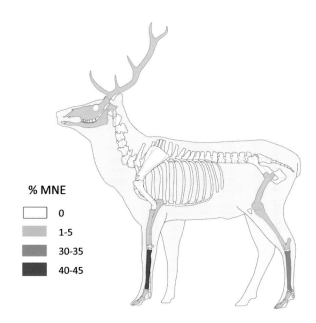

FIGURE 16.3. Proportion of the Minimum Number of anatomical Elements (MNE) present in the assemblage (image provided by Coutureau 1976).

context; (ii) the presence of animals killed out of season; (iii) an atypical representation of anatomical elements usually subject to ritualised distribution; (iv) and the rapid disposal of the remains of one episode of carcass processing in the backyard of a house with no other evidence of butchery activity.

During the Medieval period restrictions on hunting game, and particularly deer, were not rigorously adhered to (Manning 1993, 66), so that by the end of the sixteenth century poaching was rife. Poaching was prevalent amongst the upper classes that did it for sport and social engagement, and the peasant classes who supplied a prolific black market in venison and deer skin (Birrell 1996; 2001). The act of poaching itself was a potent social statement, reflecting an unwillingness of the peasant classes to accept a law that considered wild creatures as possessions of the gentry (Manning 1993, 34). Nonetheless, the risk of receiving and distributing venison was a crime punishable by fine which, if it could not be paid, would result in the perpetrator being imprisoned. Research undertaken by Jean Birrell (1996) into the legal proceedings of poaching trials has indicated that receivers and peasant poachers were often poor, although there was a thriving market for venison for those prepared to chance the consequences.

It is likely that the deer were poached from a nearby source. Wakefield was surrounded by many deer parks and the largest called the 'New Park' contained around 200 fallow deer at this time; others are recorded at Sandal Castle and Barnsdale Forest (Shirley 1867, 218) so deer would have been readily available. Indeed reference is made to a 'fellowship' of poachers who targeted Wakefield New Park (Manning 1993, 43). Poachers would have found that disposing of venison through a receiver in a nearby town (such as Wakefield) could 'provide a market and a welcome degree of anonymity' (Birrell 1996, 85–86).

From this address on Westgate, therefore, at sometime during the fifteenth or sixteenth century, the evidence suggests that someone received the poached carcasses of at least thirteen fallow deer, the skins were removed and disposed of and the carcasses cut up and sold on as meat. The highly conspicuous antlers and the lower limb bones which contained little meat or marrow were put into a cess pit and rapidly backfilled to keep the evidence hidden. Those who lived here must have been willing to risk a fine and imprisonment for this act, which suggests they were not above living outside the law.

Conclusions

Little direct archaeological evidence for poaching has previously been recorded, with the exception of the remains of large numbers of red and fallow deer from the rural site of Lyveden, Northamptonshire (Grant 1971; Grant 1975). The likely consumption of poached deer in the urban context has been inferred by Sykes (2007, 156–7), albeit from a small sample of bones from five sites. In both rural and urban cases, the presence of deer remains from poached animals has been conjectured based on the deposition of elements from all parts of the carcass; a pattern that does not conform to legitimate sources typical of high-status sites and game keepers' residences. Deer remains from the Wakefield excavation offer an important contribution to current knowledge, providing the first hard evidence in an urban context for the presence of poachers and the distribution of venison on the black market. Furthermore, the site has presented a good case for potential applications of animal bone evidence. When used in conjunction with other sources of evidence they can provide nuanced interpretations of the day to day life of people living and working in the urban environment.

Acknowledgements

Thanks to Birmingham Archaeology for the use of their site data and to Richard Thomas for access to metatarsal measurements from Dudley Castle and for commenting on an earlier draft.

References

Albarella, U. (2003) 'Tawyers, tanners, horn trade and the mystery of the missing goat', in *The Environmental Archaeology of Industry*, eds P. Murphy and P. Wiltshire, Oxbow, Oxford, 71–86.

Albarella, U. and Davis, S. (1996) 'Mammals and birds from Launceston Castle, Cornwall: decline in status and the rise of agriculture', *Circaea* 12, 1–156.

Almond, R. (2003) *Medieval Hunting*, Sutton, Stroud.

Armitage, P. (1990) 'Post-medieval cattle horn cores from the Greyfriars site, Chichester, West Sussex', *Circaea* 7, 81–90.

Armitage, P. and Butler, J. (2005) 'Medieval deerskin processing waste at the Moor House site, London EC2', *London Archaeologist* 10, 323–327.

Birrell, J. (1996) 'Peasant deer poachers in the medieval forest', in *Progress and Problems in Medieval England: Essays in Honour of Edward Miller*, eds R. Britnell and J. Hatcher, CUP, Cambridge, 68–88.

Birrell, J. (2001) 'Aristocratic poachers in the Forest of Dean: their methods, their quarry and their companions', *Transactions of the Bristol and Gloucestershire Archaeological Society* 119, 147–154.

Carden, R. F. and Hayden, T. J. (2006) 'Epiphyseal fusion in the postcranial skeleton as an indicator of age at death of European fallow deer (*Dama dama dama Linnaeus*, 1758)', in *Recent Advances in Ageing and Sexing Animal Bones*, ed. D. Ruscillo, Oxbow, Oxford, 227–236.

Chapman, D. and Chapman, N. (1997) *Fallow Deer*, Coch-Y-Bonddu, Machynlleth.

Coutureau, M. (1976) Drawings of vertebrate skeletons, http://www.archeozoo.org/fr:, accessed 16.06.2008.

Cummins, J. (1988) *The Hound and the Hawk,* Phoenix Press, London.

Grant, A. (1971) 'The animal bones', in Excavations at the Deserted Medieval Settlement at Lyveden, eds G. Bryant and J. Steane, *Northampton Museums and Art Gallery Journal* 9, 90–93.

Grant, A. (1975) 'The animal bones', in Excavations at the deserted medieval settlement at Lyveden, Northants, eds J. Steane and G. Bryant, *Journal of the Northampton Museums and Art Gallery* 12, 152–157.

Grinter, P (2009) 'Waterlogged plant macrofossil analysis', in *Excavations at Westgate, Wakefield, West Yorkshire*, eds D. McNicol and C. Hewitson, Unpublished Birmingham Archaeology Report PN1881.

Griffith, N., Halstead, P., Maclean, A. and Rowley-Conwy, P. (1983) 'Faunal remains and economy', in *Sandal Castle Excavations 1964–73*, eds P. Mages and L. Butler, Wakefield Historical Society, Wakefield, 341–349.

Harman, M. (1996) 'The mammal bones', in The excavation of a late 15th- to 17th-century tanning complex at the Green, Northampton, ed. M. Shaw, *Post-Medieval Archaeology* 30, 89–103.

Holmes, M. (2009) 'The animal bones from Upper Well Street, Coventry', unpublished Birmingham Archaeology report.

Maltby, J. M. (1982) 'The animal and bird bones', in Excavations at Okehampton Castle, Devon Part 2 – the Bailey, eds R. Higham, J. Allen and S. Blaglock, *Proceedings of the Devon Archaeological Society* 40, 114–138.

Manning, R. (1993) *Hunters and Poachers: A Social and Cultural History of Unlawful Hunting in England,* Clarendon Press, Oxford

McNicol, D. and Hewitson, C. (2009) *Excavations at Westgate, Wakefield, West Yorkshire* Unpublished Birmingham Archaeology Report PN1881.

Munsche, P. (1981) *Gentlemen and Poachers – the English Game Laws 1671–1831,* CUP, Cambridge.

Serjeantson, D. (1989) 'Animal remains and the tanning trade', in *Diet and Crafts in Towns: The Evidence of Animal Remains From the Roman to the Post-Medieval Periods,* eds D. Serjeantson and T. Waldron, BAR British Series 199, Archaeopress, Oxford, 129–146.

Shirley, E. (1867) *Some Account of English Deer Parks: With Notes on the Management of Deer,* John Murray, London.

Sykes, N. (2007) 'Taking sides: The social life of venison in Medieval England', in *Breaking and Shaping Beastly Bodies: Animals as Material Culture in the Middle Ages,* ed A. Pluskowski, Oxbow, Oxford, 149–160.

Thomas, R. (2005) *Animals, Economy and Status: Integrating Zooarchaeological and Historical Data in the Study of Dudley Castle, West Midlands (c.1100–1750),* BAR British Series 392, Archaeopress, Oxford.

Thomas, R. (2007) 'Chasing the ideal? Ritualism, pragmatism and the later medieval hunt in England', in *Breaking and Shaping Beastly Bodies: Animals as Material Culture in the Middle Ages*, ed. A. Pluskowski, Oxbow, Oxford, 125–148.

'Playing the stag' in Medieval Middlesex? A perforated antler from South Mimms Castle – parallels and possibilities

John Clark

The Norman castle at South Mimms, Hertfordshire (formerly Middlesex) lies shrouded by trees, in a copse adjacent to a now defunct chalk-pit, to the west of Potters Bar. The site was the subject of research excavation over several summer seasons in the 1960s. The excavation confirmed that the site was a motte and bailey castle of unusual construction dating to the mid-twelfth century. Documentary evidence suggests that it was erected by Geoffrey de Mandeville, first Earl of Essex, who was disgraced in 1143 and the finds confirmed the short life of the castle. Although some reports on this site have been published (Wilson and Hurst 1961, 318; Wilson and Hurst 1962–3, 322; Wilson and Hurst 1964, 255), the results of the excavation were never fully published. The project was eventually revived in 2006 and a publication of the full and updated site report has recently appeared (Kent *et al.* 2013).

Amidst the extensive demolition debris of the mid- to late twelfth century dumped in the motte ditch was a broken red deer antler (the right antler, only parts of the main beam and brow tine remaining), still attached to a fragment of the skull (Figure 17.1). A hole, about twelve millimetres in diameter, had been drilled through the main beam above the coronet, and contains the corroded remains of a square-section iron rod, perhaps part of a nail or spike. The edges of the cranial bone are broken, except along one side, which is smoothly cut. Two small semi-circular cuts in one broken edge suggest that there were originally other drilled holes.

Another unpublished antler with a similar perforation was recorded, from a City of London waterfront site: an almost complete (right) antler from a red deer (*Cervus elaphus*), the bone of the skull pierced by a round hole, six millimetres in diameter, containing the traces of a square-section iron rod or nail (Figure 17.2). A polygonal area of light green iron corrosion on the surface of the bone around the hole may reflect the head of a nail or an iron washer or plate. The evident direct contact between iron and bone suggests that the skull had

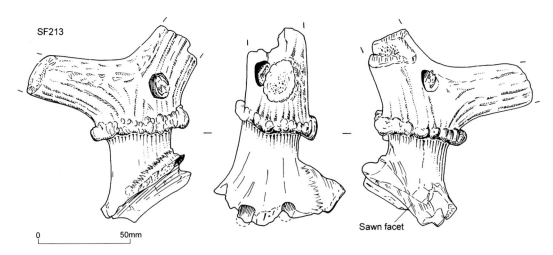

SF213

0   50mm

Sawn facet

FIGURE 17.1. *(above)* Antler fragment (SF213) from South Mimms Castle (Museum of London's London Archaeological Archive and Research Centre (LAARC) showing drilled hole containing remains of iron spike, and evidence of other modifications. (Scale 1:2. Drawing: Jon Cotton)

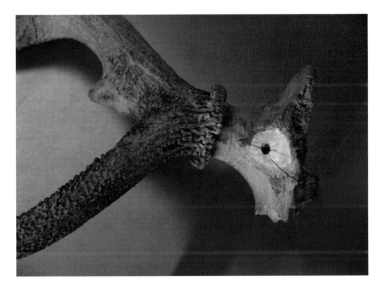

FIGURE 17.2. *(right)* Antler from excavations at St Magnus House, City of London, 1975 (Museum of London LAARC: SM75[195]<689>); close-up showing drilled hole with iron staining around. (Photo: John Clark)

been defleshed before the iron spike was inserted. This antler was mentioned in passing by Pritchard (1991, 175) in a volume on finds of the Saxo-Norman period from London: Pritchard merely noted the evidence for antler working in Late Anglo-Saxon London and that the antler was from a slaughtered animal, i.e. was not shed. It comes from a site excavated in 1974–1975 at St Magnus House/New Fresh Wharf, Lower Thames Street, City of London. The context [195] comprised river silting over the earliest Saxon structure recorded on the site, a clay, timber and rubble bank in front of the Roman timber waterfront; silting that can be dated securely to the late tenth or early eleventh century, although containing residual Roman pottery (Steedman *et al.* 1992, 12–13, 100, fig. 56; Miller *et al.* 1986, 54–5, fig. 46, 96, fig. 73).

A later parallel is from Norwich Castle, coming from a late sixteenth to mid-seventeenth century context: a fragment of antler and skull (probably fallow deer, *Dama dama*), the skull pierced by two holes, one with an iron 'rivet' (Huddle 2009, 893, fig. 10.75). Huddle suggests that this functioned as 'a crude wall hook', a practical explanation of an enigmatic object, but perhaps not the most economic or efficient way of fabricating wall furniture.

The South Mimms site notebooks record the excavators' speculation that the antler find was a 'hunting trophy'. The location of the site close to the western edges of Enfield Chase made this an attractive interpretation, although the first mention of Enfield Chase by that name dates from 1322 (Pam 1984, 10–11; Kent *et al.* 2013, 10). It is a suggestion that has been made about other modified antlers. Thus, it was concluded of a pair of roe deer (*Capreolus capreolus)* antlers from an eleventh century, but pre-Conquest, context at Norwich Castle, that 'a roe deer *trophy*… probably represented a status object rather than a specimen of any practical use' (Albarella *et al.* 2009, 185, pl. 4.42). Elsewhere it has been suggested that a fallow deer antler from a twelfth to early thirteenth century context at the Bishop's Palace at Sonning (Berkshire) was also such a trophy, because of the way the bone had been trimmed so that 'angled cuts better facilitate the hanging of the antler pair against the wall' (Hamilton-Dyer 2003, 85; Pluskowski 2007, 41, fig. 4.4).

In a Roman context at Great Chesterford (Cambridgeshire), a similar suggestion was made concerning a shed antler pierced by two nails just above the burr, which had been chopped flat on one side: 'This antler appears to have been nailed to a flat surface as a trophy' (Serjeantson 1986, 39).

However, there is no evidence for the long-term display of trophies, for which metal fixings such as those on the South Mimms and City of London finds would be appropriate, before the sixteenth century (Pluskowski 2007, 44–6: an example from 1585 illustrated fig. 4.7). Moreover, in neither the South Mimms nor the City of London find are the perforations appropriately placed for attachment to a flat surface like a wall.

What were the perforations and iron attachments for?

It is too easy to apply the modern concept of the hunting trophy to a very different Medieval milieu. When those with perforations in the bone are considered, the familiar Mesolithic pierced stag frontlets from Star Carr (Clark 1954, 168–75, pls XXII–XXIV) and a much later, but very similar, pierced frontlet from a late Roman context at Hooks Cross, Hertfordshire (Green 2005, 58) should be noted. In both cases it has been suggested they were intended to be *worn*, possibly for some symbolic or *ritual* purpose. Just because our examples are from a Medieval and Christian context, can 'ritual', as an explanation, be dismissed? There is certainly evidence for ritual surrounding Medieval deer hunting, Pluskowski (2007, 40), for example, notes the 'opportunities for impressive visual display surrounding the kill'.

At least two fourteenth century illustrations of hunting depict the slaughtered stag's head set on a pole (the Taymouth Hours, British Library MS. Yates

Thompson 13 (London *c*.1330) fol. 83v (illustrated Pluskowski 2007, fig. 4.5); the Smithfield Decretals, British Library MS. Royal 10 E IV (illustrations London *c*.1370) fol. 256r), in the latter the head is being displayed to a king who blows a hunting horn). In Rockingham Forest, Northamptonshire, in 1272, a gang of poachers killed three deer, 'and they cut off the head of one buck and put it on a stake … placing in its mouth a stick … and they made the mouth gape towards the sun, in great contempt of the lord king and his foresters' (Turner 1901, 38–9). The symbolism of the gaping mouth is unclear, but the display was certainly dramatic.

Our antler finds, however, seem to have been preserved and modified for use after the flesh had been removed from the skull, as is suggested by the iron staining on the bone of the one from St Magnus House, for example. So what other uses of antlers might there be?

There are pictorial representations of the symbolic use of antlers: a Spanish adulterer 'wearing the horns' in about 1598 (Braun and Hogenberg 1572–1617, 5: no. 7, view of Seville); the villagers of Abbots Bromley in Staffordshire, dancing through the streets in pseudo-Medieval costume with their sets of reindeer (*Rangifer tarandus*) antlers supported with heavy metal bars (in a custom first described by Robert Plot in 1686) (Cawte 1978, 65–79; Buckland 1980); 'swearing on the horns' in Highgate, Middlesex, first recorded in the eighteenth century (Roud 2008, 222–3); modern Druids processing at Avebury, led by a man carrying a set of antlers on a pole (Wallis 2009, 105, fig. 7.1).

But there is evidence for a 'ritual' symbolism of the stag's head much earlier in the Medieval period. In the sixth century Bishop Caesarius of Arles, in the south of France, (d. 542), complained in his New Year sermons about the men who go out visiting people's houses in midwinter, on the kalends of January, in the guise of *cervuli* and *anniculae* – 'little stags' and 'heifer calves':

> Whoever, therefore, on the first day of January, gives any hospitality to such wretched men, who are not just playing a game but are instead raving in a sacrilegious rite, should be aware that they are being generous not to men but to demons. And so, if you do not wish to be accomplices in their sin, do not allow a *cervulus* or an *annicula*, or any other kind of monster, to come to your houses.
>
> Arbesmann 1979, 90

This practice of *cervulum faciens* 'playing the stag', or 'putting on the head of a stag to imitate the form of a wild beast', was condemned by churchmen, and penances prescribed, on a number of occasions in Spain, Italy, and France from the fourth century to the ninth century (Arbesmann 1979). For example:

> If anyone on the kalends of January goes around as a stag or a heifer, that is, if they change themselves into the form of wild beasts and dress themselves in the skins of animals and don the heads of beasts…three years penance, because it is devilish.
>
> *Penitential of Pseudo-Theodore*, north-east France or Rhineland,
> early ninth century, *ibid*. 100–1

FIGURE 17.3. Romance of Alexander, *c.* 1344 (Bodleian Library MS. Bodl. 264, fol. 70r (James 1933, 23) (detail)): 'hobby stag', with musician playing pipe and tabor. (© The Bodleian Libraries, The University of Oxford)

This complaint may, however, have become a standard element in the clergy's condemnation of supposed pagan practices:

> The impressive sequence of prohibitions may be due not so much to the persistence of the custom itself however as to a bureaucratic inertia which persisted in repeating an originally necessary pronouncement in cultures and centuries to which it no longer applied.
>
> Pettitt 2005, 16, n. 45

But whatever its range in time and space, what did 'playing the stag' entail? Given the time of the year that it occurred, it seems likely to have been one of those traditional midwinter visiting customs, when masqueraders or mummers performed at people's doors and demanded gifts of food and drink or money (Simpson and Roud 2000, 44–5). And many such customs have involved animal disguise.

Presumably the *cervulus* was represented either by a man wearing a simple stag's head mask (as shown much later as one of a troupe of masked dancers in the margin of the well-known mid-fourteenth century Flemish manuscript of the Romance of Alexander (Bodleian Library MS. Bodl. 264, fol. 21v: James 1933, 14), or, rather more sinister, a man hidden under an animal-skin hood holding an actual or replica stag's head on a pole, as depicted elsewhere in the same manuscript (Figure 17.3).

This strange creature resembles a hobby horse with a wooden head with clacking jaws like the Kentish 'Hooden Horse' or with an actual horse's skull like the Welsh *Mari Lwyd* ('grey mare'), hobby horses of the type known as 'mast horses' (Alford 1978, 50–1, 62–5; Cawte 1978, 85–109). The form of a typical mast horse is described by Cawte (*ibid.* 8):

> A skull or carved wooden head is fixed to a pole, and the operator crouches behind, covered with a cloth attached to the base of the head. The jaw is hinged and has a handle or spring so that the operator can move it.

In the case of the Romance of Alexander stag, the performer's face can be seen peering out at the front of the figure. Additionally, not all hobby horses are horses, as is shown for example by the 'Old tup' or ram of the Sheffield area (*ibid.* 110–17) and the *Julbuk* ('Yule Goat') of Scandinavia (Alford 1978, 122–3). Indeed, some English morris sides today dance with 'hobby stags' of the mast type, like 'Eric' from the Wyre Forest and 'Harry the Hart' from St Albans (Underwood 2005). These stags, however, are all relatively recent innovations – there is nothing like them in the comprehensive survey of traditional hobby horses and other animal disguises in the British Isles published by Edwin Cawte in 1978. And 'hobby stags' seem to be rare elsewhere in Europe, although Alford (1978, 134) illustrates a mast horse with a stag's head from eighteenth century Austria.

The Romance of Alexander 'hobby stag' appears to have an artificial head, probably carved of wood, with the typical long snapping jaws of more recent mast horses; but it may be assumed that the head was surmounted by real antlers. A musician plays on pipe and tabor for the stag's performance, while on the right of the page a mother hurries her children away, with that mixture of horror and embarrassment that still greets hobby horses today!

Sadly, there is nothing to suggest continuity from the custom described in sixth century Francia by Bishop Caesarius to the hobby stag depicted by Jehan de Gris, the Flemish illustrator of the Romance of Alexander, although Flanders lies close to the area of northeast France or the Rhineland where the *Penitential of Pseudo-Theodore*, still proscribing the practice of midwinter 'playing the stag' in the ninth century (above: Arbesmann 1979, 100–1), may have been compiled (Van Rhijn and Saan 2006). Nor can we safely make the sideways leap to twelfth century Middlesex, although 'playing the stag' would have been known, at least by reputation, if only to churchmen, in Late Anglo-Saxon England: the *Pseudo-Theodore Penitential*, complete with its condemnation of the practice, was well known there, four out of the seven surviving copies being from England (*ibid.* 24–5).

There is no evidence that Middlesex peasants were 'playing the stag' on midwinter nights in the hall or at the gates of South Mimms Castle. But there is perhaps sufficient evidence that in considering the *post-mortem* use of deer antlers we should look beyond their more practical usage as raw material for the making of combs and knife handles, or the commemorative function as a trophy that we take for granted today. Sets of antlers may also have been modified and fitted with metal fastenings to facilitate their symbolic use in some ritual that is unrecorded, but which the parallels in other times and cultures certainly suggest could have once existed in Early Medieval England.

Acknowledgments

I am grateful to Dr Derek Renn and the other members of the South Mimms Castle publication project team for approving my intention to publish a separate discussion of this find in addition to that in the main site report (see Clark 2013); to Alan Pipe (Museum of London Archaeology) and Aleks Pluskowski (Reading University) for discussion of the South Mimms and City of London finds; to Isobel Thompson for information on Hooks Cross; to Dale Serjeantson for drawing my attention to the Great Chesterford find; to Jon Cotton for the drawings of the South Mimms antler; and to staff of the London Archaeological Archive and Research Centre, particularly Cath Maloney and Steve Tucker, for facilitating my access to the finds and site records from both South Mimms and St Magnus House.

References

Albarella, U., Beech, M. and Mulville, J. (2009) 'Zoological and botanical evidence: mammal and bird bone' in E. Shepherd Popescu *Norwich Castle: Excavations and Historical Survey, 1987–98, Part I: Anglo-Saxon to c.1345*, East Anglican Archaeological Report 132, Norfolk Museums Service, Norwich, 180–5..

Alford, V. (1978) *The Hobby Horse and Other Animal Masks*, Merlin, London.

Arbesmann, R. (1979) 'The "cervuli" and "anniculae" in Caesarius of Arles', *Traditio* 35, 89–119.

Braun, G. and Hogenberg, F. (1572–1617) *Civitates Orbis Terrarum*, 6 vols, Cologne; http://historic-cities.huji.ac.il/spain/seville/maps/braun_hogenberg_V_7.html, accessed 15.01.2012.

Buckland, T. (1980) 'The reindeer antlers of the Abbots Bromley Horn Dance: a re-examination', *Lore and Language* 3, 1–8.

Cawte, E. C. (1978) *Ritual Animal Disguise: A Historical and Geographical Study of Animal Disguise in the British Isles*, Brewer, Cambridge.

Clark, J. G. D. (1954) *Excavations at Star Carr: An Early Mesolithic Site at Seamer near Scarborough, Yorkshire*, CUP, Cambridge.

Clark, J. (2013) 'Appendix IV: Small finds' in J. Kent, D. Renn and A. H. Streeten *Excavations at South Mimms Castle, Hertfordshire, 1960–1991*, London and Middlesex Archaeological Society Special Paper 16, London and Middlesex Archaeological Society, London, 63–76.

Green, M. J. (2005) *The World of the Druids*, Thames and Hudson, London.

Hamilton-Dyer, S. (2003) 'The animal bones' in 'Excavation of Medieval features at St Andrew's Church vicarage, Sonning, Berkshire', eds G. Hull and M. Hall, *Berkshire Archaeological Journal* 76, 84–87.

Huddle, J. (2009) 'Antler object' in E. Shepherd Popescu *Norwich Castle: Excavations and Historical Survey, 1987–98, Part I: Anglo-Saxon to c.1345*, East Anglican Archaeological Report 132, Norfolk Museums Service, Norwich, 893.

James, M. R. (1933) *The Romance of Alexander: A Collotype Facsimile of MS Bodley 264*, Clarendon Press, Oxford.

Kent, J., Renn, D. and Streeten, A. H. (2013) *Excavations at South Mimms Castle, Hertfordshire, 1960–1991*, London and Middlesex Archaeological Society Special Paper 16, London and Middlesex Archaeological Society, London.

Miller, L., Schofield, J. and Rhodes, M. (1986) *The Roman Quay at St Magnus House, London: Excavations at New Fresh Wharf, Lower Thames Street, London 1974–78*, London and Middlesex Archaeological Society Special Paper 8, London and Middlesex Archaeological Society, London.

Pam, D. (1984) *The Story of Enfield Chase*, Enfield Preservation Society, Enfield.

Pettitt, T. (2005) 'When the Golden Bough breaks: folk drama and the theatre historian', *Nordic Journal of English Studies* 4, 1–40.

Pluskowski, A. (2007) 'Communicating through skin and bone: appropriating animal bodies in Medieval western European seigneurial culture' in *Breaking and Shaping Beastly Bodies: Animals as Material C in the Middle Ages*, ed. A. Pluskowski, Oxbow, Oxford, 32–51.

Pritchard, F. (1991) 'Small finds' in *Aspects of Saxo-Norman London: II, Finds and environmental evidence*, ed. A. G. Vince, London and Middlesex Archaeological Society Special Paper 12, London and Middlesex Archaeological Society, London, 120–278.

Roud, S. (2008) *London Lore: The Legends and Traditions of the World's Most Vibrant City*, Random House, London.

Serjeantson, D. (1986) 'The animal bones' in J. Draper 'Excavations at Great Chesterford 1953–5', *Proceedings of Cambridge Antiquarian Society* 75, 37–39.

Simpson, J. and Roud, S. (2000) *A Dictionary of English Folklore*, OUP, Oxford.

Steedman, K., Dyson, T. and Schofield, J. eds (1992) *Aspects of Saxo-Norman London: III, The Bridgehead and Billingsgate to 1200*, London and Middlesex Archaeological Society Special Paper 14, London and Middlesex Archaeological Society, London.

Turner, G. J. (1901) *Select Pleas of the Forest*, Publications of Selden Society 13. London B. Quaritch, London.

Underwood, P. (2005) *The Morris Ring Animal Archive: Morris Animals – Great Britain*, http://www.cajunmusic.co.uk/hh/mr/gb/gb_index.htm, accessed 16.01.2012.

Van Rhijn, C. and Saan, M. (2006) 'Correcting sinners, correcting texts: a context for the *Paenitentiale pseudo-Theodori*', *Early Medieval Europe* 14, 23–40.

Wallis, R. J. (2009) 'Modern antiquarians? Pagans, "sacred sites", respect and reburial' in *Antiquarians and Archaists: The Past in the Past, the Past in the Present*, eds M. Aldrich and R. J. Wallis, Spire Books, Reading, 103–21.

Wilson, D. M. and Hurst, D. G. eds (1961) 'Medieval Britain in 1960', *Medieval Archaeology* 5, 309–39.

Wilson, D. M. and Hurst, D. G. eds (1962–3) 'Medieval Britain in 1961', *Medieval Archaeology* 6–7, 306–49.

Wilson, D. M. and Hurst, D. G. eds (1964) 'Medieval Britain in 1962 and 1963', *Medieval Archaeology* 8, 231–99.

Landscapes

Forest Law in the Landscape:
Not the clearing of the woods,
but the running of the deer?

John Langton

Introduction

In a forest, wild animals were preserved for hunting exclusively by its lord. Within precisely bounded territories, stretching across land held and used by other people for all other purposes, forest law protected hunting quarry (mainly deer) known as 'venison', and the vegetation in which they fed, bred and sheltered known as 'vert'. They existed throughout Europe (Schama 1995, 237–242; Jones 2010), but it was 'in the England of the Norman kings ... that the creation of royal forests ... was most extensive and their protection most stringent' (Bloch 1961, 303): the supposedly pre-Norman Constitutions of the Forest (1016) were probably forged in the reign of Henry I (Stubbs 1921, 185), whose own forest regulations were reaffirmed in 1184 by Henry II's Assize of the Forest (Douglas and Greenaway 1953, 417–20). This code was ameliorated and amended by the Charter of the Forest in 1217 (Ruffhead 1769, 11–15), then by 160 more statutes before forest law was abrogated by the Wild Creatures and Forest Laws Act in 1971 [19&20 Eliz. II c. 52].

These laws are customarily treated as irksome impediments to the efforts of *homo economicus* to manage woods for commercial timber production, or, mainly, to clear them in order to grow crops. Therefore, in so far and for as long as they were adhered to, they ensured 'a retarded landscape' (Broad and Hoyle 1997, 48). The aim here is to examine these haunts of *homo ludens* in their own terms, as areas deliberately set aside and conserved for the running of deer (Young 1978).

Land subject to forest law

Forest law did not apply only to the 60–70 'royal forests' usually identified (Young 1979, 62, 152; Grant 1991, 221–229; Hoyle 1992, 356). It is said that forests must be royal because only kings could commission the eyres in which the system of forest justice culminated, and that forests which passed into private hands could no longer be subject to forest eyres and therefore to forest law, becoming 'chases' rather than 'forests' (Brown 1883, 13; Turner 1901, cix–cx; Bazeley 1921, 40). The granting of a forest by the crown to a subject would, therefore, be the same as disafforestation (Grant 1991, 221–29; Bond 1994, 132).

However, this distinction 'hath been the cause that readers and others have erred' (Coke 1644, 314). Of the 73 forests mentioned in the Hundred Rolls of the 1250s and 1270s, 20 did not belong the crown, and five of the 75 chases did (Illingworth 1812, 1818a); eighteen of the 40 forests mentioned in the Quo Warranto Rolls from the 1270s onwards belonged to subjects (Illingworth 1818b; English 1996, 121–282). Every earl, baron, and high cleric had forests and/or chases. Figure 18.1 illustrates that 39 of the 53 belonging to the earldom then duchy of Lancaster by 1421 originated in earldoms and other lordships rather than the crown. Forest eyres were held in some of them (Coke 1644, 314), but generally eyres were superseded by other means of governance after the Ordinances of the Forest in 1306 [34 Edw. I c. 5]. The 'vert' and 'venison' of royal and private forests and chases could be well-protected without them (Moseley 1832, 347, 356–358; Pettit 1968, 44; Birrell 1990–91; Langton 2010). Chases as well as forests extended over common waste and land belonging to people other than their lords (Coke 1644, 301). Just over half of the land in Gillingham Forest (Dorset) in the 1620s and of Alice Holt Forest (Hampshire) in 1787 belonged to the crown (Hutchins 1774 Vol. 2, 651; Anon. 1877–78, 43), but less than four percent of the deer-fed area of Cranborne Chase (Dorset, Hampshire and Wiltshire) belonged to its forest lord (Dorset County Record Office, D/PIT/M38).

'Purlieus' were areas that had been afforested by Henry II, Richard I or John, but disafforested under the Charter of the Forest (Fisher 1887, 159–61). That of the Forest of Dean (Gloucestershire) was as big as the forest itself (Hart 1945, 187); Exmoor Forest (Somerset) covered 20,000 and its purlieu 150,000 acres[1] (*c.*11,875 ha and *c.*88,060 ha) (Chanter 1907, 292–3). Notwithstanding its 'disafforestation', '*the Purlieu in some sort is Forest still*' (Manwood 1615, 170 (r); Mosley 1832, 130; Petit-Dutaillis 1915, 233–38). Chases might have equivalent areas between inner and outer bounds (Hawkins 1980, 175–81; Jackson 1885): the inbounds of Cranborne Chase contained 40,000 acres (*c.*27,749 ha), the outbounds over 700,000 (*c.*415,611 ha) (Hutchins 1774 Vol. 3, 406). Figure 18.2 shows the forests and chases so-far known, mapped to their maximum extents including 'purlieus' and outbounds.

Parks and free warrens were chartered by chief justices of the forest over the demesnes of the grantees (Turner 1901, cxxv): in parks, which must be enclosed, for 'vert' and 'venison'; in free warrens for beasts of warren (wild animals not preserved as 'game') and birds of warren (later called 'game birds'). In Medieval times free warrens were extended over other people's land and advanced to parks, chases or forests (Salzman 1941, 26; Salzman 1942–3, 49–50; Bond 1994, 144; English 1996, 58, 99, 155, 199, 207–8): Micklewood (Gloucestershire) and

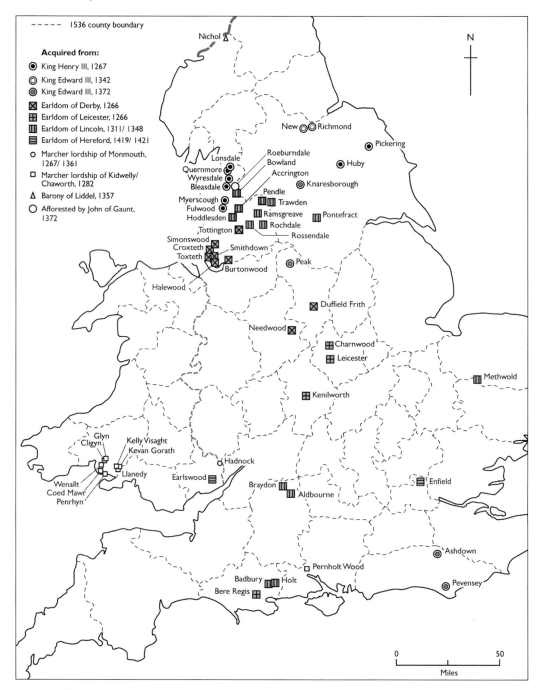

Acquired from:

◉ King Henry III, 1267
◎ King Edward III, 1342
◍ King Edward III, 1372
⊠ Earldom of Derby, 1266
⊞ Earldom of Leicester, 1266
▥ Earldom of Lincoln, 1311/ 1348
▤ Earldom of Hereford, 1419/ 1421
○ Marcher lordship of Monmouth, 1267/ 1361
□ Marcher lordship of Kidwelly/ Chaworth, 1282
△ Barony of Liddel, 1357
○ Afforested by John of Gaunt, 1372

Enfield (Middlesex) chases and Hayes and Heath forests (Shropshire) seem to have begun as free warrens (Illingworth 1818b, 256, 479, 685, 718). Purbeck (Dorset) was '*the forest, chase or warren of*' (Hutchins 1774 Vol. 2, 463); 'the chases' were also 'the king's parks' of Fulwood, Myrescough and Bleasdale (Lancashire) in 1323 (Tupling 1949, 103), and pleas for offences in parks and

FIGURE 18.1. Forests and chases belonging to the Earldom (from 1267) and Duchy (from 1351) of Lancaster by 1421, as currently known. Drawn by Ailsa Allen.

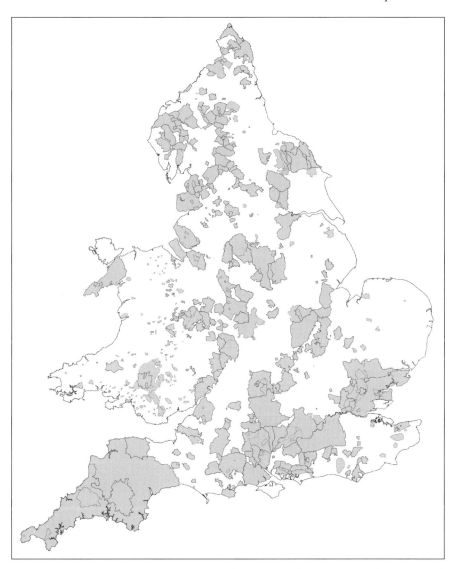

FIGURE 18.2. Forests and chases in England and Wales at their fullest but not necessarily contemporary extents, as currently known. Drawn by Graham Jones.

warrens were heard at some thirteenth century forest eyres (The National Archives, E 32/13, 14, 43–47, 152, 153, 158, 194 and 231–33), or subject to forest law through commissions of oyer and terminer (Grant 1991, 29) or at the King's Bench (Tupling 1949, 59–60, 89–92, 97–110). A statute of 1293 [21 Edw. I st. 2] applied to 'Trespassers in Forests, Chases, Parks and Warrens'; one of 1504 [19 Hen. VII c. 11] to woods of 'all the Lords and other Noblemen ... having Forests, Chases or Parks'. The abolition act of 1817 [57 and 58 Geo. III c. 61] specified 'the Chief Justices of His Majesty's Forests, Chases, Parks and Warrens'.

Needwood Forest contained ten parks (Mosley 1832, 154–157) and in Sussex 'in the main parks were enclosures in or on the edge of the seven [royal and private] forests' (Legge 1907, 297). Most parks and free warrens probably were (Turner 1901, cxvi, cxxxiii), so that initially they did not extend the area of protected hunting grounds beyond that of forests and chases, but they often survived as *lacunae* of forest law after partial or complete disafforestations (Harrison 1901, 13; Shaw 1956, 459–70).

Life under forest law

'*This law of afforesting of the lands and inheritance of other men … was thought a very extreame heavie burden … to the great hindrance and impoverishment of … subjects*' (Manwood 1615, unpaginated preface). It was not just that people could not hunt over their own ground in forests, chases and 'purlieus', no one except demesne lords over parks and warrens, and that lords of forests and chases could hunt over the land of others. Life under forest law was much more distinctive than that (Pettit 1968, 18–25; MacDermot 1973, 44–106; Stagg 1974; Birrell 1980). Every forest-dweller must swear to protect 'venison' and 'vert', serve in hunting parties, on forest court juries and at the 'drifts' which removed domestic animals during prohibited periods. They must provide board and lodging for huntsmen and their dogs ('puture'), could not carry weapons, nets or other 'engines' that might kill or catch deer, and toes must be cut from their own large dogs so that they could not chase deer. During 'fence month', from fifteen days before to fifteen days after Midsummer Day when deer were fawning, domestic animals could not be grazed and the wayleave payment of 'cheminage' was levied on anyone passing through. To erect fences preventing the free passage of deer was to commit the offence of 'purpresture'; clearing land to plant crops was to commit 'assart', and land clearance for any other purpose was 'waste'; although Henry II 'graciously allows them to take from their woods what they need … this is to be done without wasting and at the oversight of the king's foresters' (Douglas and Greenaway 1953, 418). Woodwards appointed for private woods in forests must swear to protect vert and venison. If a venison offence was detected in a private wood but not reported to forest officers by its woodward, the wood was forfeit to the forest lord.

In 'purlieus' deer could be hunted by landholders under strict conditions or not at all (Cox 1905, 9; Goodchild 1930–32, 395; Brentnall 1937–9, 379) and not by the forest lord, though escapees from his forest or chase belonged to him and were returned by his 'ranger'. The management of 'purlieu' 'vert' was not under the supervision of forest officers and woods were not subject to forfeiture to forest lords, but they could not be cut without foresters' permission, and both 'vert and venison' offences could be presented to forest courts (Chanter 1907; Brentnall 1940–42; Hart 1945; Worth 1944; Shaw 1956, 10, 15). Owners of parks had equivalent rights to forest lords, but lords of warrens could only prevent others hunting non-game animals, and warrens in forests and chases were subject to the superior rights of their lords over 'venison and vert' (Turner 1901, cxv–cxxxiv).

These regulations were policed by many more officers than the custodians of common law and manorial custom (Hackwood 1903, 34–35; Cox 1905, 17–24;

Langton 2005, 3). The most senior, including wardens, masters of the game and rangers, served for whole forests, whilst each bailiwick, ride, walk, or ward into which forests and chases were subdivided had its own numerous foresters. All had valuable perquisites (Mosley 1832, 356–359; Bazeley 1910, 193–7; Dorset County Record Office, D/PIT/L30) which could be unwarrantably extended with impunity (Mosley 1832, 129–131), and the capacity to grant gifts and favours or to oppress forest inhabitants (Fisher 1887, 142–4; Grant 1991, 112–22).

However, forest law was less severe than common law for equivalent offences (Bazeley 1910, 264, 276; Parker 1912, 30) and forest inhabitants had more rights than those held over manorial commons (St. John 1787 App. III, 7–8; Wordsworth 1907–08; Bazeley 1910, 269–70). Usually in return for customary payments, but sometimes without charge (Worth 1944, 193–4; Stagg, 1974, 33), the feeding of pigs ('pannage') and other domestic animals ('agistment'), usually *sans nombre*, was permitted when it did not affect the deer's feed (in winter) or breeding (in summer). They were also entitled to 'estovers' of 'firebote' (for firewood, turf or peat), 'heybote' (for hedging wood) and 'housebote' (for building timber), gathered nuts, fruit, edible fungi, birds' eggs and fish, and in forests with mineral deposits might be granted free mining rights by forest lords (Gough 1931; Wood 1996; Hart 2002). The first clause of the Charter of the Forest saved for ever and regularised supervision of 'the Common of Herbage, and of other things … to them which before were accustomed to have them' (Ruffhead 1769, 11), and forest common rights were exceptionally well-protected in English law (Coke 1644, 297; Linebaugh 2008, 75–6).

The forest rights of purlieu inhabitants were normally limited to grazing specified numbers of animals and lesser estovers. Those of Exmoor 'were to depasture [in the forest] 140 sheep, five mares with their young, and such cattle as they wintered, to cut and take all necessary turf, heath and fern for their households' (Chanter 1907, 294). A Dartmoor (Devon) purlieu man could graze as many cattle on the forest as his holding could over-winter, but only between sunrise and sunset, and take 'all that maye doo hym good excepte grene ocke and venyson' (Worth 1944, 202; Fox 2012, 81–107). All common rights (except that to hunt) survived over warrens, but the creators of parks usually tried to extinguish them or increase charges (Mileson 2009, 167–72).

Parks were 'devourers of the people … [containing] … nothing else than either the keeper's or warrener's lodge, or at least the manor place of the chief lord and owner of the soil' (Edelen 1968, 260), but the distinctive livelihoods in forests, chases and purlieus gathered inhabitants into communal practices, institutions, customs and seasonal rhythms different from those elsewhere, creating equally distinctive people. 'The desert forests' were prone to 'idleness, beggary and atheism … wherein infinite poor yet most idle inhabitants have thrust themselves, living covertly without law or religion' (St John 1787, 290), rioting when disafforestation and enclosure threatened (Griffin 2005; Langton 2011, 270). 'Within the memory of man' in the 1790s Selwood Forest in Somerset was still 'the notorious asylum of a desperate clan of banditti, whose depredations were a terror to the surrounding parishes' (Cox and Greswell 1911, 558), and a century later 'the inhabitants [of the forest of Dean] are ... a sort of robustic wild people, that must be civilized by good discipline and government' (J. Y. H. 1881–2, 697).

The holly and the ivy and the running of the deer

Forest law ensured the existence of large stocks of deer. Between 1234 and 1263, 2,512 bucks and does were taken from Epping and Hainault forests and Havering Park in Essex (Rackham 1989, 55); in 1315 Edward II ordered his huntsmen into 34 forests, five chases and six parks to get 322 harts, 302 bucks and 24 does (Baillie-Grohman 1904, 202–3). There were 4,977 fallow (*Dama dama*), 14 roe (*Capreolus capreolus*) and an unknown number of red (*Cervus elaphus*) deer in Windsor Forest (Berkshire and Surrey) in 1607 (Madden 1844, 181–3), and as late as 1837 'a list of those who are to claim Her Majesty's Venison in the Buck Season' included 120 for the queen plus 614 for others (Anon. 1853, 381–4). Medieval armies ate large quantities: in 1363 the Black Prince ordered that 60 bucks and twelve harts be 'salted, dressed, and sent to ... Burdeux' from Dartmoor (Deputy Keeper of the Records 1931, 204) and 100 harts and 100 bucks from Wirral Forest (Cheshire) (Deputy Keeper of the Records 1932, 458); 'during the reign of king Richard [III] at various times were slain 500 deer by the Northern men and servants of the king' in one bailiwick of the New Forest (Hampshire) (Stagg 1983, 24).

Of course, poaching was rife amongst all ranks of society, and forests served as illicit larders for rich and poor alike (Mosley 1832, 144–149; Parker 1907; Birrell 1982; Freeman 1996; Birrell 2010). But recorded poaching took fewer deer than were killed legitimately in Clarendon Forest (Wiltshire) in the thirteenth and fourteenth centuries and in the New Forest in the late fifteenth (Stagg 1983, 10–17; Richardson 2005, 53). Generally numbers killed were within the sustainable yield of wild deer (Rackham 1989, 55), and it has been suggested that winter doe hunts were mainly for culling (Stamper 1983; Richardson 2005, 26–7). After murrains (Mosley 1832, 348; Cox 1905, 28–9) and other reductions in numbers came rapid restocking (Armitage-Smith 1911, Vol. 2 206; Deputy Keeper of the Records 1932, 13, 273) and limitations on or temporary prohibitions of hunting (Cox 1905, 124; Deputy Keeper of the Records 1932, 13, 95). Censuses were taken to monitor stocks from the fourteenth century onwards (Moseley 1832, 348; Armitage-Smith 1911, Vol. 2 68, 335; Deputy Keeper of the Records 1932, 276–7; The National Archives, E 36/77), lodges were built and manned (Deputy Keeper of the Records 1931, 71; 1932, 407) and wolves were killed for the deer's safety (Booth and Carr 1991, 24, 41), boundaries were altered for their more convenient management (Moseley 1832, 14–15; Armitage-Smith 1911, 176; Deputy Keeper of the Records 1932, 4), and meadows were irrigated to ensure their feed (Armitage-Smith 1911, Vol. 2, 202).

Commercial management of woodland became more important as time passed, especially after the Act for the Preservation of Woods [35 Hen. VIII c. 17]. A 1565 crown survey of woodland in forests, chases and parks shows wide variations: the New Forest had nearly 5,000 acres of oak (*Quercus* sp.) and beech (*Fagus* sp.) in 141 woods, and nine coppices covering 447 acres set with timber standrils; in Epping Forest the 225 acres of woodland were all 'underwood', most under 20 years old, set with 'timber trees and husbands' of 60–200 years, and half the 1,132 acres of Chapel Hainault common 'is without wood or underwd except only certain short shrubby Thorns of small or no Value, the rest is thinly set with Oke & Hornbeame wch have been lopp'd & topp'd' for estovers (The National Archives, LRRO 5/39, 5–13 and 41–3). A statute of 1482 fixed the period of coppice enclosure after cutting in forests, chases and purlieus at seven years, before deer and commoners' cattle must be allowed in [22 Edw. IV c. 7]. In 1543 it was stipulated that every acre of coppice must contain 12 standard trees allowed to grow into timber [35 Hen. VIII c. 17].

However, although by 'experience it hath been found that nothing did more conduce to the raising, increase, and preservation of woods … than the execution of the forest laws' (Nisbet 1906, 451), coppice and timber management were always subservient to the preservation of vert and venison (Hammersley 1957; Pettit 1968, 26–33; Surrey History Centre, 6729/3/52; The National Archives, T 1/496/129–32). To facilitate hunting, forests and chases were 'platted with many laundes and plaines whereon groweth no timber nor underwood' (Strutt and Cox 1903, 192), which were also carefully managed for hunting. In 1351 'if any waste of the Prince's forest is enclosed in such a way as to be a nuisance to his hunting stations (*trystes*), the justices &c are to have such enclosures removed' (Deputy Keeper of the Records 1932, 9), and before James I visited Windsor Forest in 1608 holes caused by commoners' pigs were filled in 'as the damage they have caused is a danger to him riding to hunt' (Surrey History Centre, 6729/3/55). 'For the better preservation of the Game' it was forbidden to burn 'Goss, Furze or Fern, in Forests or Chaces' in 1692 and 1755 [4&5 Wm. & Mary c. 23 cl. 11; 28 Geo. II c. 19], and in 1769 'Hollies, Thorns and quicksets growing upon His Majesty's Forests and Chases' were protected for the benefit of the 'Deer and other Game therein' [9 Geo III c. 41].

These complexes of multi-purpose usage (see Figure 18.3) required intricate means of separating and combining functions (Freeman 1997; Langton 2014). Parks were cut off completely by pales and ditches, with leaps allowing deer in (Fletcher 2011, 146–9) only if they were not in or near forests or chases (Turner 1901, cxvii–cxviii; Deputy Keeper of the Records 1932, 130). Railed areas of up to 100 acres had common rights extinguished and lodges built, presumably to facilitate the winter care of deer (Madden 1844, 161–2; Wordsworth 1907–8, 303; Stewart-Brown 1910, 271). Farmers of land abutting forests and chases might be allowed to protect it with a live hedge impenetrable to deer, though 'gates' must be left for commonable animals and the return of escaped deer from purlieus (Wordsworth 1907–08, 294; Worth 1944; Rowley 1965). Within forests and chases the hedges of fields, woods and rabbit warrens could not exclude

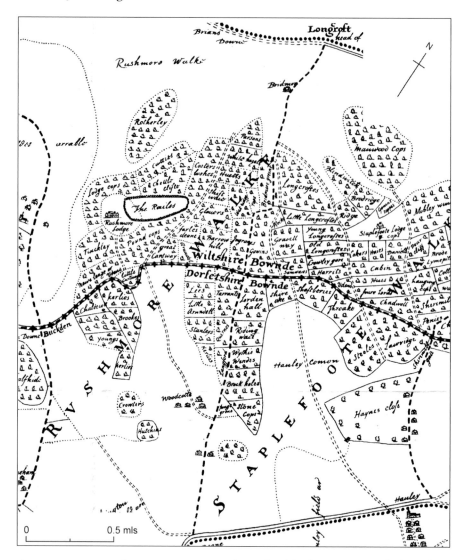

FIGURE 18.3. Parts of Rushmore and Staplefoot Walks of Cranborne Chase in *c.*1616, showing hunting lodges, rails, and the intermixture of coppices, arable, pastoral closes, and open down land. Abstracted from Thomas Aldwell's map of 1618 by Ailsa Allen.

deer, except around young coppices, where dead hedges were made by piling up the outer few yards of material from felled coppices or from hurdles, and were subject to 'leaping and creeping' to allow in deer and fawns the year before removal to admit commoners' animals (Wordsworth 1907–08, 305; Cheeseman 2009, 53). On Cranborne Chase (see Figure 18.3) the outer perimeters of groups of coppices were separated from open ground by 'borders', which 'are preserved exclusively for the Deer, & are not to be cut by the Woodmen' (Cheeseman 2009, 53), consisting of 'Holly, Ivy, Thorn and Crab and Maple, the two first of these are of greatest service to the Deer as Vert, the others are called Berrys and the Deer in the Autumn resort to them in great numbers to eat the fruit as it falls to the ground' (Dorset Record Office, D/PIT/M38). Holly was particularly prized in all forests and chases, for commoners' animals as well as deer (Jackson 1885, 170; Hackwood 1903, 77; Strutt and Cox 1903, 187).

Conclusion

In English history, land under forest law is normally abbreviated to 'royal forests' and considered mainly as a hindrance to the onward march of parliamentary control over the sovereign, the common law, and especially to the 'progress' driven by *homo economicus* through the privatisation of land to permit agriculture and other commercial enterprise. When examined from the perspective of *homo ludens*, who created and maintained them, vast tracts of territory with very distinctive lives, landscapes and ecologies come into view – the progenitors of most of the national parks and areas of outstanding natural beauty enjoyed by the successors of *homo ludens* today.

Note

1. Forest measures of length and area were bigger than customary ones. Forest perches were variable, but always larger than customary acres, most commonly of 20 rather than 16.5 feet, and forest ones were therefore multiplied by 1.466 before conversion to hectares (Brentnall 1940–42, 419; Connor 1987, 36–46).

References

Anon. (1853) *English Forests and Forest Trees: Historical Legendary and Descriptive*, Ingram Cooke, London.

Anon. (1877–78) 'Alice Holt and Woolmer forests', *Journal of Forestry and Estate Management* 1, 43–45.

Armitage-Smith, S. (1911) *John of Gaunt's Register Vols 1 and 2*, Camden Society 3rd Series 20 and 21, HMSO, London.

Baillie-Grohman, W. A. and Baillie-Grohman, F. (1904) *The Master of Game by Edward, Second Duke of York*, Ballantyne Hanson, London.

Bazeley, M. L. (1910) 'The Forest of Dean in its relations with the crown during the twelfth and thirteenth centuries', *Transactions of the Bristol and Gloucestershire Archaeological Society* 33, 153–285.

Bazeley, M. L. (1921) 'The extent of the English forest in the thirteenth century', *Transactions of the Royal Historical Society* 4th Series 4, 140–172.

Birrell, J. R. (1980) 'The medieval English forest', *Journal of Forest History* 24, 78–85.

Birrell, J. R. (1982) 'Who poached the king's deer? A study in thirteenth-century crime', *Midland History* 7, 9–25.

Birrell, J. (1990–91) 'The forest and the chase in medieval Staffordshire', *Staffordshire Studies* 3, 23–50.

Birrell, J. R. (2010) 'Families and friendships: hunting in the medieval English forest', in *Forests and Chases of Medieval England and Wales c.1000–c.1500: Towards a Survey and Analysis*, eds J. Langton and G. Jones, St. John's College, Oxford, 80–90.

Bloch, M. (1961) *Feudal Society Vol. 2*, Routledge and Kegan Paul, London.

Bond, J. (1994) 'Forests, chases, warrens and parks in medieval Wessex', in *The Medieval Landscape of Wessex*, eds T. Aston and C. Lewis, Oxbow, Oxford, 115–158.

Booth, P. H. W. and Carr A. D. (1991) *Account of Master John de Burnham the Younger, Chamberlain of Chester, of the Revenues of the Counties of Chester and Flint 1361–62*, Record Society of Lancashire and Cheshire, 125.

Brentnall, H. C. (1937–39) 'Savernake Forest in the middle ages', *Wiltshire Archaeological and Natural History Magazine* 48, 371–386.

Brentnall, H. C. (1940–42) 'The metes and bounds of Savernake Forest', *Wiltshire Archaeological and Natural History Magazine* 49, 391–434.

Broad, J. and Hoyle, R. (1997) *Bernwood: The Life and Afterlife of a Forest*, University of Central Lancashire, Preston.

Brown, J. C. (1883) *The Forests of England and the Management of them in Bye-gone Times*, Oliver and Boyd, Edinburgh.

Chanter, J. F. (1907) 'The swainmote courts of Exmoor, and the Devonshire portion of the purlieus of the forest', *Reports and Transactions of the Devonshire Association* 39, 267–301.

Cheeseman, C. (2009) 'Land and life on Cranborne Chase, 1786–1830' in *The Chase, the Hart and the Park*, ed. K. Barker, Cranborne Chase and Wiltshire Downlands Area of Outstanding Natural Beauty Occasional Papers Series 1, Cranborne, Dorset, 49–56.

Coke, E. (1644) *The Fourth Part of the Institutes of the Laws of England: Concerning the Jurisdiction of Courts*, W. Lee and D. Pakeman, London.

Connor, R. D. (1987) *The Weights and Measures of England*, HMSO, London.

Cox, J. C. (1905) *The Royal Forests of England*, Methuen, London.

Cox, J. C. and Greswell, H. P. (1911) 'Forestry', in *Victoria History of the County of Somerset Vol. 2*, ed. M. W. Page, Constable, London, 547–572.

Deputy Keeper of the Records (1931) *Register of Edward the Black Prince Preserved in the Public Record Office Part 2 (Duchy of Cornwall AD 1351–1365)*, HMSO, London.

Deputy Keeper of the Records (1932) *Register of Edward the Black Prince Preserved in the Public Record Office Part 3 (Palatinate of Chester AD 1351–1365)*, HMSO, London.

Douglas, D. C. and Greenaway, G. W. eds (1953) *English Historical Documents 1042–1189*, Eyre and Spottiswoode, London.

Edelen, G. ed. (1968) *The Description of England by William Harrison*, Cornell University Press, Ithaca.

English, B. ed. (1996) *Yorkshire Hundred and Quo Warranto Rolls 1274–1294*, Yorkshire Archaeological Society Record Series 151, Leeds.

Fisher, W. R. (1887) *The Forest of Essex*, Butterworths, London.

Fletcher, J. (2011) *Gardens of Earthly Delight: The History of Deer Parks*, Windgather Press, Oxford.

Fox, H. S. A. (2012) *Dartmoor's Alluring Uplands: Transhumance and Pastoral Management in the Middle Ages*, University of Exeter Press, Exeter.

Freeman, M. (1996) 'Plebs or predators? Deer-stealing in Whichwood Forest, Oxfordshire in the eighteenth and nineteenth centuries', *Social History* 21, 1–21.

Freeman, M. (1997) 'Whichwood Forest, Oxfordshire: an episode in its environmental history', *Agricultural History Review* 45, 137–148.

Goodchild, W. (1930–32) 'William Horder, yeoman, 1683. A story of Cranborne Chase taken from original documents', *Wiltshire Archaeological and Natural History Magazine* 45, 394–395.

Gough, J. W. (1931) *Mendip Mining Law and Forest Bounds*, Somerset Record Society 45.

Grant, R. (1991) *The Royal Forests of England*, Alan Sutton, Stroud.

Griffin, K. (2005) 'Resistance, crime and popular cultures', in *Forests and Chases of England and Wales c.1500–c.1850*, eds J. Langton and G. Jones, St John's College, Oxford, 49–52.

Hackwood, F. W. (1903) *The Chronicles of Cannock Chase*, Elliott and Stock, London.

Hammersley, G. (1957) 'The Crown woods and their exploitation in the sixteenth and seventeenth centuries', *Bulletin of the Institute of Historical Research* 30, 136–159.

Harrison, W. (1901) 'Ancient forests, chases, and deer parks in Lancashire', *Transactions of the Lancashire and Cheshire Antiquarian Society* 19, 1–37.

Hart, C. E. (1945) 'The metes and bounds of the forest of Dean', *Transactions of the Bristol and Gloucester Archaeological Society* 66, 166–207.

Hart, C. E. (2002) The *Free Miners of the Royal Forest of Dean and Hundred of St. Briavels*, Lightmoor Press, Gloucester.

Hawkins, D. (1980) *Cranborne Chase*, Victor Gollancz, London.

Hoyle, R. (1992) 'Disafforestation and drainage: the Crown as entrepreneur?', in *The Estates of the English Crown, 1558–1640*, ed. R. Hoyle, CUP, Cambridge, 297–352.

Hutchins, J. (1774) *The History and Antiquities if the County of Dorset, 3rd Edition in 4 Volumes, Introduced by Robert Douch*, EP Publishing, Wakefield, 1993.

Illingworth, W. ed. (1812 and 1818a), *Rotulorum Hundredorum Temporibus Hen. III and Ewd. I, in Turr' Lond' et in Curia Receptae Scaccarij Asservata Vols I and II*, Printed by Command of His Majesty King George III, London.

Illingworth, W. ed. (1818b) *Placita de Quo Warranto Temporibus Ewd. I. II. & III. In Curia Receptae Scaccarij Asservata*, Printed by Command of His Majesty King George III, London.

J. Y. H. (1881–82) 'The Forest of Dean', *Journal of Forestry and Estate Management* 5, 689–699.

Jackson, J. E. (1885) 'Cranborne Chase', *Wiltshire Archaeological and Natural History Society Magazine* 122, 148–173.

Jones, G. (2010) 'A common of hunting? Forests, lordship and community before and after the Conquest', in *Forests and Chases in Medieval England and Wales c.1000–c.1500*, eds J. Langton and G. Jones, St John's College, Oxford, 39–68.

Langton, J. (2005) 'Forests in early-modern England and Wales: history and historiography', in *Forests and Chases of England and Wales c.1500 to c.1850: Towards a Survey and Analysis*, eds J. Langton and G. Jones, St John's College, Oxford, 1–9.

Langton, J. (2010) 'Medieval forests and chases: another realm?', in *Forests and Chases of Medieval England and Wales c.1000–c.1500*, eds J. Langton and G. Jones, St John's College, Oxford, 15–38.

Langton, J. (2011) 'Forests and chases in Wales and the Welsh Marches: an exploration of their origins and characteristics', *Journal of Historical Geography* 37, 262–272.

Langton, J. (2014) 'Forest fences: enclosures in a pre-enclosure landscape', *Landscape History* 35, 2014, 5–30.

Legge, W. H. (1907) 'Forestry', in *Victoria History of the County of Sussex Vol. 2*, ed. W. Page, Archibald Constable, London, 291–326.

Linebaugh, P. (2008) *The Magna Carta Manifesto*, University of California Press, Berkeley.

MacDermot, E. T. (1973) *The History of the Forest of Exmoor*, David and Charles, Newton Abbot.

Madden, F. (1844) *Manuscript Maps, Charts and Plans, and Topographical Drawings in the British Museum Vol. 1*, Trustees of the British Museum, London.

Manwood, J. (1615) *A Treatise of the Lawes of the Forest*, The Society of Stationers, London.

Mileson, S. (2009) *Parks in Medieval England*, OUP, Oxford.

Mosley, O. (1832) *History of the Castle, Priory, and Town of Tutbury, in the County of Stafford*, Simpkin and Marshall, London.

Nisbet, J. (1906) 'The history of the Forest of Dean', *English Historical Review* 21, 445–59.

Parker, F. M. H. (1907) 'Inglewood Forest Part III – some stories of deer-stealers', *Transactions of the Cumberland and Westmorland Antiquarian and Archaeological Society* N. S. 7, 1–30.

Parker, F. M. H. (1912) 'The Forest Laws and the death of William Rufus', *English Historical Review* 27, 26–38.

Petit-Dutaillis, C. (1915) *Studies and Notes Supplementary to Stubbs' Constitutional History II*, Manchester University Press, Manchester.

Pettit, P. A. J. (1968) *The Royal Forests of Northamptonshire: a Study in their Economy 1558–1714*, Publications of the Northamptonshire Record Society 23.

Rackham, O. (1989) *The Last Forest: The Story of Hatfield Forest*, Dent, London.

Richardson, A. (2005) *The Forest, Park and Palace of Clarendon, c.1200–c.1650*, BAR British Series 387, Archaeopress, Oxford.

Rowley, R. T. (1965) 'The Clee forest – a study in common rights', *Transactions of the Shropshire Archaeological Society* 58, 48–67.

Ruffhead, O. (1769) *The Statutes at Large from Magna Charta to … 4 Geo. III Vols 1–9*, Baskett, Baskett, Woodfall and Strahan, London.

St John, J. (1787) *Observations on the Land Revenue of the Crown*, Debrett, London.

Salzman, L. F. (1942, 1942–43 and 1944–45) 'The Hundred Rolls for Sussex' Parts I, II and III, *Sussex Archaeological Collections* 82, 20–34, 83, 35–54 and 84, 60–81.

Schama, S. (1995) *Landscape and Memory*, HarperCollins, London.

Shaw, R. C. (1956) *The Royal Forest of Lancaster*, The Guardian Press, Preston.

Stagg, D. (1974) 'The orders and rules of the New Forest AD 1537', *Hampshire Field Club and Archaeology Society, New Forest Section Report* 13, 33–38.

Stagg, D. (1983) *A Calendar of New Forest Documents: The Fifteenth to the Seventeenth Centuries*, Hampshire Record Series 5, Hampshire County Council, Southampton.

Stamper, P. A. (1983) 'The medieval Forest of Pamber, Hampshire', *Landscape History* 5, 41–52.

Stewart-Brown, R. (1910) *Accounts of the Chamberlains and other Officers of the County of Chester 1301–60*, Record Society of Lancashire and Cheshire 59.

Strutt, F. and Cox, J. C. (1903) 'Duffield Forest in the sixteenth century', *Journal of the Derbyshire Archaeological and Natural History Society* 25, 181–216.

Stubbs, W. (1921) *Select Charters and other Illustrations of English Constitutional History 9th edition*, ed. H. W. C. Davis, Oxford, Clarendon Press.

Tupling, G. H. (1949) *South Lancashire in the Reign of Edward II*, Chetham Society, Manchester.

Turner, G. J. (1901) *Select Pleas of the Forest*, Bernard Quaritch, London.

Wood, A. (1996) 'Custom, identity and resistance: English free miners and their law, 1550–1800' in *The Experience of Authority in Early Modern England*, eds P. Griffiths, A. Fox and S. Hindle, Macmillan, Basingstoke, 249–285.

Wordsworth, C. (1907–08) 'Customs of Wishford and Barford in Grovely Forest', *Wiltshire Archaeological and Natural History Magazine* 35, 283–316.

Worth, R. H. (1944) 'The tenants and commoners of Dartmoor', *Reports and Transactions of the Devonshire Association* 74, 187–214.

Young, C. R. (1978) 'Conservation policies in the royal forests of medieval England', *Albion* 10, 95–103.

Young, C. R. (1979) *The Royal Forests of Medieval England*, Leicester University Press, Leicester.

Parks and Designed Landscapes in Medieval Wales

Spencer Gavin Smith

The creation of parks and designed landscapes in the Medieval period by English royalty and nobility is now an accepted and well-studied part of the historical and archaeological agenda. However the corresponding research agenda in Wales is still being developed, even though much of the source material is similar. This paper brings together some of the range of sources available for the study of this topic, and highlights that parks and designed landscapes were created in Wales prior to the Edwardian conquest.

In 1324, the son of notable Ruthin burgess, Cynwrig Scissor, and his wife, Isabel, was indicted for fowling by night with John Trigomide's son in the park of Ruthin and fined 40 shillings, with 30 shillings of this suspended for good behaviour (Korngiebel 2007, 20; GC1 / 2294). This misdemeanour is just one of dozens of examples recorded by the scribes for the Great Court of the lordship of Dyffryn Clwyd as happening in the parkland of the De Grey lords during the fourteenth century. The information contained in the Great Court Rolls (GC) was made available to a wider audience through an Economic and Social Research Council funded project between 1991 and 1995 to transcribe the Dyffryn Clwyd Court Roll Database 1294–1422 (Barrell n.d., 1). A section in the user manual for the database listed thirteen places which are described as parks in the Court Rolls, and indicated that further work was needed to understand the nature and extent of the parkland in the lordship of Dyffryn Clwyd in north-east Wales (Barrell *ibid.*, 162).

In addition to illicit fowling, other crimes recorded included the escape of 22 female goats in the park of Bathafarn in 1307 (GC1 / 104) and the escape of six pigs in the park of Clocaenog in 1311 (GC 1 / 309). Three men broke into the park of Clocaenog (GC 1 / 532) in 1312 and there was the theft of a cow in 1315 from the park of Bryn Cyffo (GC1 / 915).

The framework of land use in Dyffryn Clwyd, and across Wales as a whole in the fourteenth century was, in part, an inherited one. In the case of Dyffryn Clwyd the lordship was formed out one of the cantrefi of the Perfeddwlad, and contained within it the three commotes of Colion, Dogfeiling and Llanerch and had at its centre the castle of Ruthin, begun in 1277 by Welsh Prince Dafydd ap Gruffydd, but forfeited in 1282 and subsequently given to the De Grey family (Smith 1998, 428). This pattern of lordships being created out of pre-existing land units was repeated across Wales and the administrative paperwork these lordships generated contains a wealth of information. Using this in conjunction with the documents which survive from the period prior to the pre-Edwardian conquest of Wales enables a framework for understanding the transition of the landscape at this time.

The various aspects of the historical context of the information recorded in the Dyffryn Clwyd Court Rolls has been examined (Jack 1961; 1968; 1969; Pratt 1990a; Korngiebel 2007) but there has only been one project which began to explore the archaeological dimension by attempting to map the parks of Dyffryn Clwyd (Berry 1994). However this project was curtailed by the closure of the Clwyd Archaeological Service with the reorganisation of the unitary authorities in Wales in 1995 when only two parks had been mapped.

As well as inherited framework of land use, a pre-existing legal system was also in place and known as the Laws of Hywel, after the tenth century King who was thought to have codified their existence. The laws existed with regional variations across the whole of Wales and contain within them some of the most comprehensive information relating to the legal aspects of hunting and hawking to be found in any European law books. Jenkins (2000) examined their usefulness in understanding how relevant their content was to a Medieval audience and also how internal differences between the different regional variations could highlight the chronology of adoption of various types of hunting practice across Wales.

The legal system was administered from the Llys and its associated maerdref and the CADW funded Llys and Maerdref project carried out by Gwynedd Archaeological Trust between 1991 and 1995 (Johnstone 1997; Johnstone 2000) assembled the available historical information for the location of the Llysoedd and Maerdrefi of the Welsh Court in the Medieval kingdom of Gwynedd in north-west Wales. One of the sites examined was the town of Nefyn on the north coast of the Lleyn Peninsula. In late July 1284, Edward I held a Round Table there and Rees Davies, in his book 'The First English Empire' gives us an account of why the English King would chose to conduct such an event in this location at this time, drawing on themes of questing and adventure (Davies 2000, 31–32). The archaeological report prepared by Johnstone on Nefyn (2000, 180–181) mentions the field name Cae Ymryson – Tournament Field, lying to the south-west of the modern settlement, as one location which should be examined for the archaeological evidence of this particular event.

Both of these examinations of the historical and archaeological context of Nefyn provide useful evidence to further the study of Medieval Wales but are unable to ask or answer the fundamental questions of why the Round Table was held at this time of year and how was this party of a conquering king and his retinue fed in a country apparently ravaged by his army the previous year? The very fact that these questions need to be asked highlights the fact that the study of Medieval Wales has tended to concentrate on understanding the political complexities of Kingship or Lordship, and not on the economic or social use of the landscape in both pre- and post-Edwardian conquest Wales.

The questions of the cases of misdemeanour in the parks of the lordship of Dyffryn Clwyd and the logistics of the tournament held at Nefyn form part of a wider agenda than the realms of the purely academic historical and archaeological research into Medieval Wales. The geographical distribution and differing uses of the parks and designed landscapes of Medieval Wales is not well understood and appreciated by the general public who visit the castles and high status houses to which they originally belonged. The castle in Wales tends to be perceived as an overtly military symbol, and the economic and social landscape which accompanied it has a tendency not to be related through accompanying guidebooks.

The need for a concerted national research policy for the history and archaeology of the gardens, parks and designed landscapes of Wales has been highlighted in a series of articles by C. Stephen Briggs (Briggs 1991a; Briggs 1991b; Briggs 1998; Briggs 2000) and Elisabeth Whittle (Whittle 1989; 1999; 2000). The publication of 'Welsh Woods and Forests: A History' (Linnard 2000) provided a broad framework from which further avenues of study could be developed regarding the Medieval forest in Wales, and 'Houses and History in the March of Wales: Radnorshire 1400–1800'. Suggett (2005) highlighted the management of woodland resources and the consequent production of hundreds of Late Medieval houses in just one county of Wales.

The 'Register of Landscapes, Parks and Gardens of Special and Historic Interest in Wales' produced by Cadw, ICOMOS and the Countryside Council for Wales during the 1990s, (for example, the volume on Clwyd published in 1995) provided baseline information to which further examples could be added in the future. The Medieval antecedents of some of the landscapes, parks and gardens recorded in these volumes are discussed in general, although the general level of the understanding of the origins of some of these parks is still ambiguous.

Identifying examples of parks and designed landscapes in Wales is able to benefit greatly from a multidisciplinary approach. The historical and archaeological sources can be combined in the study of Medieval Wales with a rich corpus of poetry, within which there is a genre of praise poetry produced for the houses and landscapes of the Uchelwyr. Enid Roberts, in a series of articles from 1974 (Roberts 1974; Roberts 1975; Roberts 1976) highlighted the origins and development of this and closely related genres and a book (Roberts 1986) brought together edited passages from poetry composed between 1350 and 1650.

The poetry was originally written in Middle Welsh, and although some words are no longer used in Modern Welsh, the general tone, feel and meaning of the poem can be understood by a modern reader today. The perceived barrier of difficultly in learning Welsh, and the lack of translated versions of this poetry has tended to mean that this material has not been examined by historians who are unfamiliar with the Welsh language.

Between 1994 and 2009, The University of Wales Centre for Advanced Welsh and Celtic Studies produced a series of publications on the poetry of the Poets of the Nobility. All of the publications are produced through the medium of Welsh, and provide within them a critical apparatus. Two poets from this period with particularly extensive and notable collections have seen their work translated into English and made available online. The work of Dafydd ap Gwilym (*c*.1315/1320 to *c*.1350/1370) appeared online in 2007 (www.dafyddapgwilym. net) and Guto'r Glyn (*c*.1412–1493) appeared online in 2012 (http://www. wales.ac.uk/en/CentreforAdvancedWelshandCelticStudies/ResearchProjects/ CurrentProjects/PoetryofGutorGlyn/Introductiontotheproject.aspx).

The most familiar example of this genre of praise poetry, which the polymath George Borrow translated for his travelogue 'Wild Wales', first published in 1862 (Rhys 1910) and that has subsequently been quoted in part by Higham and Barker (1992, 144–146 and 300–303) Landsberg (2004, 11), Creighton (2002, 179–180; 2009, 101) and Liddiard (2005, 116–117) is the cywydd (a form of poem) composed *c*. 1390 by Iolo Goch to the Uchelwr Owain Glyn Dŵr and the llys of Sycharth (Denbighshire) and its surrounding landscape.

The cywydd is considered to be one of the finest of its genre, and is divided into two parts of unequal length. In the first part Iolo outlines that he is visiting Sycharth to sample the renowned generosity of its owner, before describing in the second part the llys buildings and their constructional details. The cywydd then provides information on the wider landscape around the llys, including an orchard, vineyard, mill, smithy and animals including peacocks and a herd of deer, which are described as being 'in another park'.

The llys at Sycharth was constructed on top of and adjacent to the earthworks of an earlier motte and bailey castle, dating from at least the early thirteenth century (XD2/1113). The site was subject to archaeological excavation in 1962–1963 (Hague and Warhurst 1966) and by the author in 2003, as well as geophysical survey in 1997 (Smith 2003) and 2009 (Gater and Adcock 2009). The research work carried out has revealed that the final phase of building could be recovered archaeologically (Hague and Warhurst 1966, 118–120) and that further buildings were present on the motte top and in the bailey (Gater and Adcock 2009, 3–7). Archaeological excavations by the author to the south of the bailey recovered what appears to be a ditch around the orchard and vineyard described in the poem and also demonstrated that the roadway into the bailey had a metalled surface which had seen at least three phases of repair.

The wider landscape surrounding the llys at Sycharth has been examined by the author. The geophysical survey conducted in 1997 revealed a probable plan of a building to the east of the castle (Smith 2003, 26) and comparison of a deed of 1382 (Anon 1849, 115–117) with the Tithe map on 1841 revealed a continuity of field names in the Cynllaith valley around the earthwork of the former llys (Smith 2003, 31–32).

A charter of 1334 for the lordship of Chirkland (Pratt 1990b, 110–111) mentions the 'boscus de Bronnedynas' as a reserved area of woodland which Pratt (1990b, 110) suggested functioned as a hay for the containment and management of deer. The 'boscus' corresponds to the prehistoric hillfort of Llwyn Bryn Dinas. This hillfort, 3.2 km (2 miles) south-west of Sycharth saw a limited rescue excavation in 1983 (Musson *et al.* 1992), the report for which did not reference the Medieval reuse of the site.

To the west of the llys at Sycharth and immediately to the east of the 'boscus' was the Forest of Mochnant or Garnau (Pratt 1990a, 24), with the Forest of Bodlith (*ibid.* 24) to the north of the llys. Within the Forest of Bodlith is Moel Iwrch 'bald hill of the roe buck' (*ibid.* 24) A house known as Moeliwrch on the south-west outskirts of the forest has two poems written to praise it during the fifteenth century, one by Ieuan ap Gruffydd Leiaf and the other by Guto'r Glyn (Roberts 1986, 12–13). The current house occupying the site is of sixteenth or seventeenth century date with nineteenth century additions (RCAHMW Coflein 2004, NPRN 36030).

The poem composed by Iolo Goch can be seen to be a more accurate reflection of the landscape surrounding Sycharth than had previously been considered, and given that the relationships between the praise poetry and the archaeological evidence is only now beginning to be explored, there remains the opportunity for the discovery of designed landscapes of equal quality and importance.

Other Uchelwyr articulated their love of the landscape through other artistic media. In 'Medieval Stone Carving' (1968) Colin Gresham illustrated six grave slabs dating from the fourteenth century which depicted different hunting scenes. The slabs are in St. Asaph Cathedral (Flintshire); Llanyblodwel church (Shropshire); Valley Crusis Abbey and the churches of Llanbedr Dyffryn Clwyd and Ysbyty Ifan (Denbighshire). Unfortunately, only one of the Grave slabs carries a name, complicating identification of the slabs to individuals, however the very fact that these Uchelwyr wished to be remembered for carrying out this activity which they so loved they had used it to depict them after their death is an important glimpse of their psyche.

As well as commissioning monuments, Uchelwyr could also commission poetry to articulate their love of hunting. The poet Dafydd ap Gwilym for example wrote poems entitled 'Yr Iwrch' (The Roebuck) and 'Serch fel Ysgyfarnog' (Love like a Hare), both of which contain detailed references to the hunt and a knowledge of the habits of these animals (www.dafyddapgwilym. net). These are just two of the many examples of this genre of poetry which have survived and provide a valuable additional dimension to the study of the Medieval park in Wales.

A love of hunting by the Royalty and Nobility of Wales can be found in some of the documents which survive from pre-Edwardian conquest Wales. In a letter from Dafydd ap Llywelyn to King Edward I in 1278, Dafydd writes 'it is hard and tedious to stay at Frodsham (Cheshire) without any solace of hunting' (Pryce and Insley 2005, 646–647). The privy seal of his Uncle Llywelyn ab Iorwerth, prince of Gwynedd during the early thirteenth century depicted a boar under a tree (Williams 1982, 19). Other seals have been found in Wales, including one of fourteenth century date recorded from Prestatyn (Flintshire) in 2010 depicting a bird riding on the back of a dog or a fox, with the inscription 'IE VOYS A BOYS – I go to the woods' (www.ukdfd.co.uk/ ukdfddata/showrecords.php?product=25760&cat=44).

One park which existed prior to the Edwardian conquest and was in the possession of the Princes of Powys Fadog serves to highlight the palimpsest of source material which relates to Medieval parks in Wales. The Medieval park at Eyton (RCAHMW Coflein 2004, NPRN 308744) south of Wrexham (Denbighshire) is recorded in a document dating from 1269 (Pryce with Insley 2005, 720) and some of the history behind this park has been examined by Cavell (2007). The park is the subject of a fourteenth century poem by Madog Benfras (Johnston 2005, 148) and in 1388 and 1389 the receiver of Bromfield and Yale accounted for empailing the park of Eyton, the whole park then belonging to the lord (Palmer and Owen 1910, 93). The drawing together of information on parks, such as the example at Eyton highlights how a more comprehensive picture of their use can be accomplished and their relationship to the wider landscape more firmly established.

The castle most probably associated with the park at Eyton is Overton Castle, but this has been lost to erosion of the riverbank of the Dee in the Late Medieval period (King in Manley *et al.* 1991, 175). The loss of the castle earthworks into the river has also meant a loss of the archaeological deposits which would have accompanied it, including the evidence for the use and consumption of deer from the park at Eyton.

The archaeological evidence of the husbandry of deer can be ascertained from the limited archaeological evidence which has been obtained from excavations in Wales. Maltby and Hambleton (this volume) have studied some of the evidence from castle excavations in south Wales. The excavation of Caergwrle Castle (Flintshire) (Manley 1994) provided important evidence for late thirteenth use and consumption of deer in north-east Wales. Other deer bones have been found during excavations directed by T. A. Glenn at Dyserth Castle in 1912 and Llys Edwin in 1934 (both Flintshire) (Glenn 1915; Glenn 1934), however these remains have yet to be re-assessed using modern techniques.

The Forests and Chases Project at St John's College Oxford (Langton, this volume) has mapped and recorded forest across Wales. The identification of a forest at Gwern Vigle (Merionethshire) was arrived at independently of the identification of an associated Medieval park to the north-west of Criccieth Castle by the author. Associated fieldwork recorded that the boundaries of the park survive in part, particularly on the western side. The relationships between forests and parks in Wales will benefit from the use of this resource and the information contained within it, providing opportunities to understand and interpret the relationship between the park and forest in Wales.

Understanding the chronology and development of Medieval parks and designed landscapes in Wales requires the view that the rulers of the different territorial units prior to the Edwardian conquest were as well connected to other countries in Europe as their English neighbours. By using a variety of source materials in parallel with one another, and understanding where the limitations of the material needs to be addressed, the production and synthesis of a searchable national resource will aid in the study and dissemination of this material.

References

XD2/1113 Madog ap Gruffudd for Strata Marcella Abbey, Gwynedd Archives Service, Caernarfon.

Anon (1849) 'Welsh Deed in Norman French', *Archaeologica Cambrensis*, 4, 115–7.

Anon, The Poetry of Guto'r Glyn. Available from: http://www.wales.ac.uk/en/CentreforAdvancedWelshandCelticStudies/ResearchProjects/CurrentProjects/PoetryofGutorGlyn/Introductiontotheproject.aspx, accessed 11.06.2011.

Anon Seal Matrix: 25760. Available from: http://www.ukdfd.co.uk/ukdfddata/showrecords.php?product=25760&cat=44, accessed 11.06.2011.

Barrell, A. D. M. (n.d.) *The Dyffryn Clwyd Court Roll Database 1294–1422 A Manual for Users*, University of Wales, Aberystwyth.

Barrell, A. D. M., Brown, M. H. and Padel, O. J. (n.d.) *Dyffryn Clwyd court roll database, 1294–1422*, UK Data Archive, Economic and Social Research Council.

Berry, A. Q. (1994) 'The parks and forests of the lordship of Dyffryn Clwyd', *Denbighshire Historical Society Transactions*, 43, 7–25.

Briggs, C. S. (1991a) 'Garden Archaeology in Wales' in *Garden Archaeology*, ed. A. E. Brown, CBA Research Report 78, 138–159.

Briggs, C. S. (1991b) 'Welsh gardens under threat: an archaeological perspective', *Journal of Garden History*, 11, 199–206.

Briggs, C. S. (1998) 'A new field of Welsh cultural heritage: inference and evidence in gardens and landscapes since *c.*1450' in *There by design: Field archaeology on parks and Gardens*, ed. P. Pattison, Royal Commission on the Historical Monuments of England, Swindon, 65–77.

Briggs C. S. (2000) 'Fields fit for Bards', *Historic Gardens Review*, Spring 2000, 35–38.

Brown, A. E. ed. (1991) *Garden archaeology*, CBA Research Report 78.

CADW–ICOMOS (1995) *Register of Landscapes, Parks and Gardens of Special Historic Interest in Wales Park I: Parks and Gardens*, Clwyd, CADW, Cardiff .

Cavell, E. (2007) 'Aristocratic widows and the Medieval Welsh frontier: The Shropshire evidence', *Transactions of the Royal Historical Society* 17, 57–82.

Charles-Edwards, T., Owen, M. E. and Russell, P. eds (2000) *The Welsh King and his Court*, University of Wales Press, Cardiff.

Creighton, O. H. (2002) *Castles and Landscapes: Power, Community and Fortification in Medieval England*, Continuum, London.

Creighton, O. H. (2009) *Designs upon the land: Elite landscapes of the Middle Ages*, Boydell, Woodbridge.

Dafydd ap Gwilym (2007) available from: http://www.dafyddapgwilym.net, accessed 11.06.2011.

Davies, R. R. (2000) *The First English empire: Power and identities in the British Isles 1093–1343*, OUP, Oxford.

Edwards, N. ed (1997) *Landscape and Settlement in Medieval Wales*, Oxbow, Oxford.

Gater, J. and Adcock, J. (2009) *Sycharth Motte and Bailey, Powys*, GSB Geophysical Survey Report 2009/12 [unpublished report for CADW].

Glenn, T. A. (1915) 'Prehistoric and historic remains at Dyserth Castle', *Archaeologica Cambrensis* 70, 47–86 and 249–252.

Glenn, T. A. (1934) *Excavations at Llys Edwin Celyn Farm, Northop, Flintshire*, Harrison and Sons, London.

Gresham C. A. (1968) *Medieval stone carving in North Wales: Sepulchral slabs and effigies of the thirteenth and fourteenth centuries*, University of Wales Press, Cardiff.

Hague, D. B. and Warhurst, C. (1966) 'Excavations at Sycharth Castle, Denbighshire, 1962–1963', *Archaeologica Cambrensis* 115, 108–127.

Higham, R. and Barker, P. (1992) *Timber Castles*, Batsford, London.

Jack, R. I. (1961) *The Lords Grey of Ruthin 1325 to 1490: A Study in the lesser baronage*. Unpublished PhD dissertation. University of London.

Jack, R. I. (1968) 'Records of Denbighshire Lordships, II, the lordship of Dyffryn Clwyd in 1324', *Denbighshire Historical Society Transactions* 17, 13–18.

Jack, R. I. (1969) 'Welsh and English in the medieval lordship of Ruthin', *Denbighshire Historical Society Transactions* 18, 23–49.

Jenkins, D. (2000) 'Hawk and Hound' in *The Welsh King and his Court*, eds T. Charles-Edwards, M. E. Owen and P. Russell, University of Wales Press, Cardiff, 255–280.

Johnston, D. (2005) *Llen Yr Uchelwyr: Hanes Beirniadol Llenyddiaeth Gymraeg 1300–1525*, University of Wales Press, Cardiff.

Johnstone, N. (1997) 'An Investigation into the Location of the Royal Courts of Thirteenth-Century Gwynedd' in *Landscape and settlement in Medieval Wales*, ed. N. Edwards, Oxbow Monograph 81, Oxford, 55–69.

Johnstone, N. (2000) 'Llys and Maerdref: The royal courts of the princes of Gwynedd', *Studia Celtica* 34, 167–210.

King, D. C. (1991) 'The stone castles' in *Archaeology of Clwyd*, eds J. Manley, S. Grenter and F. Gale, Clwyd County Council, Mold, 173–185.

Korngiebel, D. M. (2007) 'English colonial ethnic discrimination in the lordship of Dyffryn Clwyd: Segregation and integration, 1282–c.1340', *The Welsh History Review* 23, 1–24.

Landsberg, S. (2004) *The Medieval garden*, British Museum Press, London.

Liddiard, R. (2005) *Castles in context, power, symbolism and the landscape, 1066 to 1500*, Windgather Press, Macclesfield.

Linnard, W. (2000) *Welsh woods and forests: A history*, Gomer Press, Llandysul.

Manley, J., Grenter, S. and Gale F. eds (1991) *Archaeology of Clwyd*, Clwyd County Council, Mold.

Manley, J. (1994) 'Excavations at Caergwrle Castle, Clwyd, North Wales: 1988–1990', *Medieval Archaeology* 38, 83–133.

Musson, C. R., Britnell, W. J., Northover, J. P., and Salter, C. (1992) 'Excavations and Metal-working at Llwyn Bryn-dinas Hillfort, Llangedwyn, Clwyd', *Proceedings of the Prehistoric Society* 58, 265–283.

Palmer, A. N. and Owen, E. (1910) *A history of ancient tenures of land in North Wales and the Marches*, 2nd edition, Printed by the authors.

Pattison, P. ed. (1998) *There by design: Field archaeology on parks and gardens*, Royal Commission on the Historical Monuments of England, Swindon.

Pratt, D. (1990a) 'Parkers of Clocaenog', *Denbighshire Historical Society Transactions* 39, 116–120.

Pratt, D. (1990b) 'A Chirk charter', 1334, *Denbighshire Historical Society Transactions* 39, 109–115.

Pryce, H. and Insley, C. (2005) *The acts of Welsh rulers: 1120–1283*, University of Wales Press, Cardiff.

RCAHMW (2004) Moeliwrch, Llansilin NPRN 36030. Available from: http://www.coflein.gov.uk/, accessed 11.06.2011.

RCAHMW (2004) [online] Park Eyton, Park, Ruabon NPRN 308744. Available from: http://www.coflein.gov.uk/, accessed 11.06.2011.

Rhys, W. ed (1910) *Wild Wales: The people, language and scenery by George Borrow*, J. M. Dent and Company, London.

Roberts, E. (1974) 'Llys Ieuan, Esgob Llanelwy', *Transactions of the Denbighshire Historical Society*, 23, 70–103.

Roberts, E. (1975) 'Uchelwyr y Beirdd', *Transactions of the Denbighshire Historical Society*, 24, 38–73.

Roberts, E. (1976) 'Uchelwyr y Beirdd', *Transactions of the Denbighshire Historical Society*, 25, 51–91.

Roberts, E. (1986) *Tai Uchelwyr y Beirdd 1350–1650*, Cyhoeddiadau Barddas, Swansea.

Smith, J. B. (1998) *Llywelyn ap Gruffydd*, University of Wales Press, Cardiff.

Smith, S. (2003) 'Report on the Geophysical and Historical Survey at Sycharth Castle', *Transactions of the Denbighshire Historical Society*, 52, 17–36.

Suggett, R. F. (2005) *Houses and history in the March of Wales: Radnorshire 1400–1800*, Royal Commission on the Ancient and Historical Monuments of Wales, Aberystwyth.

Whittle, E. (1989) 'The Renaissance gardens of Raglan Castle', *Garden History*, 17, 83–94.

Whittle, E. (1999) 'The Tudor Gardens of St Donat's Castle, Glamorgan, South Wales', *Garden History*, 27, 109–126.

Whittle, E. (2000) 'The historic parks and gardens of Carmarthenshire', *The Carmarthenshire Antiquary*, 36, 87–102.

Williams, D. H. (1982) *Welsh seals through history*, National Museum of Wales, Cardiff.

Preliminary Fieldwork and Analysis of Three Scottish Medieval 'Deer Parks'

Derek Hall, Kevin Malloy and Richard Oram

Introduction

The idea of deer parks is understood to have reached Scotland from mainland Europe via England in the early twelfth century as part of a new style of 'feudal' lordship introduced by Scottish kings (Dixon 2009, 27–46). Part of that new-style lordship was a privatisation of hunting – particularly for venison. This was regarded as a reserved privilege of royalty and nobility in extensive tracts of landscape designated as 'forest' (Gilbert 1979, 19). Parks have long been viewed as elements of that package, usually as enclosed reserves to contain managed deer herds, but this narrow view of their function has undergone much recent revision in England and Wales as a result of recent interdisciplinary research (Rotherham 2006; Liddiard 2007; Mileson 2009; Fletcher 2011). While still viewed as reflections of status, English parks are now considered to have a wide range of socio-economic functions, of which deer management is only one. No similarly focussed interdisciplinary research on Scottish parks has been undertaken. This means that our understanding of these nationally significant monuments and landscapes, of the distinct ecologies which they contain, and of the human and environmental pressures and threats which confront them, has advanced little in the last 30 years. This study presents results from trial excavations taken from the sites of three potential Medieval Scottish 'deer parks' (for locations of parks see Figure 20.1).

In recent years there has been a marked increase in publications dealing with the function and operation of the Medieval park but little comparable increase in fieldwork or surveys (Rotherham 2006; Liddiard 2007; Mileson 2009; Fletcher 2011). This is especially so in Scotland where at least 80 deer parks are known. The purpose of this study was to carry out the first preliminary archaeological excavations of this site type to examine evidence for Medieval deer parks at three Scottish park sites.

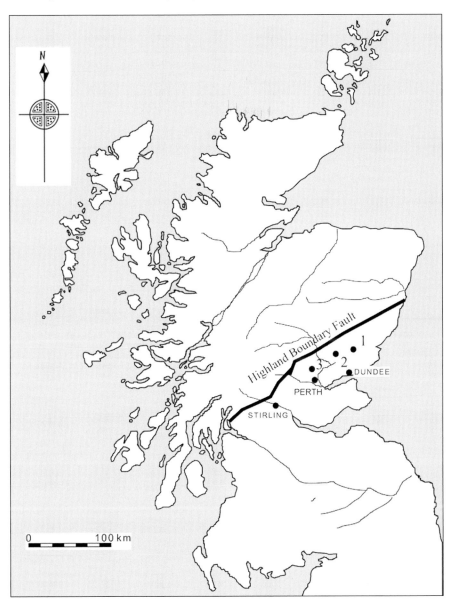

FIGURE 20.1. Locations of the three proposed Medieval Scottish deer parks: 1. Kincardine; 2. Durward's Dyke; 3. Buzzart Dykes.

Sites

The first park presented in this study, Kincardine in Aberdeenshire, was once a royal hunting park. This park was described by John Gilbert (1979, 356) as 'the most outstanding remains of a park pale to be found in Scotland'. The park may have been created initially by King William the Lion (1165–1214), who issued several charters there. The park was later extended by King Alexander III in 1266. The boundary of this park falls into two parts, the main pale and a northerly extension (Gilbert 1979, 82). It is not clear whether the park was associated with nearby Kincardine Castle (2 km south) or the potential nearby early castle site at Green Castle (<2 km).

The second park presented in this study, Durward's Dyke in Angus, was once a baronial park. The Medieval deer park of Lintrathen was probably bound on the east by the Loch of Lintrathen and elsewhere by a bank and ditch, now discontinuous, but best preserved on the east slope of the Knock of Formal between NO 2579 5450 and NO 2598 5486. The bank and ditch are known as Durward's Dyke and the park is said to have belonged to Alan Durward in the thirteenth century (Gilbert 1979, 85–86, 220). The associated hunting lodge for this park is thought to have been located at Peel Farm, located towards the assumed southern boundary of the park.

The function of the third park, Buzzart Dykes' in Perth and Kinross, is poorly understood. Originally interpreted on the First Edition Ordnance Survey map of 1860 as a 'Caledonian camp' the earthworks at Buzzart Dykes were re-identified as a possible deer park in the 1940s (Childe and Graham 1942–3, 45). Despite possessing all the recognisable hallmarks of such a monument, research, which has been limited, has been unable to confirm this identification. One basic difficulty is that the original, historic name of the park is unknown. It has been suggested that this site may have been associated either with Glasclune castle, Drumlochy castle or the royal forest of Clunie (R. Oram *pers. comm.*). Documentary evidence for a link with Drumlochy is provided by six entries in 'The Exchequer Rolls of Scotland' for the years 1464 to 1469 which record the *redditus* (rent) payment of 'eight broad arrows' (*octo amplarum saggitarum*) presumably for hunting (*ER* VII, 231, 340, 394, 473, 532 and 616).

Excavation methodology

Following initial site visits to all three parks in October 2009, trench locations were decided upon. Scheduled Monument Consents were then sought for the necessary sites (Kincardine and Buzzart Dykes). Excavations took place by hand and samples were taken from the banks and the buried ground surface with the aim of analysing their soil profiles and isolating any datable material. Cross sections of each of the excavated trenches are presented in Figure 20.2.

Excavation results

The excavation at Kincardine was targeted at an easily accessible section of park bank and ditch located at NO 663 780. A single trench measuring 7.5 m by 2 m was opened by hand across the bank and ditch line. The external park bank measured 2.10 m wide by 0.80 m high and comprised of a substantial stone core sealed by a layer of silt clay and stone. The depth of the ditch was 0.30 m. From field observation the section of park bank which ran up to 70 m north of the excavation site had a visible stone core. The section of park bank that was trenched appeared to be constructed of large unbonded stones (*c.*0.20 m by 0.10 m) covered with a layer of soil. There was no evidence of either a palisade or hedge line along the top of the bank. An adjacent deep natural gorge was also located towards Clatterin' Brig on the north eastern side of the park.

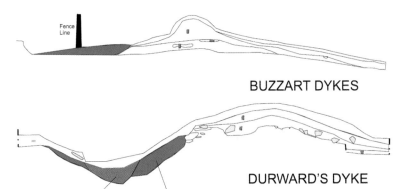

KINCARDINE

BUZZART DYKES

DURWARD'S DYKE

FIGURE 20.2. Sections of the excavated trenches at the three parks of Kincardine, Buzzart Dykes and Durward's Dyke, green indicates ditches, small red boxes show sample locations, and yellow on the Durward's Dyke drawing shows the bank.

FIGURE 20.3. Excavated stone base to bank at Buzzart Dykes.

The excavation at Durward's Dyke was targeted at a well-preserved section of park bank and ditch located at NO 257 456. A trench measuring 10.3 m by 2 m was opened across the external bank and internal ditch line. The park bank was represented by a 0.48 m thick layer of silt and small stones that had two small postholes dug into its top, these features were 0.20 m and 0.15 m deep respectively. The upstanding bank never seems to have been any higher than 0.48 m high and may have had a palisade running along its top. Evidence for such a palisade line is represented by the two small postholes that were located on the top of the bank.

The excavation at Buzzart Dykes was targeted at a section of park bank and ditch located at NO 119 476 which lay towards the western edge of the park at a point where there appeared to be either an entrance or break in the bank. A trench measuring 9.5 m by 1.5 m was opened by hand across the bank and ditch line. The external western bank at Buzzart proved to have a stone base that appears to have been constructed using glacial erratic stones found in the vicinity (Figure 20.3). The upcast from the ditch was thrown up onto this base to create a substantial upstanding earthen bank. The gap between the bank and ditch appeared to be a deliberately built entrance, although there was no sign of any gate structure. The ditch had been recut at least once. In its original form it was 3.5 m wide and 1.20 m deep but its recut was much narrower being only 1.34 m wide. The lack of an obvious eastern bank and ditch to this park is the subject to some debate (RCAHMS 1990, 93); however, analysis of the First Edition Ordnance Survey map (1860) revealed evidence for such a boundary; as a small section of bank and ditch ran along the expected alignment. The RCAHMS survey of North East Perth also indicates upstanding elements of short stretches of a bank at the eastern ends of both the northern and southern boundaries to the park (RCAHMS 1990, 93, 216).

Discussion

This study provides some of the first preliminary results regarding the existence of Medieval deer parks in Scotland.

The strongest evidence for the existence of a genuine hunting deer park came from Buzzart Dykes, the least documented and smallest of the three sites examined. The excavation of this site revealed a substantial western bank and ditch which may have served as a barrier to the movement of deer. This bank, which consisted of a stone base (Figure 20.3) and appeared to have been built in separate sections, would have required a large work force, possibly with more than one work group for its construction. Although cartographic and survey evidence require further examination there is also some evidence for an eastern boundary to the park. Several other features, which would be expected to be attributed to a deer park, were also found when additional parts of the park were examined. First, a potential enclosure for livestock (corral) may have existed in the north-west corner of the park as revealed when aerial survey of Buzzart

Dykes was examined (see Grid reference: NO 121121 48157). Secondly, the internal landscape appears not to have been cultivated and retains upstanding prehistoric hut circles and burial cairns which may hint towards an intention to create a 'designed' landscape. Finally, the existence of two buildings (one inside and one outside of the park boundary) may have been associated with the park and even functioned as keepers and/or hunting lodges. It is clear that further research is required to examine this site as a whole.

For the other two parks, Kincardine and Durward's Dyke, it is more difficult to ascertain their role as a Medieval deer park based on our excavation alone, which covered only a small area of both sites. Our small excavation revealed very shallow ditch lines for both Kincardine and Durward's Dyke (0.30 and 0.15 m respectively). Ditches of these depths could not have acted to enclose deer. However, for at least Kincardine there was some evidence for an additional stone built bank, similar to that described at Buzzart Dykes, which may have inhibited deer movement. Furthermore, the existence of a deep natural gorge which ran adjacent to Kincardine could have acted as a 'kill zone' for deer. Such areas are known from other parks, especially in Europe, and were designed to allow those not taking part in the hunt a grandstand view of the proceedings (Arigoni-Martelli 2012).

Taken together, there is great potential to further investigate the role of the three Scottish sites presented in this study as deer parks in the Medieval landscape. Further excavations, along with careful environmental analysis of samples retrieved from all the sites in this study, are required to provide more clues about former land use. The archaeology of Medieval rural Scotland is still largely unexplored and continued multi-disciplinary study of parks, forest and other hunting landscapes is vital to take research and understanding forward.

Acknowledgements

The authors would like to thank Stirling students Ben Fanstone and Simon Parkin for their help in the excavations. We also thank the landowners Mr Ewan Berkeley of Savills at Kincardine, Mr Rick Knight at Buzzart Dykes and Mr David Bryce of Formal Farm at Durward's Dyke, for allowing access for the various pieces of work to be carried out. Finally the help and interest of Alasdair Ross is gratefully acknowledged.

Author contributions

This work took place as part of a Masters degree at Stirling University by K. M. The archaeological excavation was supervised by D. H. and the entire exercise was directed by R. O.

References

Arrigoni-Martelli, C. (2012) *The Prince, the Park and the Prey: Hunting in and around Milan in the Fourteenth and Fifteenth Centuries* http://www.medievalists.net/2012/05/15/the-prince-the-park-and-the-prey-hunting-in-and-around-milan-in-the-fourteenth-and-fifteenth-centuries/, accessed 30.03.2013.

Burnett, G. ed. (1884) *The Exchequer Rolls of Scotland; Volume 7 AD 1460–1469*; *Rotuli Scaccarii Regum Scottorum*, Edinburgh.

Childe, G. and Graham, A. (1943) 'Some notable Prehistoric and Medieval monuments recently examined by the Royal Commission on Ancient and Historical Monuments of Scotland', *Proceedings of the Society of Antiquaries of Scotland* 77, 31–49.

Dixon, P. (2009) 'Hunting, summer grazing and settlement: competing land use in the uplands of Scotland', *Medieval Settlement in Marginal Landscape, Ruralia* VII, 27–46.

Fletcher, J. (2011) *Gardens of Earthly Delight: The History of Deer Parks*, Windgather Press, Macclesfield.

Gilbert, J. (1979) *Hunting and hunting reserves in Medieval Scotland*, John Donald, Edinburgh.

Liddiard, R. ed. (2007) *The Medieval Park: New Perspectives*, Windgather Press, Macclesfield.

Mileson, S. (2009) *Parks in Medieval England*, OUP, Oxford.

RCAHMS (Royal Commission on Ancient and Historical Monuments of Scotland) (1990) *North-East Perth an archaeological landscape*, HMSO, Edinburgh

Rotherham, I. D. ed. (2006) *The History, Ecology and Archaeology of Medieval parks and Parklands,* Wildtrack Publishing, Sheffield.

Post-Medieval Hunting in the UK

English Icons:
the deer and the horse

Mandy de Belin

Introduction

In Medieval and Early Modern England the pursuit of deer was considered to be the highest form of the sport of hunting. This is reflected in the literature of the time: early modern authors described the hunting of a wide variety of beasts, but there was consensus that the pursuit of the red deer (*Cervus elaphus*) stag represented the pinnacle of the sport (e.g. Cox 1674; Blome 1686). In modern England, however, the term 'hunting' is equated with foxhunting, and foxhunting is regarded as being primarily an equestrian sport. How was it that the deer ceased to be the iconic animal of the hunt, and why was it replaced by the horse (*Equus* sp.) rather than by the fox (*Vulpes* sp.) or the hound (*Canis* sp.)?

The transition from one form of hunting to another happened across the course of the eighteenth century. During this time the deer was supplanted by the fox as the favoured quarry. The well-established explanation of this transition has tied it to change in the landscape, and a consequent depredation of the deer population. The argument suggests that as the woodland cover of the forests disappeared, so the deer declined. Hunters looked around for an alternative prey and fixed upon the fox. The earliest rehearsal of this argument is found within the 1912 work *Hunting in the Olden Days* by William Scarth Dixon, and it has remained unchallenged since then, most recently repeated in Griffin's general history of hunting with dogs (Scarth Dixon 1912; Griffin 2007).

The purpose here is to challenge this account, based on research into hunting in Northamptonshire between 1600 and 1850. This county was chosen because it contained both the archetypal deer-hunting landscape of royal forest, and some of the prime landscape of the sport of modern foxhunting.

Three royal forests of Northamptonshire, Whittlewood, Salcey and Rockingham, were part of a forest block that ran originally from Oxford to Stamford (see Figure 21.1). By 1600 these forests were diminished from their greatest expanse, but still considerable in size, and there is evidence of their continued suitability as deer habitat. Of the venison supplied to Charles

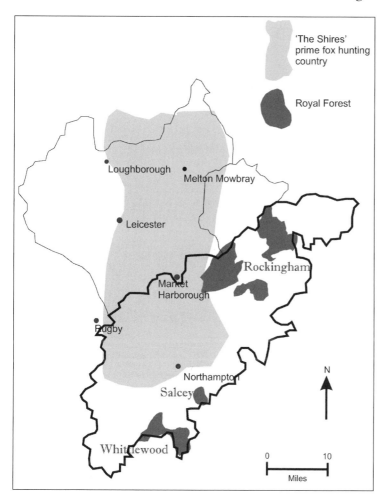

'The Shires' prime fox hunting country

Royal Forest

Loughborough

Melton Mowbray

Leicester

Rockingham

Market Harborough

Rugby

Northampton

Salcey

N

Whittlewood

0   10
Miles

FIGURE 21.1. Royal forests of Northamptonshire and the 'shires' foxhunting area.

I for Christmas 1640, by far the largest consignment came from Rockingham forest; the next largest came from Whittlewood, which tied for second place with the New Forest (Cox 1905, 78–9).

In the nineteenth century, another area of Northamptonshire gained fame as part of the 'Shires': the quintessential foxhunting country (see Figure 21.1). The elite hunted the fox in Northamptonshire, Leicestershire and Rutland; to hunt anywhere else was to hunt in 'the provinces'. The importance of foxhunting in this place and in this period should not be underestimated; it was the height of fashion to spend the winter in Melton Mowbray or Market Harborough hunting with the various shire packs (de Belin 2013, 63).

Northamptonshire contained the landscape of the old and the new, so if evidence of landscape change driving the hunting transition existed anywhere, it should surely be found in Northamptonshire.

The Northamptonshire forest

Can a decline in the forest and a corresponding decline in the deer population be detected? In short, the answer is no. As regards to the existence of habitat, map evidence for all three royal forests shows very little change between around 1600 and 1850. With Whittlewood it is possible to compare a forest map from 1608 with one from 1787, and with Bryant's large scale map of the county from 1827 (NRO, map 4210; NA, MR1/359). There was no notable decrease in the amount of woodland visible in these maps.

Whittlewood was a royal demesne – the crown owned the land as well as having various rights over it. At the end of the seventeenth century, Whittlewood was granted to the Dukes of Grafton. Under their stewardship the forest continued to be managed largely as coppiced compartments. These were enclosed immediately after the coppice was harvested to provide protection from deer and the other commonable animals (cattle and horses in the case of Whittlewood). The compartment was opened when the trees were sufficiently regrown to withstand browsing. The coppice compartments were separated by wide grassy 'rides'. There were also open grazing areas known as plains, and enclosed ones – usually dedicated to the deer – known as 'lawns'. The forest landscape did not significantly change until after disafforestation and enclosure in the 1850s, which can be seen in the Ordnance Survey maps from the late nineteenth century (de Belin 2013, 27–33).

Salcey, although much smaller than Whittlewood, was managed in the same way. A forest map from 1787 shows the similar coppice compartments, opened and closed in turn, together with a large open lawn at the centre of the forest (NA, MPE 1/938). Salcey was not much altered by disafforestation and enclosure, and remains much the same size and shape to present day.

Rockingham was the largest of Northamptonshire's royal forests. In contrast to Whittlewood, the forest was held by a much greater number of (usually squabbling) nobles and gentlemen. Portions of it were disafforested across the Early Modern period, but even here the woodland remained fairly consistent. Map evidence must be pieced together from a broader range of sources, as maps survive from different dates for various portions of the forest. A seventeenth century map covering the area between Northampton and Stamford shows the entire expanse of Rockingham forest, but not in the same detail as the 1608 Whittlewood map (NA, MPE 459). Other maps depict smaller areas, such as that commissioned by Sir Christopher Hatton in the 1580s to show his estates (NRO, FH272; NRO, BRU Map 126). These maps tell much the same story as the Whittlewood ones: there is little diminution in woodland cover until after enclosure in the mid-nineteenth century. Coppicing was found to be a good use for what was otherwise fairly marginal land.

Deer population

The woodland decline at the heart of the traditional explanation of the change in hunting practice is not apparent in Northamptonshire over the period of the transition. What about the accompanying decline of the deer population? There certainly is evidence of the impact of the English Civil War in the mid-seventeenth century. When Charles II was restored to the throne in 1660, he made the state of the nation's deer one of his foremost concerns. He repeatedly issued restraints on warrants for taking deer because of low numbers. Rockingham seems to have suffered less than other areas, however, as he started to issue warrants there while hunting was still suspended elsewhere. The Earl of Exeter was annually granted a license to hunt in Rockingham forest in the summer months (the buck season) (CSPD 1661–1662, 627). In October 1662 Sir John Robinson was granted a warrant to kill deer in Farming Woods (in Brigstock Bailiwick), 'provided he leave sufficient of the Royal disport' (CSPD 1661–1662, 530).

Evidence about the deer population in Northamptonshire is eclectic and tends towards the qualitative rather than the quantitative. There were few attempts to assess the actual deer population, but information retrieved from a variety of sources suggests that the Northamptonshire deer population had been restored to healthy numbers by the late seventeenth and certainly the eighteenth centuries. In 1714 the Earl of Westmoreland reported that, in Cliffe Bailiwick in Rockingham, 172 bucks and does could be 'safely' killed per year plus 20 brace for the queen and forest officers. Accounts of deer in Whittlewood were equally favourable (NRO, W(A)VI 2/25; W(A)VI 2/23). Morton, writing in 1712, reported that Wakefield Lawn 'shews sometimes seven or eight hundred deer, generally three or four hundred in any fine day' (Morton 1712, 11). The parliamentary reports of the 1790s made some attempt at estimating the strength of the deer population. In Whittlewood there were calculated to be some 1,800 deer 'of all sorts' in the forest. In Salcey the figure was given as 1,000 (Commons Journal 47, 143, 99).

An examination of woodland landscape and deer population in the period of the hunting transition has suggested that, for Northamptonshire at least, the traditional explanation of the change does not stand up to scrutiny. Some other account must be found, and it is here that the history of the horse, and more specifically to the appearance of the thoroughbred horse, is examined.

The thoroughbred horse

The foundations of the English thoroughbred were laid at the end of the seventeenth and beginning of the eighteenth centuries. The animal was created in response to the burgeoning popularity of the sport of horseracing. People had long pitted one horse against another in competitions of speed, but the sport began to emerge in a more organised form in the seventeenth century. Initially, horseracing was quite closely related to hunting. A popular form of the competition was the 'hunting match'. This provided a way of delimiting a race course over an expanse of open ground. A dead animal, most often a cat (*Felis* sp.), was dragged across the desired course of the race to lay a 'train scent'. Hounds then followed this scent, and two horses pitched against each other in the match followed the hounds. Other forms of competition included running for a plate, where there were four or more horses running in up to four heats of up to four miles each (Fairfax 1758, 62–4). This was clearly an endurance event compared to modern racing.

Horse racing grew in popularity at a time when hunting itself was going out of fashion. The 'polite society' that arose at the beginning of the eighteenth century was metropolitan in orientation. The hunting-obsessed Tory squire, locked away in the countryside, became a figure of fun and derision (e.g. *Spectator* magazine had Sir Roger de Coverley as the archetypal hunting squire). Racing, however, was considered a 'polite' pastime with the opportunities it provided for socialising and display. It was also popular because of the gambling opportunities it afforded.

The popularity of racing led to changes in the type of horse bred, and in the style of riding it. Arabian and Barb stallions were imported from the middle east and used on native English mares (themselves with various degrees of eastern blood) to produce the English thoroughbred. All modern English thoroughbreds have one of three foundation sires in their pedigrees: the Byerley Turk, the Darley Arabian, and the Godolphin Arabian. The breeding programme resulted in the production of the ultimate equine athlete (de Belin 2013, 111–119).

The lessons learned in the selective breeding of horses were more widely implemented in general stock improvement, a mania for which gripped the agriculturists of the eighteenth century. These same techniques were applied to the breeding of hounds, also producing a faster and more nimble animal that came to be used for the pursuit of the fox (Thirsk 1985, 578).

To ride the new thoroughbred horses fast required their riders to adopt a new position, and a new design of saddle. The deep saddle, long stirrups, and upright position gave way to a flat-seated saddle, short stirrups and a forward riding position. This came to be known as 'the English hunting seat' (Landry 2009). As foxhunting developed in the eighteenth century, and rose to an unprecedented level of popularity in the nineteenth, the thoroughbred horse came to be the aspirational mount for the rider to hounds. One of the major motivations in participating in the sport was the search for a good fast gallop across country.

To understand the role of the horse in modern foxhunting, it is necessary to compare the methods of hunting the fox with those of hunting the deer.

Hunting methods

The hunting of deer was heterogeneous in its methods. The 'highest' form was hunting *par force des chiens* ('by force of dogs'). Superficially, this conforms to our modern notions of hunting. The deer was pursued by a pack of hounds accompanied by people on horseback. But there were many significant differences with modern foxhunting. First was the overwhelming importance of selecting the animal to hunt – a suitably prestigious mature red deer stag – and of hunting that animal alone. To switch to another deer was considered a failure. Unlike in modern foxhunting, men on foot played a vital role, in both selecting the animal to hunt and in handling the hounds. The hounds were deployed in relays: they were not laid on the prey *en masse*, but rather stationed along the route it was predicted that the stag would take and let loose as the stag passed. The horse had a supporting role in all this. The principals of the *par force* hunt were the stag and the hounds (Blome 1686, 82–4). The horse played an important part in allowing people to keep up with the hounds, and this might require skilled horsemanship, but that was not the focus of the sport (and this is reflected in the lack of attention given to the horse by the Medieval hunting sources; for example, Edward, Duke of York makes no recommendations about the horse to use for hunting) (Ballie-Grohman 2005).

There were other forms of hunting the deer too. The foremost of these was 'bow and stable' hunting. This was most often staged in a deer park. Large numbers of (usually fallow, *Dama dama*) deer were driven past standings where a collection of hunters waited with crossbows to dispatch them as they passed. The deer were driven by dogs, and held to course by men (known as 'the stable') stationed along the route to the standings. This form of hunting clearly demanded less exertion from the occupants of the standings than did hunting *par force*, and was consequently popular as a form of entertainment. The horse played no role other than carrying away the slaughtered deer (Ballie-Grohman 2005, 188–96).

Deer coursing was another popular form of hunting. Here the deer was pursued over a set course by greyhounds (who hunted by sight rather than scent). A pair of dogs pursued a single deer with money put on the outcome. Again the participation of the horse was limited; in this case doing no more than providing a 'viewing platform' (Blome 1686, 96–7).

Foxhunting, in contrast to deer hunting, came to be homogenous in its methods. The modern form of the sport had it origins in the 1750s, when the breeding of faster, specialist foxhounds coincided with the idea of starting the hunt at 11 am or midday. By then the rested fox had digested the night's meal and would run fast. The entire pack was sent into a covert to 'draw' for the fox: to flush it out and get it running. The mounted followers would then set off in hot pursuit. The hunt prized short sharp runs of twenty minutes to an hour in duration.

Foxhunting adopted some of the social innovations of the eighteenth century. Early foxhunts, such as the Charlton or the Pytchley, were organised as clubs. Hunting came to be financed partly by subscriptions paid by the hunt followers, rather than relying totally on the largesse of a lord or gentleman. The new sport rescued hunting from its position in the provincial backwaters. By the end of the eighteenth century, foxhunting was being viewed as a smart and fashionable winter pastime for gentlemen. In the nineteenth century, the sport became central to the nation's cultural identity.

One of the biggest transformations wrought by foxhunting was in the number and the interests of the hunt followers, known as 'the field'. For popular and fashionable hunts in the nineteenth century, fields could number from three to five hundred. Most of these followers were there for the riding. Nineteenth-century hunting literature is dominated by tales of hard riders and hard riding. The hunting horse came to be identified as 'the hunter', and the animal recommended above all for this pursuit was the English thoroughbred. (Modern notions of a hunter are a larger, much more substantial horse, but the literature of the nineteenth century universally recommends a thoroughbred, or near thoroughbred, horse.)

There was a certain tension between the hunt personnel – the Master, the Huntsman, and the Whippers-in – and the field, who might override the hounds and generally get in the way. The majority of the hunt followers hunted in order to ride, as opposed to those who rode in order to keep up with the hounds. Hunting had become an equestrian sport, and its iconic animal was not the prey or the hound, it was the horse.

Later deer hunting

Deer hunting did not cease altogether with the ascendancy of foxhunting. Historians have, in fact, used the changing nature of the sport as evidence to support the argument that pursuit of the fox supplanted pursuit of the deer because the former were plentiful and the latter were scarce. Deer hunting was transformed by the nineteenth century, and the most significant development was the growth of the practice of hunting the carted deer. This involved loading a captured animal into a cart and transporting it to the appointed place of the meet. The deer was then set loose and given a small head start before the hounds were released, *en masse*, to start the pursuit. Initially the deer would be killed once caught, but by the nineteenth century the practice was to recapture the deer and transport it home once more, when, after sufficient rest and recuperation, it could be hunted again.

This was the form of the sport pursued by the royal buckhounds. The meetings of the buckhounds could be extremely popular, although this seems to have been more for the opportunity of viewing members of the royal family at leisure rather than for the sheer pleasure of the chase. The buckhounds' carted deer could obtain a celebrity status. *The Sporting Magazine* talked of two of them, High Flyer and Moonshine, having 'blood and bottom', using the type of language typically employed to describe racehorses (Sabretache 1948, 116).

The hunting of deer elsewhere in the country followed the pattern seen for the royal hunt. There was a concentration of activity within reach of London, and in East Anglia. Few packs of deer hounds are recorded as existing within the venerable foxhunting country of the shires. Neither do the packs generally coincide with the existence of royal forests (with the notable exception of Exmoor) (Whitehead 1980, 206–52). Apart from the North Devon hunt, these packs all pursued carted deer. It was generally regarded as being a somewhat inferior sport to foxhunting. William Angerstein, a late resident of Northamptonshire and follower of the Pytchley, established a pack of stag hounds when he moved to Norfolk, on the theory that 'half a loaf' was 'better than no bread'. But Angerstein was not long in discovering that 'the pursuit of the deer in an essentially non-hunting country, and that of the fox over the big pastures in the neighbourhood of Crick or Market Harborough are enjoyments as distinct in their character as light from darkness' (Nethercote 1888, 255).

Hunting the carted deer mimicked foxhunting in its methods. They used foxhounds, rather than the older breed of slower staghounds, and hunted in the winter, rather than in the traditional summer stag and buck season. The hunt's main virtues seem to have lain in the certainty of the sport and its comparatively short duration. The foxhunting writer Nimrod observed that 'turning out deer before fox-hounds in the neighbourhood of the metropolis' had the 'advantage of affording a certainty of something in the shape of a run' which was most useful to 'persons whose time is precious' (Nimrod 1843, 424). No one seemed to have expected the sport to match the excitement offered by a fox hunt. Even advocates of hunting the wild deer on Exmoor acknowledged that their sport would only satisfy 'a firstflight Melton Man' if 'he is not merely a rider, but a sportsman to boot' (Palk Collyns 1902, 171).

Historians have viewed the ascendancy of carted-deer hunting as evidence supporting the traditional explanation of the hunting transition. The switch was made because the traditional haunts of deer had disappeared, and it was no longer possible to pursue the wild animal. Carr summed it up thus: 'fewer forests and fewer deer parks meant fewer wild deer. The hunting of carted deer – was one answer' (Carr 1986, 24). There is an alternative argument to be made that the practice of hunting carted deer was intended to, and largely succeeded in, bringing the new style of hunting to a population who might otherwise not be able to enjoy it on a regular basis. Those who followed these hunts valued fast runs and good quality horses, just as foxhunters did. Carr suggested that hunting the carted deer was 'a tame substitute for the real thing', but, for enthusiasts of the fast horseback pursuit, the reverse seems to have been true (*ibid.*). Nineteenth century stag hunters were not 'making do' with some pale imitation of an ancient and noble sport, rather they were making the best of a rather 'watered down' version of foxhunting.

The Northamptonshire landscape

Returning to the Northamptonshire landscape once more, it has been argued here that the large-scale disappearance of woodland in the county did not coincide with the hunting transition; rather it happened in the second half of the nineteenth century when foxhunting was well established. But there was a major change in the midlands landscape that did coincide with the period of transition: this was the enclosure of the Medieval open fields, and large-scale conversion to pasture. This change gave rise to grass fields, good scenting, and the fast going that characterised foxhunting in the shires. The prime foxhunting country was associated with those areas of Northamptonshire and Leicestershire that had been converted to pasture early – in the sixteenth and seventeenth centuries – so providing acres and acres of well-established and comparatively open grassland.

It is not intended to argue here that conversion to pasture itself caused the hunting transition, however. A prey animal requires habitat to produce a sustainable population, and grass fields are not a habitat for the fox. As foxhunting developed and gained popularity, fox coverts were planted or preserved to maintain the fox population. But this 'grassed down' landscape did become the aspirational landscape over which to hunt fox. An argument can be made that it did so because it arrived with the English thoroughbred, and provided the ideal arena in which the horse could perform.

Conclusion

For much of the history of hunting in England, the deer was the iconic animal of the sport. When the role of deer hunting was supplanted by foxhunting, it was neither the prey animal nor the hound that became the icon, but rather the horse. In the nineteenth century the sport involved unprecedented numbers of followers, but the vast majority participated for the thrill of the ride, rather than the 'science' of hound pursuing scent. Hunting had become essentially an equestrian sport.

Abbreviations

NRO   Northamptonshire Record Office
NA   National Archives
CSPD   Calendar of State Papers Domestic

References

Blome, R. (1686) *The Gentleman's Recreation*, London.

Carr, R. (1986) *English Foxhunting: A History*, Weidenfeld and Nicolson, London.

Cox, J. C. (1905) *The Royal Forests of England*, Methuen, London.

Cox, N. (1674) *The Gentleman's Recreation*, London.

Baillie-Grohman, W. A. and Baillie-Grohman, F. N. eds (2005) *The Master of Game by Edward of Norwich, 1406–1413*, University of Pennsylvania Press, Philadelphia.

de Belin, M. (2013) *From the Deer to the Fox: The Hunting Transition and the Landscape, 1600–1850*, University of Hertfordshire Press, Hatfield.

Edward of Norwich (2005) *The Master of Game*, eds A. William and F. N. Ballie-Grohman, University of Pensylvania Press, Philadelphia.

Fairfax, T. (1758) *The Complete Sportsman; or Country Gentleman's Recreation*, London.

Griffin, E. (2007) *Blood Sport: Hunting in Britain Since 1066*, Yale University Press, New Haven and London.

Landry, D. (2009) *Noble Brutes: How Eastern Horses Transformed English Culture*, John Hopkins University Press, Baltimore.

Morton, J. (1712) *Natural History of Northamptonshire*, London.

Nethercote, H. O. (1888) *The Pytchley Hunt Past and Present*, Sampson Low Marston Searle and Rivington, London.

Nimrod (C. Apperley) (1843) *The Horse and the Hound*, Adam and Charles Black, Edinburgh.

Palk Collyns, C. (1902) *Notes on the Chase of the Wild Red Deer in the Counties of Devon and Somerset*, Lawrence and Bullen, London.

Sabretache (Barrow) (1948) *Monarchy and the Chase*, Eyre and Spottiswoode, London.

Scarth Dixon, W. (1912) *Hunting in the Olden Days*, Constable, London.

Thirsk, J. (1985) 'Agricultural innovations and their diffusion' in *The Agrarian History of England and Wales*, ed. J. Thirsk, CUP, Cambridge, 533–89.

Whitehead, G. K. (1980) *Hunting and Stalking Deer in Britain through the Ages*, Batsford, London.

Femmes Fatale:
Iconography and the courtly huntress in the Later Middle Ages and Renaissance

Richard Almond

The idea that Medieval and Renaissance courtly women were hunting like men, perhaps in some cases better than their male counterparts, has generated increasing academic interest over recent years (Almond 2009, 147–157).[1] For the majority of male hunters, historians and writers this has been a controversial notion and to a significant extent remains so. Yet prior to the Later Middle Ages and the Renaissance there is no lack of precedent in European Classical myth and legend for divine and heroic women actively hunting, and its recognition, acknowledgement, admiration and even worship by mortal men. Ancient Greek and Roman mythology was a rich and versatile source of allegorical imagery, and the huntress, divine or otherwise, was a significant element of this iconography (Almond 2009, 17). In the concluding chapter to his *Cynegetica, On Hunting,* Xenophon states that the young huntsman is 'dear to the gods and pious, believing that these activities [hunting] are overseen by one of the gods [Artemis]' (Phillips and Willock 1999, 129). Surprisingly, Xenophon continues 'And not only men who have had a passion for the hunt have become virtuous, but women also, Atlanta or Procris or whoever else, who have been given this pursuit by the goddess.' (Phillips and Willock 1999, 13–18). Artemis, the goddess of hunting, later called Diana, was worshipped and consulted by hunters for many centuries for her divine powers of intercession before and during the hunt and the placation of the slain quarry's spirits at the conclusion and afterwards (Figure 22.1). It seems, therefore, that the notion of some women as capable and successful hunters, albeit 'special women', was acceptable to men from early times.

Amanda Richardson points out that the Renaissance 'idealisation' of hunting was based around the images and exploits of the Classical goddess of the chase (Richardson 2012, 253). Artemis, her Roman equivalent Diana and other mythological huntresses thus provided popular subjects for late medieval and Renaissance artists in various art media. An illustration by Robinet Testard in a French manuscript, *Les echecs amoureux* (*c.*1496–1498), shows Diana, attended by her nymphs, standing in an open space near a castle with woodland as a backdrop, about to use a longbow to shoot a fine hart.[2] A similar French manuscript of *Le Livre des echecs amoureux* (1495) depicts Diana holding a short bow and wearing a quiver of arrows, seated in an open space enclosed by woodland with her master of game, huntsmen and hounds in attendance.[3] Both these illustrations can be likened to the reality of Medieval park hunting, enclosed and private hunting spaces, usually artificially stocked with fallow deer (*Dama dama*), adjacent to castles or lodges in which courtly ladies could hunt decorously and in relative safety. Various versions of the story of Diana and Actaeon were made by artists and that of Matteo Balducci illustrates one particularly favoured them, the horrific transformation of the unfortunate hunter into a stag being torn to pieces by his own hounds.[4]

It is now accepted that there were skilled, knowledgeable and courageous royal huntresses in the Middle Ages, Renaissance, and Early Modern period. There were many notable examples of such high-status women in Medieval France and Burgundy, including Valentine de Milan (1368–1408), wife of Louis of Orleans; Isabeau of Bavaria (*c.*1370–1435); Agnes Sorel (1422–1450), mistress of Charles VII of France; Anne de Beaujeu (1461–1522), daughter of Louis XI; Mary of Burgundy (1457–1482), wife of Holy Roman Emperor Maximilian I, and her daughter Margaret of Austria (1480–1530); Diane de Poitiers (1499–1566), mistress of Henry II of France; Gabrielle d'Estrées (1573–1599), mistress of Henry IV of France; and Marie-Adélaide de Savoie (1685–1712), duchese de Bourgogne. Other regal huntresses were native to Hainault. Many English queens enjoyed hunting and hawking, including Eleanor of Castile (1241–1290), first queen consort of Edward I; Margaret of France (1279?–1318), second wife of Edward I; Isabella of France (1295–1358), wife of Edward II; and Philippa of Hainault (1314?–1369), wife of Edward III (Almond 2009, 58–59, 83; Richardson 2012, 253–254). Richardson puts forward the possibility that these foreign-born English hunting queens of the late thirteenth and early fourteenth centuries, brought up in a Continental courtly culture which included active hunting roles, influenced the greater acceptance of such participation by the end of the Middle Ages in England (Richardson 2012, 266). However, the brief and tantalising references in contemporary documents to female elites hunting and hawking are largely confined to household and wardrobe account books. Such books tell us basically how much money was expended on hawks and falcons, occasionally on hunting costumes and very rarely on the weaponry of ladies[5] but little directly about what these high-status women did in the hunting field and how they did it. These *femmes fatale* are sometimes referred to as 'Medieval Dianas' by historians but interestingly a number of these royal and aristocratic women perceived themselves as latter-day personifications of the goddess of hunting and were depicted as such in various forms of iconography, particularly paintings and statuary. King Henry II's mistress, Diane de Poitiers was portrayed as Diana the huntress several times, both in oils and bronze. In the Louvre

FIGURE 22.1. Royal courtesan Diane de Poitiers. Diana the Huntress (oil on panel), Fontainebleau School, (1550–60). Louvre, Paris, France, The Bridgeman Art Library.

oil painting of her, naked except for a gauzy cloak, she regards the viewer coyly whilst striding through the woods carrying her attributes of a bow and quiver, accompanied by her greyhound. The Louvre bronze depicts her as a nude Diana, as usual accompanied by her faithful hounds (Almond 2009, 56),[6] reclining languidly beneath the figures of a hart, his harem of hinds and two wild boar (*Sus scrofa*) (Figure 22.1).[7]

Textual and iconographic sources show clearly that Queen Elizabeth I (1533–1603) was a knowledgeable, pragmatic and enthusiastic huntress. Although Susan Doran concludes that in some ways 'Elizabeth was no feminist icon', it is probable that her position as English sovereign, strong sense of her own queenship and innate intelligence allowed her to rise above the gender roles normally ascribed to even royal women (Doran 2003, 29–35; Doran 2003, conclusions 35). In a letter to Robert Cecil (*c*.1593), Raleigh described her as 'hunting like Diana', a personification which doubtless pleased her (Richardson 2012, 253).[8] The Dutch artist Cornelius Vroom famously painted Queen Elizabeth I as Diana, wearing her attribute of a crescent moon on her brow, holding a bow in her left hand and a leashed hound in her right (Almond 2009, 90).[9] An anonymous engraving with a political-religious message dating from the sixteenth century shows Elizabeth as Diana, sitting in judgement over Pope Gregory XIII (Almond 2009, 90).[10] Elizabeth is portrayed out hunting in the queen's Forest in several woodcuts of *Turbervile's Booke of Hunting 1576*. She is the central figure at three key events of the royal hart hunt: presiding at the *assemblée* (hunt breakfast); examining the *fewmets* (deer dung) in order to choose the best hart to hunt; and making the first cut in the *breaking up* (ritual butchering) of the slain hart in order to *assaye* or ascertain the thickness of its fat (Figure 22.2).[11]

To the present-day researcher, Medieval women as participators are conspicuous by their absence in the literature of Medieval hunting. The didactic so-called *hunting books* and treatises in their chapters on quarry species, methodology and procedures do not mention women, with the notable exception of *The*

Master of Game. Edward of Norwich, grandson of the ardent hunter Edward III, describes 'Queen and ladies, and gentlewomen' being accommodated at the tryst (Baillie-Grohman and Baillie-Grohman 1909, 263: the place or stand where the hunter took up position to shoot driven game from) and several references to the queen using her bow at the royal stand, in his chapter on bow and stable hunting in Forests and parks (*ibid.*, 189–196). Neither do women feature in the many and varied dedications in the Medieval hunting books except in *The Master of Game* (Almond 2009, 5). In the concluding passage of the Shirley Manuscript of *The Master of Game,* Edward ends his dedication with the hope that all his readers: 'that hath herde or rude [sic] this lytell tretys', have approved of and corrected it as necessary according to their own knowledge and he presents it 'openly to the knowledge of all lords, *ladies,* gentlemen and *women,* according to the customs and manners used in the high noble court of this realm of England [author's italics]' (Baillie-Grohman and Baillie-Grohman 1909, 200; Cummins 1988, 249).[12]

FIGURE 22.2. Royal expertise in the hunting field: Queen Elizabeth about to assaye the thickness of the fat of the slain hart. Queen Elizabeth I at a stag hunt, from the *Noble Art of Venerie and Hunting,* by George Turberville (1575). British Library, London/The Bridgeman Art Library.

Edward is not only indicating the active presence of the queen and ladies at the stable stand, he is also tacitly acknowledging that courtly women may read, perhaps have read out to them, or certainly have some knowledge of his didactic text. This is a very rare textual indication that aristocratic women were *lerned* in hunting vocabulary and specialised knowledge in the same manner as courtly men.

The main reason for this gender exclusion is that the contemporary illustrative and textual material on hunting was commissioned and written by educated gentlemen for a male audience of the ruling elite. In addition, social mores may have restricted or perhaps even have disapproved of an active female role in the chase. Furthermore, Medieval women had historically no voice, especially in what were claimed as 'manly' and 'warrior' activities such as hunting (Almond 2009, 5–6).

This tradition of female exclusion from English hunting literature continued well into the late seventeenth century and beyond. *The Gentleman's Recreation*

(London, 1677) by Nicholas Cox, a didactic text in four parts on hunting, hawking, fowling and fishing, states in typical chauvinistic manner: 'Hunting is a Game and Recreation commendable not onely [sic] for Kings, Princes, and the Nobility, but likewise for private Gentlemen' (Cox 1973, 1).

We can therefore dismiss the hunting books and treatises as positive evidence for women hunting except insofar that their very negativity on this aspect is a clear indication of aristocratic male pre-occupation with exclusivity, encompassing knowledge, methodology, procedure, ritual and, above all, the *lexis* of hunting which marks a man as noble and gently-born.

For positive indications of women's participation and roles in hunting, an alternative source of evidence proved to be iconography. Here was a more fertile field, relatively unexplored and deconstructed. The general impression provided by Late Medieval and Renaissance art, including illuminated manuscript miniatures, paintings, engravings and tapestries, is that high-ranking women were undoubtedly to be found in the hunting-field and in various roles. Most modern historians admit that for aristocratic women, hunting was part of what was expected of them and what they did in the public gaze and in private. Peter Coss comments that female landowners were barred from many of the public roles undertaken by their male counterparts. They could not sit in parliament, nor could they occupy posts such as sheriff, coroner or justice of the peace. Nevertheless they could participate in many aspects of noble life. Women were sometimes found hunting with their households. From time to time they, too, were fined for poaching deer in the king's Forests (Coss 1998, 67).

In *Picturing Women in Late Medieval and Renaissance Art,* Christa Grössinger remarks similarly 'One of the sports in which ladies could participate was hunting. They would ride side-saddle or astride, and hunt with small hawks and merlins.' As part of her evidence she cites *The Devonshire Hunting Tapestries* (*c.*1430) where deer, wild boar, otters (*Lutra lutra*), swans (*Cygnus* sp.) and wild duck are depicted being hunted, slain and broken up ritually, courtly ladies sharing the experiences of hunting with their menfolk (Grössinger 2005, 68). Apropos women's style of riding, Giraldus Cambrensis in *The Topography of Ireland* (*c.*1188) wrote 'The women, also, as well as the men, ride astride, with their legs stuck out on each side of the horse' (Wright 1863). However, pictorial evidence shows clearly that women rode not only astride but also 'on-the-side', pillion and probably after around 1500, true side-saddle. Out hunting, potentially a dangerous activity, many were undoubtedly riding astride and later side-saddle (Almond 2009, 75–78; Almond 2012, 36–39).

Yet the active role of aristocratic women as hunters is not always clear in iconography. At first glance, it often appears that women are present in a purely passive role, apparently as decorative audience, admiring and applauding the hearty exploits of their husbands, lovers and male members of their family. Thus, an illustration in a Bodleian manuscript, MS Douce 336, *Miroir du Monde* (before 1463), shows hunters and their ladies viewing a red deer (*Cervus elaphus*) hart which has taken to water (*foillyng*) being attacked by swimming

FIGURE 22.3. Young lady, first-year entry hound and brocket learning how to hunt: 'Stag Hunting' (engraving, 1609), Antonio Tempesta (1555–1630). Oxford, Ashmolean Museum, Print Room, WA2003, Douce 244.

hounds. Quite possibly these ladies have just ridden up from their safe viewing place to be present at the *dénouement* of the stag hunt (Almond 2009, 55).[13] Less usually are women to be pictured actively participating in the dangerous, sweaty and perhaps unseemly excitement of the *par force de chiens* chase or in the safer and more physically remote but bloody *bow and stabley* game drive. John Cummins comments 'for women to take part in the rigours of the classic "par force" hunting, as opposed to its social preliminaries and aftermath, must have been a rarity' (Cummins 1988, 8). However, there are such illustrations of active female participation. The lively *bas-de-page* illustration hunting scene in *The Hours of Marguerite D'Orléans* shows both courtly men and their women engaged in *a force* stag-hunting.[14] The illustration for the month of August in the Grimani Breviary (1510–20) features a noble lady accompanied by courtiers and hunt servants carrying spears going stag-hunting and hawking.[15] Lucas Cranach the Younger's painting of the *grand battue* hunting party given by Elector John Frederick of Saxony for Emperor Charles V at the castle of Torgau (1544), a politico-religious painting now in Vienna,[16] includes an elaborately attired lady armed with a loaded crossbow waiting at a hidden stand by a lake into which red deer are being driven. She is John Frederick's wife, the Electress Sybille. As an active representative of her sex she is the exception and so has been allocated little picture space, yet she is fulfilling the same hunting function as her many male counterparts. Three male assistants have two more crossbows ready-loaded to quickly hand to her, an indication, perhaps, of her skill and dexterity with this specially-made light weapon. She is accompanied by six similarly attired ladies-in-waiting, who are providing her with company and support in this male-dominated venue. Cranach has made a clear distinction between Sybille, the active royal huntress, and her passive courtly female audience. In addition, she is looking directly out of the picture at the viewer, not at where she is shooting, thus arresting our attention and emphasising her distinction as the lone woman hunter (Almond 2009, 90–91 for discussion). Certainly women do feature in hunting illustrations less than men but this does not exclude them as active or semi-active participants in the theatre of courtly hunting.

Women are often depicted in hunting illustrations alongside men and there can be difficulties in reading such images correctly. A fifteenth-century tapestry, *Le Retour de la chasse,* in the Musée National du Moyen Age (Musée de Cluny) demonstrates this ambiguity.[17] Here, a blonde-haired woman on foot, holding a hunting-spear and wearing a hunting-knife on her belt, is presenting a hare (*Lepus* sp.) to a mounted noble; alternatively, she is being handed the hare by the young man. It is impossible to say. Probably the gift is also charged erotically so there are at least two meanings to this tapestry image. Sometimes, as in the *Hunting Scenes Series* of engravings by Antonio Tempesta (1555–1630), women are clearly apprentices or 'learners' receiving instruction from male hunters and professionals (Almond 2009, 62–65). *Stag Hunting* (1609) features a gentleman hunter and a young lady galloping in pursuit of an immature male red deer, termed a *brocket* (Baillie-Grohman 1919, 226). A greyhound runs alongside the couple, controlled by a long leash attached to the hunter's left arm or belt. The lady is riding astride, probably the first illustration of a woman in clothing designed for that purpose (Baillie-Grohman 1919, 163 and 167). The theme here is clearly one of control and instruction within relatively safe boundaries: the young lady is learning how to hunt on horseback; the brocket will not be slain, having learned how to run to escape hunter and hound; and the greyhound will not be slipped from its leash but has learned how to pursue a deer. All three 'apprentices' are prepared for the real hunt in the future, as we are reminded by the mature hunt in the background (Figure 22.3).[18]

Tempesta's *Boar Hunting* (1609) provides a similar visual apprentice/master theme. Here, the young lady sitting side-saddle is being briefed on the adjacent boar-hunting action by a mounted gentleman with a falcon on his fist, both courtiers guarded by a hunt servant carrying a cross-hilted boar spear, indicating their shared vulnerability whilst immobile.[19] Noblewomen are also portrayed hunting or hawking alone or with female companions, thus lacking the controlling presence of gentlemen or male hunt servants. Several well-known manuscripts show such independent activities. *MS Yates Thompson 13* is a personal prayer book, more usually known as *The Taymouth Hours*, dating from *c*.1325–35. It was probably produced in London for a female member of the Neville family (Sandler 1986, 107–9) who may be seen on folio 7. If the subjects of the manuscript were, as is likely, specified by or for the patron, then they clearly indicate the first owner's love of sport and her fondness for romantic stories (Thompson 1914, 34). The margins of the Hours of the Virgin in *The Taymouth Hours* include over 30 coloured illustrations of noblewomen actively participating in the courtly pastimes of hawking and hunting. These images may be interpreted as being representative of fourteenth-century aristocratic female skills and they include women flying a peregrine (*Falco peregrinus*) at wild duck; boar hunting on foot; and stag hunting with hounds, both on horseback and dismounted. In addition, the traditional and gory *après chasse* rituals are shown being carried out by the lady hunters, including feeding hare guts to a hound; presenting a boar's head on a spear; and breaking up a hart.[20] In this latter *bas-de-page* illustration, four ladies are busily engaged in the process. On the far-left, the first lady fixes the beast's head by pushing the antlers into the ground; on the mid-left, the second lady slits open the belly; on the mid-right, the third lady pulls the hide over the legs; on the far-right, the fourth lady winds (blows) the hunting horn to alert those of the field not present and as a finale to the hunt procedures, then hoists the antlered head aloft for all to see. Four of the procedures of breaking up are thus neatly depicted in one miniature, and all are being correctly performed by lady hunters.[21]

It has been correctly observed that this particular cycle of pictures is unrealistic in that only women are portrayed, yet this Book of Hours was intended for a young woman who probably loved such activities. The presence of males does not need to be acknowledged in the illustrations, particularly if they are only the professional hunters or servants. As 'ungentlemen' they are 'unseen', although necessary to the organisation and success of the hunt. There is no practical reason why properly trained and skilled courtly women could not hunt 'alone' or in small groups, despite deeply entrenched gender prejudice and male concerns over the safety of 'weak women'. Again, there is no practical reason why breaking up a deer or other large quarry could not have been performed by women who were 'lerned' in the correct procedures (Almond 2009, 82–83).

No men are included in a *bas-de-page* illustration in a Bodleian Flemish manuscript of *the Romance of Alexander,* dated 1338–1344, which shows on the left two ladies on foot hunting the hart using leashed hounds and on the right two ladies on foot using a lure and hawking 'at the brook' for wild duck.[22] A vigorous miniature in a version of Christine de Pisan's *L'epitre d'Othea* (*c.*1400) in the Bibliothèque nationale de France is also women only, featuring a blonde-haired Diana-like huntress, standing in a clearing bounded by woodland, expertly using a longbow to shoot a deer, possibly a fallow buck. She is accompanied by her three female attendants and several hounds.[23] In the late seventeenth-century engraving *A noble lady on a deer hunt*, the royal huntress not only stars as the central Diana-like figure, sitting side-saddle and performing an *haute-école levade*[24] movement (Summerhays 1962, 142 and 149) on her grey horse, she is also shown leading the mounted stag hunt along a park fence in the background.[25] One cannot help wondering whether in such iconographical contexts in which authoritative single huntresses are deliberately featured there is a covert message that men of any class were not always necessary (or even welcome) to the enjoyment and successful climax of aristocratic hunting forays.

There is one illuminated manuscript which provides a surprisingly different view of aristocratic women, showing them not only hunting and hawking but also being regarded by men as capable of having opinions and enjoying discussion upon such learned matters (Almond 2009, 65–67 for discussion). This textually incomplete manuscript consists of a debate between two noblewomen written by a court poet, Guillaume Crétin (d. 1525). *Le Débat entre deux dames sur le passetemps des chiens et oyseaux* considers the relative merits of hunting with dogs or birds. This was a subject of enduring debate in the Late Medieval hunting books and was still being aired in *The Compleat Angler* by Izaak Walton in 1653 (Walton 2010, 11–33).[26] The poem narrates how a noble couple out hunting with hounds and hawks meet another couple whom they invite back to their castle for a banquet. A discussion quickly changes into a quarrel, at which point their husbands propose the arguments be presented formally to the Constable of France, the count of Tancarville, a renowned and expert hunter. The formal debate, recorded by a clerk, takes place after the banquet. Each woman describes the ideal hunt, together with the merits of each creature: the cleanliness of hawks and falcons versus the loyalty of hounds. The clerk sends his written report to the count who returns his considered judgement that his hounds provide better sport.

  Only one surviving manuscript is illuminated but it is textually incomplete. Its four miniatures illustrate the major divisions of the poem and form a neat sequence summarising the content. Fragments of the poem appear as *trompe-l'oeil* placards across miniatures two, three and four which were painted by a French artist, the Master of Francis I, probably in Paris, *c.* 1530–40 (Hindman 1988, 69–73).[27] The cycle comprises:

  (i)   The poet composing his poem with the two court ladies who are his inspiration. The woman to the left holds a falcon on her gloved right hand and the woman to the right has a leashed hound at her feet. The base of the frame is supported by two hounds, foretelling the eventual winners of the debate. The scene is filled with quarry species.

  (ii) A hawking party on horseback, made up of two nobles and two ladies using hawks and falcons, plus their two female attendants and a male servant on foot. Below the text placard, a plainly dressed man, possibly a poacher, uses a crossbow to hunt wild duck. This *bas-de-page* illustration is reminiscent of the unlawful duck hunting marginal scene in *The Hours of Marguerite D'Orléans* (*ibid.*, 70).[28]

  (iii) Hunting with hounds, including mounted stag hunting in woodland and in water. A lady riding a grey horse side-saddle, accompanied by her husband, look on while a hunter on foot spears a black wild boar which is being harassed by hounds (Hindman 1988, 71).

  (iv) The final judgement has been made and the count's letter is delivered to a lady waiting at the castle gatehouse (*ibid.*, 72).

Although this is a piece of sixteenth century romantic literature with an exceptional theme, it suggests that not only was it potentially possible that aristocratic women could debate the merits of hunting and hawking methodology because they, too, were 'lerned' and educated in such matters, but also that such a conundrum could be written about for reading aloud to courtly men and women. This is positive evidence demonstrating that at least some courtly ladies were taught the lore and language of hunting, probably after puberty or in early womanhood, and equates favourably with the didactic theme of the *Hunting Scenes Series* of engravings by Antonio Tempesta (Almond 2009, 67).

This work has presented some of the iconographic evidence from the Middle Ages and Renaissance which attempts to clarify the involvement and various roles of royal and aristocratic women in the socially exclusive theatre of hunting 'noble' quarry, primarily red and fallow deer. The reading, deconstruction and interpretation of such elitist images is not conclusive in itself but used in conjunction with the available literary evidence provides a better understanding of noblewomen's lives and of the gender dynamics within upper class society. It is surely significant that hunting on horseback with hounds and shooting driven game continued to be an important part of courtly and aristocratic life into the Early Modern Age and beyond, women being acknowledged increasingly as active and expert participants.

Notes

1. Almond 2009, 147–157; 'Men must face the facts, however indigestible: women are better at most things than men, including hunting.'
2. Paris, Bibliothèque nationale de France, département des Manuscrits, MS Fr. 143, *Les echecs amoureux,* Diana the Huntress, fol. 116.
3. Paris, Bibliothèque nationale de France, MS Fr. 9197, *Le Livre des echecs amoureux,* Diana Goddess of Hunting, fol. 202.
4. *Diana and Actaeon,* Matteo Balducci, private collection.
5. Two ladies' weapons at the Royal Armouries, Leeds, are rare survivals: Queen Sophia Amelia of Rosenborg's enamelled and gold-hilted hunting sword, *c.*1650; a 'bullet crossbow', *c.*1720, traditionally the property of Queen Maria Leczinska, wife of Louis XV.
6. *Diana at the Bath,* Master of the Fontainebleau School, (*c.*1590), oil on wood, Musée des Beaux-Arts, Dijon; *Diana the Huntress,* Master of the Fontainebleau School, (1550–60), oil on canvas, Musée du Louvre, Paris; *Nymph of Fontainebleau,* Benvenuto Cellini, (1542–44), bronze, Musée du Louvre, Paris.
7. The red deer hart (a stag of at least six years), hind, hare, boar, and wolf were called Beasts of Venery (or correctly Venary) in the hunting books. The fallow buck and doe, roebuck and doe, fox, and marten were termed Beasts of the Field or Chase. Legally, both classes were considered Beasts of the Forest. See Dryden on Twici 1844/1908, 19–20 and Note 5, 30–31.
8. Walter Raleigh's letter quoted in David Hume, (1830) *The History of England from the Invasion of Julius Caesar to the Revolution in 1688,* Cadell, London, vol. 4, 545.
9. *Queen Elizabeth I as Diana the Huntress,* Cornelius Vroom, Collection of the Marquis of Salisbury, Hatfield House.
10. The engraving *Queen Elizabeth I of England as the Goddess Diana, Sitting in Judgement over Pope Gregory XIII* is in a private collection.
11. Turbervile, George 1908, *Assemblée* 90; *Choice* 95; *Breaking up* 133; Baillie-Grohman and Baillie-Grohman 1909, 282.
12. *The Master of Game,* British Library, Add. MS 16, fol. 65.
13. Oxford, Bodleian Library, MS Douce 336, *Miroir du Monde,* detail: hunters and their ladies, hart foillyng and swimming hounds, fol. 58r.

14. *The Hours of Marguerite D'Orléans*, Bibliothèque nationale de France, MS Lat. 1156B, 'The Trinity', fol. 163.
15. Venice, Biblioteca Nazionale Marciana, *Grimani Breviary*, August, hunting and hawking.
16. *The Stag Hunt Given by the Elector John Frederick of Saxony for Emperor Charles V at the Castle of Torgau*, (1544), Lucas Cranach the Younger, Vienna, Kunsthistoriches Museum.
17. Paris, la Musée National du Moyen Age, *Le Retour de la chasse*.
18. *Stag Hunting*, Antonio Tempesta, Oxford, Ashmolean Museum, Print Room, WA2003, Douce 244.
19. *Boar Hunting*, Antonio Tempesta, Oxford, Ashmolean Museum, Print Room, WA2003, Douce 245.
20. *The Taymouth Hours*, British Library, MS Yates Thompson 13, fols 68–83v.
21. *Ibid.*, fol. 83v.
22. Oxford, Bodleian Library, MS. Bodl. 264, pt 1, *Romance of Alexander*, fol. 77r.
23. Paris, Bibliothèque nationale de France, MS. Fr. 606, Christine de Pisan, *L'epitre d'Othea*, Lady and attendants hunting deer, fol. 30r.
24. 'haut école' is the 'High School' Art of Riding based upon the classical traditions of the past; 185 the 'levade' is a High School movement in which the horse raises itself from the ground with its forefeet and then draws them in, the entire weight of the body being supported by the deeply bent hind quarters.
25. *A Noble Lady on a Deer Hunt*, Unknown artist. Oxford, Ashmolean Museum, Print Room, WA2003, Douce 258.
26. This particular and lengthy discourse is between *Piscator* (Angler), *Venator* (Hunter) and *Auceps* (Falconer).
27. Being the catalogue of an exhibition held in London, 1988 and Japan in 1989. MS 34 in the catalogue was formerly in a private collection.
28. *Ibid.*, 70; *The Hours of Marguerite D'Orléans*, Bibliothèque nationale de France, MS Lat. 1156B, 'The Trinity', fol. 163.

References

Almond, R. (2009) *Daughters of Artemis: The Huntress in the Middle Ages and Renaissance*, D. S. Brewer, Cambridge.
Almond, R. (2012) 'The Way the Ladies Ride', *History Today* 62, 36–39.
Baillie-Grohman, W. A. (1919) *Sport in Art: An Iconography of Sport*, Simpkin, Marshall, Hamilton, Kent and Co., London.
Baillie-Grohman, W. A. and Baillie-Grohman, F. eds (1909) Edward, Duke of York, *The Master of Game,* Chatto and Windus, London.
Coss, P. (1998) *The Lady in Medieval England 1000–1500,* Sutton Publishing, Stroud.
Cox, N. (1677 facsimile edn 1973) *The Gentleman's Recreation*, EP Publishing, Wakefield.
Cummins, J. (1988) *The Hound and the Hawk: The Art of Medieval Hunting,* Weidenfeld and Nicolson, London.
Doran, S. (2003) 'Elizabeth I: gender, power and politics', *History Today* 53, 29–35.
Dryden, H. ed. (1844 revised edn 1908) William Twici, *The Art of Hunting,* William Mark Printers, Northampton.
Grössinger, C. (2005) *Picturing Women in Late Medieval and Renaissance Art,* Manchester University Press, Manchester.
Hindman, S. (1988) *Medieval and Renaissance Miniature Painting,* Akron, London.

Phillips, A. A. and Willock, M. M. eds (1999) *Xenophon and Arrian, On Hunting,* Aris and Phillips, Warminster.

Richardson, A. (2012) 'Riding like Alexander, hunting like Diana: gendered aspects of the Medieval hunt and its landscape settings in England and France', *Gender and History* 24, 253–270.

Sandler, L. F. (1986) *Gothic Manuscripts, 1285–1385,* OUP, Oxford.

Summerhays, R. S. (1952 revised 1962) *Summerhays' Encyclopaedia for Horsemen,* Frederick Warne and Co. Ltd, London and New York.

Thompson, H. Y. (1914) *Illustrations from One Hundred Manuscripts in the Library of Henry Yates Thompson, vol. IV,* Chiswick Press, London.

Turbervile, G. (1908) *The Noble Arte of Venerie or Hunting,* reprinted as *Turbervile's Booke of Hunting 1576,* Clarendon Press, Oxford.

Walton, I. (1653 reprinted 2010) *The Compleat Angler, Or: The Contemplative Man's Recreation,* Arcturus Publishing Limited, London.

Wright, T. ed. (1863) *The Historical Works of Giraldus Cambrensis,* H. G. Bohn, London.

Deer Management

Supplemental Feeding and our Attitude Towards Red Deer and Natural Mortality

Karoline T. Schmidt

Historical basis

The hunting system in Austria and Germany was considerably restructured as a consequence of the revolution in 1848. The aristocracy lost their hunting privileges when the revier system, which is still in place today, was established: this system incorporates hunting and landownership rights although the right to actually carry out hunting is restricted to a minimum property size (1.15 km2 or 3 km2 in the different Austrian federal states) and this right may be leased (Prossinagg 1999; Bode and Emmert 2000; Zentralstelle der Oesterreichischen Landesjagdverbaende 2012). Since the established hunting society feared that the untitled middle-class would use their newly-acquired hunting rights too much and over-exploit game populations, hunting authorities encouraged hunters to not only hunt but rather foster game populations (Bode and Emmert 2000). The new attitude is best summarised by the catchphrase, 'Kein Heger, kein Jäger' (Silva-Tarouca, 1899). 'No gamekeeping, no hunt' – this slogan became the new belief and is still the ethical foundation of Austrian and German hunters.

In deer, this attitude was reinforced by putting the focus on antlers as a trophy. At the First International Hunting Exhibition in Vienna in 1910 the term antler was replaced with trophy and competitive trophy measurements were standardised in 1927 (Prossinagg 1999). As far as red (*Cervus elaphus*) and roe deer (*Capreolus capreolus*) are concerned, the moral obligation to foster game was understood not just as an appeal for sustainable use, habitat amelioration or to set aside safe areas, but as a call to trophy production planning. The crucial and indispensable management expedient was supplemental feeding. Feed or salt-licks have been used in the past to entice red deer into locations resulting in immediate hunting success (Stahl 1979), but when in the early twentieth century feeding trials demonstrated a formidable effect of feed on antler size (Vogt 1936), feeding was used to this avail in several reviers (Geist 1986; Peek *et al.* 2002; Schmidt 2010). In the larger reviers, intensive feeding successfully impeded and ceased red deer migrations, and where feeding alone did not succeed to inhibit the urge to migrate, mile-long fences across valleys, were erected to retain red deer in the hunting ground (Schmidt 2010). During the 1960s, in the economic upturn after World War II, easy availability of cheap feed allowed supplemental feeding to become a widespread management measure in several, though not all, Central European countries (Peek *et al.* 2002; Putman and Staines 2004). Purposes were manifold according to the specific local situation: to boost numbers by attracting and retaining deer or by largely reducing levels of winter mortality, to increase body condition and trophy quality as well as to increase hunting success (Gossow and Dieberger 1989; Zeiler *et al.* 1990; Peek *et al.* 2002; Putman and Staines 2004). By the late 1970s winter feeding was regarded as a *sine qua non* for the survival of alpine red deer populations (Schmidt 2010).

Feeding goals were not always attained, but in each instance it quickly became clear that supplemental feeding is a fundamental encroachment on the mode of life of wild red deer. Red deer abandoned migrations to wintering areas (even where such areas were still available and accessible). Deer were lured into habitats with unfavourable thermal conditions (e.g. northern exposure) and with a low range of visibility in afforested valley bottoms because these sites were easily accessible to those who provided the supplemental feed (Zeiler *et al.* 1990; Peek *et al.* 2002; Schmidt 2010). At most feeding stations there is no sexual segregation and a high level of aggressive interactions (Linn 1987; Schmidt 2005) may ensue which may result in deer-to-deer disease transmission (Peek *et al.* 2002; Putman and Staines 2004). The most noticeable negative effect, already observable in the 1960/70s, was severe forest damage, especially bark-stripping, caused by large concentrations of deer (Voelk 1997; Gossow 2004). Because of these environmental costs, the negative effects on the deer, as well as a management approach to keep wildlife natural, the Swiss province of Graubuenden and the state of Liechtenstein, where red deer inhabit areas and environmental conditions similar to those found in Austria, abandoned feeding in 1989 and 2004, respectively. Instead, the installation of a network of safe, undisturbed wintering areas, with no admittance for humans, including hunters was created (Kersting and Naescher 2008; Imesch-Bebié *et al.* 2010; Jenny *et al.* 2011). On the same grounds, several German states prohibited winter feeding in recent years on state-owned hunting grounds, though feeding is still prevalent on private lands (Wotschikowsky 2010). Given undisturbed wintering areas, red deer may exploit behavioural (Schmidt 1993; Schmidt 2005), as well as physiological, mechanisms to cope with alpine winters, such as nocturnal hypometabolism associated with peripheral cooling (Arnold *et al.* 2004). Game management in Graubuenden and Liechtenstein proved that even in alpine habitats, red deer not only survive but thrive without supplemental feeding and that feeding is definitively not a prerequisite to maintaining annual hunting bags (Kersting and Naescher 2008; Imesch-Bebié *et al.* 2010; Jenny *et al.* 2011).

The Austrian situation

In Austria, the revier system and its related economic aspects channelled red deer management into a system inconceivable without supplemental feeding (Zeiler 2005). Hunting is an important economic sector with a turnover of 475 million Euros in 2008 (Lebersorger 2012). As the landowner may let out his hunting rights on lease or sell individual culls, usually trophy stags, hunting provides predictable economic returns (Lebersorger 2012): the more game, the higher the profit. While this is in principle a good incentive for habitat and species conservation ('use it or lose it'), supplemental feeding perverts wildlife management into a highly productive economic enterprise (Gossow 2004; Zeiler 2010; Schmidt 2010).

Today, more than 95 percent of the deer are provided with supplemental feed (Schmidt 2010). For red deer, game keepers calculate 5 kg feed/individual/day (consisting of corn-silage, grass-silage, grain, forage beets, apple pomace and hay; Zandl 2004). This feed mix is supposed to provide 100 percent of the nutritional requirements (Peek *et al.* 2002; Zandl 2004). The amount of feed provided has to ensure that even the lowest ranking deer get their share. Feeding stations have to be refilled daily, so large barns are built in the middle of the forests, sometimes up to timberline elevation and roads are constructed and ploughed to guarantee daily access (Schmidt 2005; 2010).

Intensive feeding allows red deer densities to be high where, from an ecological and societal point of view, they should be low. Hunting revenues are an especially valuable source of income to landowners where income from forest land use and timber sale is low: higher elevation alpine slopes on poor and shallow soil. However, these forests are especially important for water retention and stabilisation of the slope to protect settlements in the valleys below from snow and mud avalanches. One third of the Austrian forests are so-called 'protective forests'. These protective forests are ecologically and societally economically important: if they are absent in part or whole, government-financed expensive artificial measures for certain protective measures (e.g. avalanche barriers) are required. For the landowner these forests are of little financial profit, but the attitudes to such areas change (to an asset) when red deer densities and trophy stag harvest rates are high thereby allowing for high monetary returns on a hunting lease (i.e. the revier system). This can be achieved only by intensive artificial feeding, which permits hunters to keep red deer densities far exceeding natural carrying capacity (Gossow 2004; Zeiler 2010). As a consequence rejuvenation of the young trees and regeneration are severely impaired due to browsing and bark stripping for two thirds of the Austrian forests (WEM 2010) even though artificial feeding is explicitly implemented to prevent forest damage (Voelk 1997; Peek *et al.* 2002; Gossow 2004).

High intensity feeding not only has an adverse impact on forest rejuvenation and regeneration but likewise has a harmful effect on red deer health. For example, in the province of Tyrol, intensive feeding and a high concentration of deer at feeding sites has lead to an eruption of tuberculosis (with 40 percent of the tested deer infected) which affected livestock in the area. To keep the infection at bay, hunters were obliged by authorities to reduce deer densities for more than a decade, but the hunters were unwilling or unable to comply. Therefore a culling brigade was despatched to cull deer inside purpose built reduction-enclosures (Koessler 2011).

Rather than to question the concept when the negative consequences were obvious, implementation was and still is continuously modified, intensified and refined, and feeding was and still is seen as a panacea for all problems it had caused. Forest damage is attributed to inadequate feed stuff, unsuitable feeding sites, and wrong feeding times (Schmidt 2010). Supplemental feeding is mandatory during times of food scarcity, but scarcity is an elastic term and on the grounds that if implemented as an emergency survival tool, then feeding needs an introductory period to allow rumen adaptability; the onset of feeding was increasingly antedated to early autumn and continued into late spring or early summer (Schmidt 2010). In most areas red deer are supplied with feed for 200 days/year, so that feed is not supplemental but rather the staple food for up to two thirds of the year (Peek *et al.* 2002; Schmidt 2005). Additionally feeding stations are often spaciously fenced-in, 10–50 ha with 40–200 animals inside (Reimoser and Reimoser 2010), and gates are closed until late spring to restrict winter bark-stripping to small areas and to hinder deer to range too early in spring to areas that are silviculturally and agriculturally sensitive (Peek *et al.* 2002; Reimoser and Reimoser 2010). These winter feeding enclosures are in some places used to cull a large percentage of the non-trophy individuals (i.e. calves and hinds) within the enclosure (Winkelmayer 2010). Where winter enclosures are adopted as a management strategy, red deer are selected for amenability, as all deer that do not enter these feeding enclosures have to be tracked and killed without exception to effectively restrict forest damage (Pirker 2010).

Since the mid-1970s, when feeding was implemented nation-wide, red deer populations have been at a continuous high level, with a calculated maximum density of ten red deer/km² (Reimoser and Reimoser 2010). Throughout Austria, supplemental feed is provided to the extent it reduces natural selection by eliminating density-dependent, as well as density-independent, winter mortality (Schmidt and Hoi 2002) thereby reducing considerable fluctuations in population density and enabling a sustainable annual cull of *c.*50,000 red deer (Reimoser and Reimoser 2006). Furthermore, the maintenance of populations at high enough levels allows the calculation of the yield of a certain number of high quality stags (trophy antlers) – the commercial basis for the current red deer management (Zandl 2004; Zeiler 2010; Schatz 2011). By eliminating winter mortality and controlling deer movements, supplemental feeding allows for highly selective culling which enables hunters within Austria to largely control population density as well as population characteristics, such as age structure, sex ratio and culling of antlered adult males for trophy hunting

(Schmidt 2010). In addition, feeding retains red deer within a limited annual home-range which not only prevents free ranging behaviour of red deer within an area, but also ensures a high success rate when hunting trophy stags due to the predictable presence of such deer within the landscape (Schmidt 2005). This is why Austrian hunters not only adhere to the feeding management but it is carried out to an extreme.

Repercussion of winter feeding deer on societal perception of hunters and wildlife

Intensive supplemental feeding changes hunting from an appropriative land use through culling (comparable to mushroom or berry picking) to a productive land use by investing enormous expenditures and employment of labour for feeding (Arnold 2010), which until recently included medical treatment. It is only since 2003, that the use of pharmaceuticals such as wormers is outlawed for wild game (Deutz 2006). Within this system of deer management, wild red deer are literally reared from calf to trophy stag age.

By maintaining a food-supplemented population without year-to-year fluctuations in numbers and quality to achieve a sustainable harvest of high-quality stags, red deer management in Austria is based on domestication and breeding: the provision of feed for up to nine months, selection based on antler criteria and manageability and control of deer movement (Clutton-Brock 1999). Furthermore, this hunting and game management system that relies on intensive supplemental feeding created all the three conditions, which the anthropologist and economist Samuel Bowles (2011) considers responsible for farming to prevail over foraging: sedentary lifestyle, the consequent emergence of private property and the military prowess needed to defend it.

For instance, in a revier system, hunters are sedentary (in terms of hunting within a restricted area) and by restricting deer home ranges and halting migrations by means of supplemental feeding, hunters force their main game species to a sedentary lifestyle too. With hindsight this seems as a stringent necessity in a revier-based hunting system where average revier size is less than 5 km² (Reimoser and Reimoser 2006) and hunters are encouraged to foster game on their land.

Additionally, hunters feed 'their' deer and not ownerless game. Hunters do not legally own the live animal (it is ownerless), they only have the right to shoot a certain percentage of game within their revier and they own an animal after they have killed it. But intensive feeding of sedentary game by sedentary hunters provokes the assumption of ownership. This is why hunters may allow for individual trophy stags which they feed throughout winter in their management plans, and why hunters will bluntly say: 'my stag is no snack for a bear' (Mueller and Rohrhofer 2009).

Within Austria, hunters do defend what they consider their deer against any source of mortality other than culling: winter as well as predators. Through winter feeding hunters eliminate natural winter mortality, which by 'saving' deer, they aim to societally justify their activities. In a democratic society, hunters have to explain themselves; after all, they exclusively exploit a public good. To gain public acceptance, hunters present and view themselves as custodians of the interest of wildlife and the wellbeing of the game. Throughout the last century, hunters were obliged to meet ethical standards for hunting, in accordance with animal welfare: for example, the use of traps which avoided animal suffering (Prossinagg 1999). Consequently hunters presented themselves as wardens of wildlife, not shooters, and hunting was considered and increasingly marketed as wildlife euthanasia: the rifle enables a quick death for the sick, the injured and the old and winter starvation is prevented and averted by supplemental feeding.

Since Palaeolithic times red deer have been a source of meat (Jarman 1972), a game species with a high net meat gain, 82 percent of the body weight compared to 70 percent in boar (*Sus scrofa*) (Richter, 2005), and any emaciated deer is considered wastage, either of venison or the wastage of a trophy stag (e.g. Ruhdorfer 2010). Therefore it is the hunters who most vigorously defend the perception of winter feeding from an animal welfare necessity viewpoint and argue that feeding is an obligation in accordance with animal welfare while accusing those who think otherwise of animal cruelty (Hespeler 2007; Ruhdorfer 2010; Schmidt 2010; Wild and Hund 2011). For example, after authorities abolished all kind of supplemental feeding in 2004 in Liechtenstein and Grisons, it was not the public, but the older hunters that accused the authorities of letting deer starve to death (Brosi *pers. comm.*). Or when in August 2010, Cornelia Behm from the German Green Party suggested to abandon supplemental feeding (Behm 2010), based on the German Forestry Commission statement that high deer densities prevent forest rejuvenation (of young trees) and cause severe ecological and economical damage (Deutscher Forstwirtschaftsrat 2011), it was not animal rights groups, but the hunting community who stressed animal welfare arguments and accused the Green Party of cruelty against game (Wild and Hund 2011). In February 2011 the aforementioned politician was forced to withdraw her statement (Behm 2011).

One hundred years ago, Ferdinand von Raesfeld, the foremost expert on game management in Germany at that time, regarded winter starvation as a positive selective force (Raesfeld 1920). Lettow-Vorbeck, in his revision of Raesfeld's classical book (Raesfeld/Lettow-Vorbeck 1965) stresses that winter death is not selective and considers this argument be an excuse by those who do not wish to supplementally feed game. The argument that wild game should be exposed to the hardships of nature is no longer considered valid by mainstream Austrian or German hunters.

Hunters have a bad reputation (Beutelmeyer 2010). The only aspect the non-hunting society appreciates is the care for game (Lebersorger *pers. comm.*). For the public, hunters who feed are well received because they support and enable life, contrary to their usual activity of ending life, and hunters do everything to reinforce this image of being the red deer's life preserver, to establish winter feeding of red deer as an indispensable deer management measure. Throughout Austria, hunters take primary school children to feeding stations to observe red deer and impart an understanding of hunting and wildlife. There is a whole generation growing up with the image of hunters inextricably linked with feeding deer, and, more worryingly, the image of red deer, the epitome of wild game, inextricably linked with feeding, unable to fend for itself. Likewise, for a whole generation it is becoming increasingly inconceivable that winter death is neither cruel nor a loss, but a natural process, a necessity and an important link in the food chain (Hespeler 2007; Schmidt 2010; Winkelmayer 2010). After all, the Scottish government pays for deer carcasses in supplemental upland feeding stations to boost eagle populations (Scottish Government 2012).

Predator control is integral to game management (Manhart 2007). Hunting officials declare predator control more important than ever and vital for many game species (Lebersorger 2007). Large predators (e.g. brown bear) were exterminated throughout central Europe by the end of the nineteenth century. In the late 1990s a population of *c.*30 brown bears had resettled in Austria but was exterminated once again by 2011 (Mohl 2011), with evidence pointing towards illegal hunting (Rauer and Kaczensky 2009). Within the last five years, wolves (*Canis lupus*) have sporadically moved into Austria each year (Pichler *et al.* 2011). Hunting representatives declare that immigrating wolves have to be respected as they are protected by national and international law, but at the same time the hunters' official representative states that for the next 50 years, wolves will certainly not settle in Austria (Lebersorger pers. comm).

Hunters arguing against the presence of wolves use mainly economic rationales: wolves will reduce red deer densities to a level, where hunting is no longer feasible, and landowners will consequently lose their income from hunting leases (Wadl and Rudigier 2010). Likewise hunters know and hunting officials clearly state that the red deer feeding system will create problems which will be difficult to counter measure (Gach 2009): wolves will use red deer feeding stations. Wolves are also indirectly incompatible with artificial winter feeding, because they cause irregularities in population structure plans (von Eggeling 2010) and the feeding-based predictability of red deer management. With a wolf in a hunters' revier, it becomes impossible to know which red deer (in terms of sex, age and quality), according to the officially sanctioned administrative culling plan, is available to hunt and disturbances to the deer daily ranging behaviour by wolves may impair the predicatablity of the hunting success. In short, wolves are 'a real handicap' for hunting (von Eggeling 2010).

Artificial feeding has put Austrian hunters in a quandary. If they adhere to the current feeding management and inhibit the settlement of wolves, then hunting is far from species conservation, thereby further damaging their reputation and their standing in increasingly more frequent conflicts with tourism over land use. If hunters accept wolves they will have to abolish feeding as the cornerstone of their ideology – which is tantamount to a paradigm shift to a less interventional approach to wildlife management.

References

Arnold, W., Ruf, T., Reimoser, S., Tataruch, F., Onderscheka, K. and Schober, F. (2004) 'Nocturnal hypometabolism as an overwintering strategy of red deer (*Cervus elaphus*)', *American Journal of Physiology: Regulatory, Integrative and Comparative Physiology* 286, 174–181.

Arnold, W. (2010) 'Wider die pauschale Diffamierung der Jägerschaft. Gastkommentar', in *Die Presse* (printed date: 27.10.2010).

Behm, C. (2010) http://www.corneliabehm.de/cms/presse/dok/350/350745.behm_novelle_des_landesjagdgesetzes_notw.html, accessed 28.01.2012.

Behm, C. (2011) http://www.cornelia-behm.de/cms/presse/dok/372/372003.die_wildtiere_vor_den_widrigkeiten_der_n.html, accessed 28.01.2012.

Beutelmeyer, W. (2010) 'Das Image der Jagd aus Perspektive der Bevölkerung – Ergebnisse einer aktuellen market Studie 2009', *Österreichische Jägertagung* 16, 73–82.

Bode, W. and Emmert, E. (2000) *Jagdwende. Vom Edelhobby zum oekologischen Handwerk*. Beck, Muenchen.

Bowles, S. (2011) 'History lesson from the first farmers', *New Scientist* 2837, 26–27.

Clutton-Brock, J. (1999) *A Natural History of Domesticated Animals*. CUP, Cambridge; The Natural History Museum, London.

Deutscher Forstwirtschaftsrat (2011) *Positionsapier des Deutschen Forstwircharfsrates. Für eine zeitgemäße Jagd: Wald und Schalenwild in Einklang bringen!*, Berlin.

Deutz, A. (2006) 'Darf es heute noch Fallwild geben?', *Der Anblick* 3/06, 28–30

Eggeling, F. K. von, (2010) 'Wenn der Wolf jagt', *Jäger* 09/2010, 48–51

Gach, H. (2009) 'Mit Wölfen ist zu rechnen', *Der Anblick* 12/2009, 10–11.

Geist, V. (1986) 'Super antlers and pre-world war II European research', *Wildlife Society Bulletin* 14, 91–94.

Gossow, H. and Dieberger, J. (1989) *Gutachten zur Behandlung der Wildtiere im Bereich der Sonderschutzgebiete des Nationalparks Hohe Tauern*, Institut für Wildbiologie und Jagdwirtschaft der Universitaet für Bodenkultur, Wien.

Gossow, H. (2004) 'Unter welchen Rahmenbedingungen ist die Winterfuetterung von Rot- und Rehwild im Ostalpenraum entbehrlich?' in Emährung des Rot-, Reh- und Gamswildes – Grundlagen, Probleme und Lösungsansätze: 16.–17. Februar 2004, BAL, Gumpenstein. Bundesanstalt für alpenländische Landwirtschaft Gumpenstein, A-8952 Irdning, Austria, 10.

Hespeler, B. (2007) Fuetterung ja oder nein. Lecture manuscript http://www.wildundhund.de/438,6623/, accessed 28.01.2012.

Imesch-Bebié, N., Gander, H. and Schnidrig-Petrig, R. (2010) 'Ungulates and their management in Switzerland', in *European Ungulates and their Management in the 21st century*, eds M. Apollonio, R. Andersen and R. Putman, CUP, Cambridge, 357–391.

Jarman, M. R. (1972) 'European red deer economies and the advent of the Neolithic', in *Papers in Economic Prehistory*, ed. E. S Higgs, CUP, Cambridge, 125–149.

Jenny, H., Gadient, R., Plozza, A. and Brosi, G. (2011) 'Der Umgang mit dem Rothirsch – faszinierend aber anspruchsvoll'. *Info pic Jagd broschure* 01/2011, Amt für Jagd und Fischerei Graubünden, Switzerland

Kersting, W. and Näscher, F. (2008) '*Der Rothirsch im Winter*', Amt für Wald, Natur und Landschaft, Fürstentum Liechtenstein.

Koessler, J. (2011) 'Bekämpfung der Tuberkulose beim Rotwild im Oberen Lechtal – aktueller Stand', *Jagd in Tirol* 11/2011, 12–15.

Lebersorger, P. (2007) 'Bejagung der Beutegreifer – rechtliche Grindlagen und Blick in die Zukunft', Österreichischer Jägertagung Gumpenberg, 13, 9–11.

Lebersorger P. (2012) Wirtschaftsfaktor Jagd. http://www.bljv.at/pr_m_wbj.htm, accessed 29.08.2012.

Linn, S. (1987) Zum sozialen Verhalten eines Rotwildrudels (*Cervus elaphus*) am Winterfuetterungsplatz unter besonderer Beruecksichtigung soziobiologischer Hypothesen. Unpublished PhD dissertation. University of Geneva.

Manhart, M. (2007) 'Umgang mit Beutegreifern in Vorarlberg', Österreichischer Jägertagung Gumpenberg, 13, 63–64.

Mohl, C. (2011) cited in: Braunbaeren zum zweiten Mal in Oesterreich ausgerottet. *Die Presse* (printed date: 02.12.2011).

Mueller, W. and Rohrhofer, M. (2009) '"Mei Hirsch is ka Jausn"', *derStandard* (printed date 26.3.2009).

Peek, J. M., Schmidt, K. T., Dorrance, M. J. and Smith, B. L. (2002) 'Supplemental feeding and framing of elk', in *North American Elk: Ecology and Management*, Wildlife Management Institute eds D. E. Toweill and J. W. Thomas, Smithsonian Institution Press, Washington DC, 617–647.

Pichler, C., Walder, C. and Kohler, B. (2011) *Strategische Überlegungen zur Einwanderung von Wölfen nach Österreich*, WWF Positionspapier. WWF Austria.

Pirker, H. (2010) *Abschussplanung. Der steirische Berufsjäger.* Juni 2010, 5–9.

Prossinagg, H. (1999) *Österreichs Jagd im 20. Jahrhundert*, Österreichischer Jagd- und Fischerei-Verlag, Austria.

Putman, R. J. and Staines, B. W. (2004) 'Supplementary winter feeding of wild red deer *Cervus elaphus* in Europe and North America: justifications, feeding practice and effectiveness', *Mammal Review* 34, 285–306.

Raesfeld, F. von (1920) *Die Hege in der freien Wildbahn*, Verlag Paul Parey, Berlin.

Raesfeld, F. von and Lettow-Vorbeck, G. (1965) *Die hege in der freien wildbahn*, Verlag Paul Parey, Berlin.

Rauer, J. and Gutleb, B. (1997) *Der Braunbär in Österreich*, Monographien Band 88, Umweltbundesamt, Wien, Austria.

Rauer, J. and Kaczensky, P. (2009) 'Rückkehr der Beutegreifer – Chance oder Last?' Nationalpark Akademie, Nationalpark Hohe Tauern, 15–16. Oktober 2009 (conference), abstract booklet 18–20.

Reimoser, S. and Reimoser, F. (2006) 'Lebensraum und Abschuß', *Weidwerk* 12/2006, 16–18.

Reimoser, F. and Reimoser, S. (2010) 'Ungulates and their management in Austria', in *European Ungulates and their Management in the 21st century*, eds M. Apollonio, R. Andersen and R. Putman, CUP, Cambridge, 338–356.

Richter, M. (2005) Wildbretvermarktung in der Rhön – Möglichkeiten der Gewinnoptimierung. Diploma thesis, Resource Management University of Applied Sciences and Arts, Goettingen.

Ruhdorfer, W. (2010) 'Gedanken zur Jaegertagung', *Der Steirische Berufsjaeger* June 2010, 4.

Schatz, H. (2011) 'Wieder mehr G'spuer als Jaeger kriegen', *Jagd in Tirol* 05/2011, 5–7.

Scottish Government http://www.scotland.gov.uk/Topics/farmingrural/SRDP/RuralPriorities/Packages/GoldenEagle/Supplementaryfoodraptors, accessed 28.01.2012.

Schmidt, K. (1993) 'Winter ecology of nonmigratory Alpine red deer', *Oecologia* 95, 226–233.

Schmidt, K. T. and Hoi, H. (2002) 'Supplemental feeding reduces natural selection in juvenile red deer', *Ecography* 25, 265–272.

Schmidt, K. (2005) 'Land of plenty natural history', *Natural History Magazine* 114, 44–49.

Schmidt, K. (2010) 'Wie man wild erntet', *Die Presse* (printed date 15.10.2010).

Stahl, D. (1979) *Wild, lebendige Umwelt: Problem von Jagd, Tierschutz u. Ökologie, geschichtlich dargestellt und dokumentiert, (Orbis academicus 2/1)*, Freiburg/München.

Silva-Tarouca, E. (1899) *Kein Heger, kein Jäger. Handbuch der Wildhege für weidgerechte Jagdherren und Jäger*, Paul Parey, Berlin.

Vogt, F. (1936) *Neue Wege der Hege*, Verlag J. Neumann-Neudamm AG, Melsungen, Germany.

Voelk, F. (1997) *Schaelschaeden und Rotwildmanagement in Abhaengigkeit von Jagdgesetz und Waldaufbau in Österreich*. Unpublished PhD dissertation. Universität für Bodenkultur, Wien.

Wadl, B. and Rudigier, E. (2010) 'Der Wolf in Kärnten', *Kärntner Jagdaufseher*. June 2010, 13–15.

WEM (2010) Wildeinflussmonitoring 2004–2009. http://www.wildeinflussmonitoring.at/, accessed 28.01.2012.

Wild and Hund (2011) http://www.wildundhund.de/438,6623/, accessed 28.02.2011.

Winkelmayer, R. (2010) 'Notwendigkeit und Sinnhaftigkeit der Winterfütterung – was ist vermittelbar?', *Österreichische Jägertagung, Gumpenstein* 16, 35–42.

Wotschikowsky, U. (2010) 'Ungulates and their management in Germany' in *European Ungulates and their Management in the 21st century*, eds M. Apollonio, R. Andersen and R. Putman, CUP, Cambridge, 201–222.

Zandl, J. (2004) 'Was kostet die Rotwildfütterung', *Tagung für die Jägerschaft, 16 und 17 Februar 2004*, BAL Gumpenstein, 33–38.

Zeiler, H., Preleuthner, M., Grinner M. and Gossow, H. (1990) *Zur Bewirtschaftung und Regulierung des Schalenwildes im Hegerung Fusch*, Gutachtliche Stellungnahme und Empfehlungen, Anhang, IWJ, Universität für Bodenkultur, Wien.

Zeiler, H. (2005) *Rotwild in den Bergen*, Österreichischer Jagd- und Fischerei-Verlag, Wien, Austria.

Zeiler, H. (2010) 'Rotwild-'Bewirtschaftung' in der Zukunft- Vom Hirschvater zum Weltrekord', *Jagd in Tirol* May 2010, 7–9.

Zentralstelle Österreichischer Landesjagdverbände (2012) http://www.ljv.at/, accessed 30.08.2012.

Estimating the Relative Abundance of the Last Rhodian Fallow Deer, *Dama dama dama*, Greece, through Spotlight Counts: a pilot study

Marco Masseti, Anna M. De Marinis, Nikos Theodoridis and Konstantinia Papastergiou

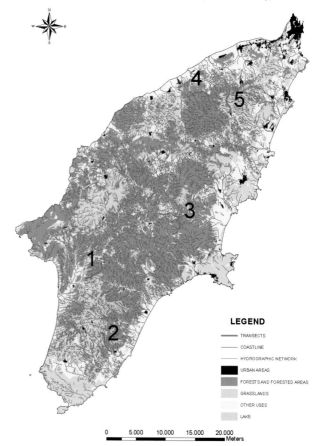

FIGURE 24.1. Map of Rhodes (study area) showing the location of the survey transects (numbered 1–5).

Introduction

The occurrence of European fallow deer, *Dama dama dama*, on the island of Rhodes, Dodecanese (Greece), has been documented since the Neolithic (sixth millennium BC). This population can be regarded as the oldest still surviving on any Mediterranean island. The latest techniques for investigating population genetics have shown that Rhodian fallow deer are an ancient lineage, distinct even from the relic population that still survives in Anatolia, the likely source of the stock introduced onto Rhodes in Neolithic times (Masseti *et al.* 2008). The Rhodian fallow deer established by humans has resulted in the chance preservation of a significant portion of the mitochondrial genetic variability of the species up to the present. Biology, biogeography, archaeozoology, history, eco-ethology, and genetics of the Rhodian fallow deer have been mainly studied only over the last decades (cf. Masseti 1999; Masseti 2002a; Masseti *et al.* 2008; Masseti 2009;

LEGEND

— TRANSECTS
— COASTLINE
— HYDROGRAPHIC NETWORK
▪ URBAN AREAS
▪ FORESTS AND FORESTED AREAS
GRASSLANDS
OTHER USES
LAKE

0   5.000   10.000   15.000   20.000
Meters

Masseti and Vernesi this volume). The current challenge is how to use this specific knowledge to manage and conserve this anthropochorous population.

Today it is reputed that Rhodian fallow deer do not exceed a few hundred in the wild, and appear to be seriously threatened by illegal shooting and by the continual reduction of the natural areas suitable for its diffusion (Theodoridis 2002; Masseti *et al.* 2008). The fact that Rhodian fallow deer can be considered as already virtually extinct implies that all necessary actions aimed at its protection and conservation must be planned and implemented with the greatest caution, in order to avert the definitive disappearance of the last fallow deer of insular Asia Minor.

Monitoring trends in animal populations is a key aspect in the conservation strategies and programs (Thompson *et al.* 1998; Nichols and Williams 2006). Abundance indices have often been used to track trends in population size of large mammals, because this indirect approach offers an attractive, low-cost method to managers, especially at large spatial scales (cf. Lancia *et al.* 1994; Engeman 2005). If sufficiently rigorous sampling protocols are employed, these indices can successfully monitor population changes over time (Engeman 2003). The kilometric index revealed a reliable index of relative abundance of deer (Whipple *et al.* 1994; Vincent *et al.* 1996; cf. Garel *et al.* 2010; Putman *et al.* 2011), even with low densities such as those occurring in Mediterranean habitats (Acevedo *et al.* 2008). We conducted spotlight surveys to evaluate the kilometric index for the Rhodian fallow deer population (McCullough 1982). This index could become a helpful tool to develop appropriate management strategies. In fact, these preliminary data can establish baseline information to start monitoring the population.

Interspecific competition occurs when different species, such as deer and domestic livestock, compete for habitat resources that are in short supply, such as food or shelter (Feldhamer *et al.* 2007). The impact of sympatric livestock on deer can be very different in relation to different factors, as well as the spatio-temporal scale considered. Therefore such competition is often misunderstood and/or overlooked in the management of deer and livestock. To start a study of resource partitioning between the Rhodian fallow deer and the local livestock, we conducted spotlight counts on domestic sheep and goats along the same transects as the deer surveys.

Material and methods

The study area (1,126.012 ha) is characterised by cultivated areas (winter cereals, forage, row crops and fruit crops such as orchards, vineyards, olive and orange groves) and woodlands.

Spotlight counts of Rhodian fallow deer were carried out from November 2010 to April 2011, taking into account the social behaviour of the species (rutting and fawning periods). The Rhodian fallow deer birth season spans from the last ten days of May to the first ten days of June, in agreement with what was found in other Mediterranean areas (San José and Braza 1992) and the rut is related to the autumnal rainy season in accord with the known photoperiodic control of the sex cycle in cervids (Jaczewski 1954; Goss 1969). On Rhodes, the rut takes place from the first substantial rains which fall in October, after a five to six months dry period (Masseti 2002b).

Two transects were selected in the south of the study area and three in the north central part, along dirt roads and throughout cultivated areas and woodlands (Figure 24.1). During the survey period, livestock occurred mainly in the north central areas of the island. The transects were surveyed once a month in similar visual and weather conditions, and in the absence of rainfall and/or fog, using four-wheel drive vehicles at a driving average speed of 10 km/hour. Surveys began approximately one hour after sunset (cf. Progulske and Duerre 1964) and required approximately four to six hours to be completed, for the northern and southern parts respectively. On each route, one observer searched for deer using 1,000,000 candle power spotlights and 10 × 40 binoculars whilst another observer recorded the number of encountered animals. Deer, goats and sheep were observed and counted at a maximum distance of 100 m from the moving vehicle. We attempted to maintain consistency among observers by having one observer participate in all spotlight surveys. The total sampling effort was 60 hours per five transects (mean length 6.16 km; total length 30.8 km).

Following Vincent *et al.* (1991) and Whipple *et al.* (1994) the kilometric abundance index (hereafter KAI) was obtained as the average number of animals seen per kilometre surveyed in each month.

Results

Spotlight counts led to an average of 2.9±0.92 and 1.1±0.68 animals/km in the southern and north central transects respectively (Figure 24.2). The higher detectability was recorded in November (Figure 24.2). Goats and sheep were not observed along the southern transects, while an average of 4.7±2.6 animals/km were observed along the north central transects (Figure 24.3). Livestock was continuously moved from one pasture to another, therefore KAI values changed between recording months. These preliminary results, however, cannot allow a deeper analysis of the KAI variation per month for deer and livestock.

Discussion

Published data in the scientific literature on the density of the Mediterranean fallow deer population are few and inter-population comparisons are biased by several factors, such as those related to the habitat use. Density estimated based upon pellet group counts, for example, in the same study area changes from less than ten individuals/100 ha in cultivated fields to 45 individuals/100 ha in abandoned olive groves and pastures (Ferretti *et al.* 2011). Furthermore, census techniques and methods adopted change according to different authors. To our knowledge, KAI is not available for Mediterranean free-ranging populations

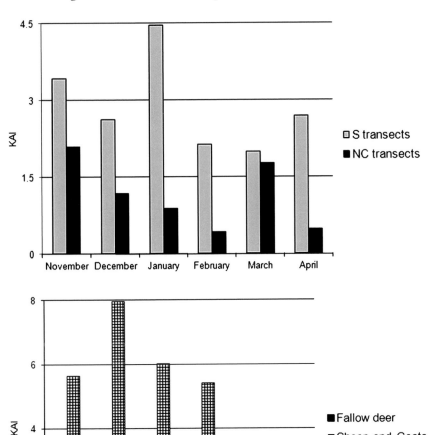

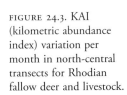

FIGURE 24.2. KAI (kilometric abundance index) variation per month in southern (S) and north-central (NC) transects for Rhodian fallow deer.

FIGURE 24.3. KAI (kilometric abundance index) variation per month in north-central transects for Rhodian fallow deer and livestock.

of fallow deer. It was only recorded for an enclosed stock of known size, and it was variable between years from 6.8 to 9.14 (Vincent *et al.* 1996). These values corresponded to population densities of 60 individuals/100 ha and 73 individuals/100 ha, respectively.

Our spotlight counts provide, for the first time, baseline data to quantify the fallow deer population of Rhodes at a given point in time, against which future data can be compared. We suggest a future monitoring program should be planned for Rhodian fallow deer, using a standardised protocol. Data should not be collected during rutting and fawning period or when animals may be clumped in their distribution in relation to food availability. Thereby allowing the overall detection of any changes of the population size relative to whether the population is increasing, stable or in decline. Moreover, data collected from these transects can estimate the population structure with respect to age and sex for management purposes. The mating system of fallow deer is supposed to be correlated with several factors such as density and habitat heterogeneity. The Rhodian population, which is characterised by occurring at low densities within the heterogeneous Mediterranean environment, has potential for future research and may be an interesting case study of a lekking versus a non-lekking fallow deer population.

Spotlight surveys are also suggested to be one of the most informative and logistically simple methods for examining relationships between deer and livestock (Lindeman and Forsyth 2008). The low relative abundance of fallow deer recorded in the north central part of Rhodes may be related to the high relative abundance of livestock. It would be interesting to determine the overlap in habitat use and dietary niche between sheep/goats and deer. In fact, densities of livestock can negatively affect those of wild ungulates through the depletion of food resources (Baldi *et al.* 2001; Madhusudan 2004; Mishra *et al.* 2004) and could lead to changes in habitat use and the distribution of wild ungulates both at fine and broad scales (Ferretti *et al.* 2011).

On Rhodes, farmers traditionally (see Chalahiris 2000), and even today, come in to conflict with fallow deer due to negative impacts (damage) inflicted upon their cultivations, such as olives and melons. Farmers perceive that these negative impacts are exclusively caused by the feeding activity of fallow deer. Consequently, it will be appropriate to study the interactions between the local agricultural land use and the habitat/food selection by this cervid species. According however to Rutter and Langbein (2005) and Wilson *et al.* (2009), impacts of deer on agricultural crops are generally localised, occurring at the level of individual farms, or even individual fields. Therefore, an investigation could be carried out in areas where frequent complaints on the part of the farmers are recorded, to assess the actual significance of cultivation damage. This could potentially alleviate future problems arising between this species and humans. Moreover, data shown in Figure 24.3 reveal how heavy the presence of the free-ranging livestock could be in the same areas where also deer occur. Future studies should aim to evaluate the extent to which damage is caused by deer and/or livestock. Finally, it is important to assess the impact of illegal shooting on the deer population, which shows no signs of abating, even today (see Theodoridis 2002).

Conclusion

The management of fallow deer of Rhodes should not focus simply on assessing the population trend, but should invest resources in collecting additional data on the ungulate–habitat system. For monitoring the interaction between a deer population and its habitat, a set of indicators of animal performance and habitat use has been proposed (cf. Morellet *et al.* 2007). This integrated approach could improve future decisions on the conservation planning according to an adaptive resource management system (Walters 1986). This approach is also needed because there is a conflict of interest between the deer population and other land use management objectives.

References

Acevedo, P., Ruiz-Fons, F., Vicente, J., Reyes-García, A. R., Alzaga, V. and Gortázar, C. (2008) 'Estimating red deer abundance in a wide range of management situations in Mediterranean habitats', *Journal of Zoology* 276, 37–47.

Baldi, R., Albon, S. D. and Elston. D. A. (2001) 'Guanacos and sheep: evidence for continuing competition in arid Patagonia', *Oecologia* 129, 561– 570.

Chalahiris, K. (2000) *Η οικονομικη πολιτικη της Ιταλιας στα Δωδεκανησα*, Trochalia Publishing, Athens.

Engeman, R. M. (2003) 'More on the need to get the basics right: population indices', *Wildlife Society Bulletin* 31, 286–287.

Engeman, R. M. (2005) 'Indexing principles and a widely applicable paradigm for indexing animal populations', *Wildlife Research* 32, 203–210.

Feldhamer, G. A., Drickamer, L. C., Vessey, S. H. and Krajewski, C. (2007) *Mammalogy, Adaptation, Diversity, Ecology*, 3rd edition, The Johns Hopkins University Press, Baltimore, Maryland, USA.

Ferretti, F., Bertoldi, G., Sforzi, A. and Fattorini, L. (2011) 'Roe and fallow deer: Are they compatible neighbours?', *European Journal of Wildlife Research* 57, 775–783.

Garel, M., Bonenfant, C., Hamann, J.-L., Klein, F. and Gaillard, J.-M. (2010) 'Are abundance indices derived from spotlight counts reliable to monitor red deer *Cervus elaphus* populations?', *Wildlife Biology* 16, 77–84.

Jaczewski, Z. (1954) 'The effect of changes in length of daylight on the growth of antlers in deer (*Cervus elaphus* L.)', *Folia Biologica* 2, 133–143.

Lancia, R. A., Nichols, J. D. and Pollock, K. H. (1994) 'Estimating the number of animals in wildlife populations', in *Research and Management Techniques for Wildlife and Habitats*, ed. T. A. Bookhout, 5th edition, The Wildlife Society, Bethesda, Maryland, USA, 215–253.

Lindeman, M. J. and Forsyth, D. M. (2008) *Agricultural Impacts of Wild Deer in Victoria*, Arthur Rylah Institute for Environmental Research Technical Report Series No. 182, Department of Sustainability and Environment, Heidelberg, Victoria.

Madhusudan, M. D. (2004) 'Recovery of wild large herbivores following livestock decline in a tropical Indian wildlife reserve', *Journal of Applied Ecology* 41, 858–869.

Masseti, M. (1999) 'The fallow deer, *Dama dama* L., 1758, in the Aegean region', *Contributions to the Zoogeography and Ecology of the Eastern Mediterranean Region* 1 Supplement, 17–30.

Masseti M. and Theodoridis N. (2002) 'Recording the data on the former and present distribution of the free-ranging deer populations on Rhodes', in *Island of Deer. Natural history of the Fallow Deer of Rhodes and of the Vertebrates of the Dodecanese (Greece)*, ed. M. Masseti, City of Rhodes, Environmental Organization, Rhodes, 169–180.

Masseti, M. ed. (2002a) *Island of Deer. Natural History of the Fallow Deer of Rhodes and of the Vertebrates of the Dodecanese (Greece)*, City of Rhodes, Environmental Organization, Rhodes.

Masseti, M. (2002b) 'The fallow deer on Rhodes', in *Island of Deer. Natural History of the Fallow Deer of Rhodes and of the Vertebrates of the Dodecanese (Greece)*, ed. M. Masseti, City of Rhodes, Environmental Organization, Rhodes, 139–158.

Masseti, M., Pecchioli, E. and Vernesi, C. (2008) 'Phylogeography of the last surviving populations of Rhodian and Anatolian fallow deer (*Dama dama dama* L., 1758)', *Biological Journal of the Linnean Society* 93, 835–844.

Masseti, M. (2009) 'A possible approach to the "conservation" of the mammalian populations of ancient anthropochorous origin of the Mediterranean islands', *Folia Zoologica* 58, 303–308.

McCullough, D. R. (1982) 'Evaluation of night spotlighting as a deer study technique', *Journal of Wildlife Management* 46, 963–973.

Mishra, C., Van Wieren, S. E., Ketner, P., Heitkong, I. M. A. and Prins, H. T. (2004) 'Competition between domestic livestock and wild bharal *Pseudois nayaur* in the Indian Trans-Himalaya', *Journal of Applied Ecology* 41, 344–354.

Morellet, N., Gaillard, J.-M., Hewison, A. J. M., Ballon, P., Boscardin, Y., Duncan, P., Klein, F. and Maillard, D. (2007) 'Indicators of ecological change: New tools for managing populations of large herbivores', *Journal of Applied Ecology* 44, 634–643.

Nichols, J. D. and Williams, B. K. (2006) 'Monitoring for conservation', *Trends in Ecology and Evolution* 21, 668–673.

Progulske, D. R. and Duerre, D. C. (1964) 'Factors influencing spotlighting counts of deer', *Journal of Wildlife Management* 28, 27–34.

Putman, R., Watson, P. and Langbein, J. (2011) 'Assessing deer densities and impacts at the appropriate level for management: a review of methodologies for use beyond the site scale', *Mammal Review* 41, 197–219.

Rutter, S. M and Langbein, J. (2005) *Quantifying the Damage Wild Deer cause to Agricultural Crops and Pastures*, Contract report VC0327 to the Department of Environment, Food and Rural Affairs, London.

San José, C. and Braza, F. (1992) 'Antipredator aspects of fallow deer behaviour during calving season at Doñana National Park (Spain)', *Ethology, Ecology and Evolution* 4, 139–149.

Theodoridis, N. (2002) 'The illegal hunting of deer on Rhodes' in *Island of Deer. Natural history of the Fallow Deer of Rhodes and of the Vertebrates of the Dodecanese (Greece)*, ed. M. Masseti, City of Rhodes, Environmental Organization, Rhodes, 181–184.

Thompson, W. L., White, G. C. and Gowan, C. (1998) *Monitoring Vertebrate Populations*, Academic Press, San Diego, California, USA.

Vincent, J.-P., Gaillard, J.-M. and Bideau, E. (1991) 'Kilometric index as biological indicator for monitoring forest roe deer populations', *Acta Theriologica* 36, 315–328.

Vincent, J.-P., Hewison, A. J. M., Angibault, J.-M. and Cagnelutti, B. (1996) 'Testing density estimators on a fallow deer population of known size', *Journal of Wildlife Management* 60, 18–28.

Walters, C. J. (1986) *Adaptive Management of Renewable Resources*, MacMillan, New York, NY.

Whipple, D. J., Rollins, D. and Schacht, W. H. (1994) 'A field simulation for assessing accuracy of spotlight deer surveys', *Wildlife Society* B22, 667–673.

Wilson, C. J., Britton, A. M. and Symes, R. G. (2009) 'An assessment of agricultural damage caused by red deer (*Cervus elaphus* L.) and fallow deer (*Dama dama* L.) in south-west England', *Wildlife Biology in Practice* 2, 104–114.